# Mathematics of Keno and Lotteries

# Mathematics of Keno and Lotteries

By

Mark Bollman

Albion College
Albion, Michigan, USA

CRC Press is an imprint of the
Taylor & Francis Group, an **informa** business

CRC Press
Taylor & Francis Group
6000 Broken Sound Parkway NW, Suite 300
Boca Raton, FL 33487-2742



Printed on acid-free paper
Version Date: 20180308

International Standard Book Number-13: 978-1-138-72380-1 (Hardback)
International Standard Book Number-13: 978-1-138-72372-6 (Paperback)



**Visit the Taylor & Francis Web site at**
**http://www.taylorandfrancis.com**

**and the CRC Press Web site at**
**http://www.crcpress.com**

*For Laura, again, and for my parents,*
*Ed and Linda Bollman.*

# Contents

**Preface** **ix**
Acknowledgments . . . x

**1 Historical Background** **1**
1.1 History of Keno . . . 1
1.2 History of Lotteries . . . 9

**2 Mathematical Foundations** **17**
2.1 Elementary Probability . . . 17
2.2 Addition Rules . . . 21
2.3 Conditional Probability . . . 24
2.4 Random Variables and Expected Value . . . 27
2.5 Combinatorics . . . 37
2.6 Binomial Distribution . . . 50
2.7 Exercises . . . 55

**3 Keno** **59**
3.1 Standard Keno Wagers . . . 59
3.2 Game Variations . . . 116
3.3 The Big Picture . . . 153
3.4 Video Keno . . . 156
3.5 Keno Side Bets . . . 168
3.6 Keno Strategies: Do They Work? . . . 186
3.7 Exercises . . . 192

**4 Lotteries** **203**
4.1 Return of the Numbers Game . . . 203
4.2 Passive Lottery Tickets . . . 213
4.3 Lotto Games . . . 233
4.4 Games Where Order Matters . . . 264
4.5 Powerball . . . 277
4.6 Lotto Strategies: Do They Work? . . . 285
4.7 Exercises . . . 294

**Answers to Selected Exercises** **303**

**References** **307**

**Index** **321**

# *Preface*

Why write a mathematics book on keno and lotteries?

Using the gambler's adage that "the simpler a bet is, the higher the house advantage", keno doesn't seem worthy of serious attention. It's fairly easy to show that the house advantage (HA) in these games is among the highest in any legitimate gambling game, often on the order of 25–35%, and the combinatorics involved, while certainly interesting, doesn't appear to vary all that much.

That's a gambler's perspective. However, a deeper look at the mathematics reveals some fascinating complexity in these related and fairly simple games. Game designers over the decades have examined "draw 20 numbers in the range 1–80" inside and out, creating a lot of variations that lead to interesting mathematical questions. A case can be made that keno has changed more with the advent of computer and video technology than any other casino game, with the exception of slot machines, and these changes have led to meaningful differences in how keno can be played. For example, one popular video keno game, Caveman Keno (page 162), cuts the HA to under 5%, and relies on increased game speed to generate the casino's profit.

Computerized game operations have made it possible for a wider variety of keno games and creative wagers to reach the casino floor, and the mathematics involved here is sometimes more intricate and more interesting than elementary combinatorics. A look at Penny Keno leads to an excursion into applied mathematics where counting is only the start of the story.

Additionally, many state and provincial lotteries run keno-like games among their offerings, which bring this game, with its history spanning many centuries, to a wider audience. This affords a neat transition to lottery mathematics. There can be no denying the hold that multi-million dollar lotteries have on the public; we need only consider the excitement when the Powerball jackpot topped $1.5 billion in January 2016 to see that lotteries, despite their long odds of winning, can still capture a large share of media attention. Moreover, there are or were innovative lottery games offered around the world that are mathematically different—in an interesting way—from Powerball and traditional daily drawings: City Picks in Wisconsin and Lucky Lines in Oregon, for example.

Some of the exercises included in Chapters 2–4 extend ideas explained in the text; some of them are dedicated to exploring different games or lottery options and can be attempted independently. Answers, though not solutions, to most exercises with numerical results are provided.

## Acknowledgments

Parts of this book were class-tested in Math 287: Mathematics of the Gaming Industry, at Albion College during the spring semester of 2017. Thanks to the students in that class: Michael Augugliaro, Henry Carnick, Alessio Gardi, Justin Kraft, Angela Morrison, Shannon Murphy, Nicole Straley, and Ethan Sutton, for their feedback on early versions of the text.

I would like to thank everyone at CRC Press/Taylor & Francis who have guided this project through to publication: Sunil Nair, who was my original editor on the project; Sarfraz (Saf) Khan, who took over as editor early in the writing and led the book through to the end; Alex Edwards and Callum Fraser, Saf's assistants and two important points of contact with the press; and Robin Lloyd-Starkes, back for a second tour moving a book of mine through the production process.

Some material in Sections 2.5 and 4.6 is reprinted from my previous book, *Basic Gambling Mathematics: The Numbers Behind The Neon*, © 2014, and is reproduced by permission of Taylor and Francis Group, LLC, a division of Informa plc. Permission was conveyed through Copyright Clearance Center, Inc.

This work was supported by a grant from the Hewlett-Mellon Fund for Faculty Development at Albion College, Albion, MI. This grant supported my travel to Las Vegas for three weeks of study at the Center for Gaming Research at the University of Nevada, Las Vegas. I am very grateful for the assistance of the staff at the Center during those three weeks.

# Chapter 1

# Historical Background

## 1.1 History of Keno

> *Suppose that 20 monkeys are loose on 80 mountains, with no more than one monkey per mountain. If we send 10 soldiers out to catch the monkeys, and each soldier looks on only one mountain, how many monkeys do we expect them to catch?*

That, in a nutshell, is keno. Over the course of time, the monkeys, mountains, and soldiers have been replaced by numbers, but the essence of modern keno is the same: The casino draws 20 numbers [monkeys] in the range 1–80 [mountains], either electronically or from a set of numbered ping-pong balls. Players select some numbers [soldiers]—not always 10 anymore—in the same range, and are paid based on how many of their numbers are among the casino's drawn set of 20. In a nod to this fable, the language of keno still speaks of "catching" numbers, or how many numbers are "caught" by the player.

This story, of Chinese origin, is said to date back 1500 years, giving keno one of the longest histories of any casino game [91]. Legend holds that Chéung Léung (205–187 BCE), of the Han Dynasty, devised an early ancestor of keno to raise money for the army and for the defense of the capital, at a time when funds were low [27]. This guessing game called for players to choose 8 Chinese characters from 120; the risk to the players was small and the rewards for a correct selection were great. Gamblers could buy a chance for 3 lí, and a winning ticket paid 10 taels, where 1 tael = 1000 lí, for payoff odds of 10,000 for 3—about 3333 for 1. The game was a success from the start; within a few decades, the operators' wealth was said to be boundless [27].

### Pák kòp piú

*Pák kòp piú* or *White Pigeon Ticket* is an ancestor of keno that was brought to America by Chinese immigrants. The game derives its alternate name from the fact that in China, where lotteries were illegal in the 19th century, homing pigeons were frequently used to convey wagers and winnings between gamblers and game operators. Pák kòp piú tickets (Figure 1.1) bore 80 different Chinese characters rather than numbers, and drawings were typically conducted once daily, in the evening. The characters were the first 80 characters in the sixth-

FIGURE 1.1: Pák kòp piú ticket.

century book *Ts'in Tsz' Man*, or the *Thousand Character Classic.* This book consists of a poem using exactly 1000 different Chinese characters, no two alike [27].

The mechanics of a pák kòp piú drawing were very different from modern keno. The operator would, carefully and in full view of patrons, separate 80 pieces of paper, one bearing each character and rolled up against detection, into four bowls of 20 numbers each. A player would then be selected to choose one of the bowls, which was designated the set of winning characters. These steps were a confidence-building measure: if gamblers could see the mechanics of the drawing taking place, there was less reason to suspect that the game was rigged [27]. Another tamper-proof method of choosing the winning bowl was to number them from 1–4, then roll 3 standard dice, divide the sum by 4, and take the remainder as the number of the lucky bowl [92].

The primary wager at pák kòp piú called for players to choose 10 of the 80 characters. Its payoff table is shown in Table 1.1.

Payoffs in Table 1.1 are quoted as "for 1" rather than "to 1", as is common in other casino games. The difference lies in how the player's original wager is handled. A payoff of 2 for 1 includes the original wager as part of the 2-unit

TABLE 1.1: Pák kòp piú pay table, without commissions deducted [27]

| **Catch** | **Payoff (for 1)** |
|---|---|
| 5 | 2 |
| 6 | 20 |
| 7 | 200 |
| 8 | 1000 |
| 9 | 1500 |
| 10 | 3000 |

payoff, so the net profit is only 1 unit; if a game pays off a winning bet at 2 to 1, the player receives 3 units for every unit wagered: the original stake plus the prize of two units. This is standard practice when paying winners at keno and lotteries, where there is often a significant time lag between when the wager is placed and when the outcome is determined. It follows then that a bet paying off at "$x$ for 1" is equivalent to a payoff of "$(x-1)$ to 1".

Game operators charged a 5% commission on all winning bets, which reduced these payoffs. If the ticket was sold through an outside agent rather than directly by the operator, a further 10% was deducted from the payoff and paid to the agent.

Pák kòp piú spread to New Zealand and Australia with the entry of Chinese mine workers, where the name was soon Anglicized to *pakapoo*. Two pay tables for pakapoo are shown in Table 1.2. The wager was 6 pence (d); this was in pre-decimal British currency, with 12 pence per shilling (s) and 20 shillings to the pound (£).

TABLE 1.2: Pakapoo pay tables

| | **Payoff (for 1)** | |
|---|---|---|
| **Catch** | **Game A** [61] | **Game B** [13] |
| 5 | 1s | 0 |
| 6 | 8s 6d | 0 |
| 7 | £3 10s | £4 |
| 8 | £19 2s 6d | £20 |
| 9 | £35 | £40 |
| 10 | £70 | £80 |

As the game spread, the Chinese characters were replaced by numbers in order to allow non-Chinese to gamble without the need to understand the language. In the USA, pák kòp piú moved eastward, to Massachusetts first, then to Pennsylvania, Maine, Connecticut, and New York City [92]. The game found its strongest American market in Montana, when Chinese mining workers brought the game to their new country [34]. The Crown Cigar Store in

Butte was home to what was called the Chinese lottery. Proprietors Joseph and Francis Lyden took possession of the game from Chinese agents in a deal with local officials.

The Chinese lottery provided for tickets containing anywhere from 9–20 spots; these tickets were processed by lottery officials by transforming them, in various ways, to collections of 10-spot tickets [16].

- A *9-spot ticket* was interpreted as 71 different 10-spot tickets by combining the player's 9 numbers with each of the 71 unmarked numbers to form a different 10-spot ticket. This ticket was sold for 71¢: 1¢ per virtual ticket. The pay table for a 1¢ ticket is shown in Table 1.3.

TABLE 1.3: Chinese lottery pay table: 1¢ ticket [16]

| Catch | Payoff (for 1) |
|---|---|
| 5 | $.01875 |
| 6 | $.1875 |
| 7 | $1.875 |
| 8 | $9.375 |
| 9 | $18.75 |
| 10 | $37.50 |

- The *11-spot ticket* was drawn as 11 different virtual 10-spot tickets; each one formed by striking out one of the player's 11 numbers. This ticket was also offered starting at 1¢ per ticket, though it could be purchased for any number of cents per ticket.

- A *12-spot ticket* could be interpreted in several ways. The simplest led to 6 10-spot tickets. The player's choices were numbered from 1–12, and a 6 × 10 array of numbers was formed by writing each number 5 times, proceeding vertically and filling the columns from left to right, as in Figure 1.2.

| | | | | | | | | | |
|---|---|---|---|---|---|---|---|---|---|
| 1 | 2 | 3 | 4 | 5 | 7 | 8 | 9 | 10 | 11 |
| 1 | 2 | 3 | 4 | 6 | 7 | 8 | 9 | 10 | 12 |
| 1 | 2 | 3 | 5 | 6 | 7 | 8 | 9 | 11 | 12 |
| 1 | 2 | 4 | 5 | 6 | 7 | 8 | 10 | 11 | 12 |
| 1 | 3 | 4 | 5 | 6 | 7 | 9 | 10 | 11 | 12 |
| 2 | 3 | 4 | 5 | 6 | 8 | 9 | 10 | 11 | 12 |

FIGURE 1.2: Resolution of a 12-spot Chinese lottery ticket into 6 10-spot tickets [16].

The rows of this table were then used to identify the numbers on each of the six tickets. This ticket could be had for 6¢.

- An alternate approach to the 12-spot ticket involved grouping the 12 numbers into four groups of three each. Twelve different 10-spot tickets could then be constructed by choosing one number from each group and combining it with the 9 numbers in the other three groups to make 10. If the numbers were grouped {1,2,3}, {4,5,6}, {7,8,9}, and {10,11,12}, then these 12 tickets would be those shown as the rows in Figure 1.3.

| | | | | | | | | | |
|---|---|---|---|---|---|---|---|---|---|
| 1 | 4 | 5 | 6 | 7 | 8 | 9 | 10 | 11 | 12 |
| 2 | 4 | 5 | 6 | 7 | 8 | 9 | 10 | 11 | 12 |
| 3 | 4 | 5 | 6 | 7 | 8 | 9 | 10 | 11 | 12 |
| 1 | 2 | 3 | 4 | 7 | 8 | 9 | 10 | 11 | 12 |
| 1 | 2 | 3 | 5 | 7 | 8 | 9 | 10 | 11 | 12 |
| 1 | 2 | 3 | 6 | 7 | 8 | 9 | 10 | 11 | 12 |
| 1 | 2 | 3 | 4 | 5 | 6 | 7 | 10 | 11 | 12 |
| 1 | 2 | 3 | 4 | 5 | 6 | 8 | 10 | 11 | 12 |
| 1 | 2 | 3 | 4 | 5 | 6 | 9 | 10 | 11 | 12 |
| 1 | 2 | 3 | 4 | 5 | 6 | 7 | 8 | 9 | 10 |
| 1 | 2 | 3 | 4 | 5 | 6 | 7 | 8 | 9 | 11 |
| 1 | 2 | 3 | 4 | 5 | 6 | 7 | 8 | 9 | 12 |

FIGURE 1.3: 12 10-spot tickets derived from a single 12-spot ticket.

- *13-spot tickets* were processed much like 12-spot tickets. The array of numbers, with each of the 13 choices repeated 4 times, is given in Figure 1.4.

| | | | | | | | | | | |
|---|---|---|---|---|---|---|---|---|---|---|
| 1 | 2 | 3 | 4 | 6 | 7 | 8 | 9 | 11 | 12 | 13 |
| 1 | 2 | 3 | 5 | 6 | 7 | 8 | 10 | 11 | 12 | 13 |
| 1 | 2 | 4 | 5 | 6 | 7 | 9 | 10 | 11 | 12 | |
| 1 | 3 | 4 | 5 | 6 | 8 | 9 | 10 | 11 | 13 | |
| 2 | 3 | 4 | 5 | 7 | 8 | 9 | 10 | 12 | 13 | |

FIGURE 1.4: Interpretation of a 13-spot Chinese lottery ticket as 3 10-spot and 2 11-spot tickets [16].

Once again, each row of this figure determines a ticket. This numbering scheme led to 25 tickets in all: the three 10-spot tickets, which were priced at 10¢ apiece, and 22 additional 1¢ 10-spot tickets arising from the two 11-spot tickets.

Similar schemes, all of which involved reducing a ticket down to some combination of 10-, 11-, or 12-spot tickets, were in common use for players wishing to mark 14–20 different numbers.

Keno evolved as it spread from Montana, eventually flourishing in Nevada after that state legalized gambling in 1931. Francis Lyden took the game from Montana to the Palace Club in Reno [34]. Initially, the game was called *racehorse keno* in Nevada in order to evade anti-lottery laws which still remain on the books, and the Chinese characters were replaced by the names of 80 fictional racehorses. Each horse was tagged with a number from 1–80, and by 1951, with a new tax on off-track betting introduced in Nevada, the horses' names were dropped to avoid any possible connection with racing and players wagered on numbers chosen in that range [92, 141].

The game came to southern Nevada in 1939, with a game opening at the Las Vegas Club [156]. Joseph Lyden moved to Las Vegas in 1956, at which time he increased the drawing frequency from once a day to several times per hour, which became the standard frequency [34]. Then as now, the casino draws 20 of the 80 numbers, and pay tables were issued that paid out increasing amounts as the number of catches rose. A current pay table from the Orleans Casino in Las Vegas may be found in Table 1.4.

TABLE 1.4: Keno pay table issued by the Orleans Casino in Las Vegas. All payoffs are "for 1" and assume a $1 wager.

| **Pick 1** | |
|---|---|
| **Catch** | **Payoff** |
| 1 | 3 |

| **Pick 2** | |
|---|---|
| **Catch** | **Payoff** |
| 2 | 12 |

| **Pick 3** | |
|---|---|
| **Catch** | **Payoff** |
| 2 | 1 |
| 3 | 45 |

| **Pick 4** | |
|---|---|
| **Catch** | **Payoff** |
| 2 | 1 |
| 3 | 2 |
| 4 | 160 |

| **Pick 5** | |
|---|---|
| **Catch** | **Payoff** |
| 3 | 1 |
| 4 | 15 |
| 5 | 750 |

| **Pick 6** | |
|---|---|
| **Catch** | **Payoff** |
| 3 | 1 |
| 4 | 4 |
| 5 | 80 |
| 6 | 2000 |

| **Pick 7** | |
|---|---|
| **Catch** | **Payoff** |
| 4 | 1 |
| 5 | 15 |
| 6 | 400 |
| 7 | 10,000 |

| **Pick 8** | |
|---|---|
| **Catch** | **Payoff** |
| 5 | 5 |
| 6 | 75 |
| 7 | 1500 |
| 8 | 50,000 |

| **Pick 9** | |
|---|---|
| **Catch** | **Payoff** |
| 5 | 6 |
| 6 | 30 |
| 7 | 300 |
| 8 | 5000 |
| 9 | 50,000 |

| **Pick 10** | |
|---|---|
| **Catch** | **Payoff** |
| 5 | 1 |
| 6 | 25 |
| 7 | 125 |
| 8 | 1000 |
| 9 | 10,000 |
| 10 | 100,000 |

| **Pick 12** | |
|---|---|
| **Catch** | **Payoff** |
| 6 | 5 |
| 7 | 20 |
| 8 | 200 |
| 9 | 2000 |
| 10 | 7500 |
| 11 | 50,000 |
| 12 | 250,000 |

| **Pick 15** | |
|---|---|
| **Catch** | **Payoff** |
| 7 | 5 |
| 8 | 25 |
| 9 | 150 |
| 10 | 1000 |
| 11 | 5000 |
| 12 | 25,000 |
| 13 | 100,000 |
| 14 | 250,000 |
| 15 | 500,000 |

## Keno Mechanics

In the earliest days of keno, players used brushes and ink to mark their chosen numbers on their bet slips. Keno ticket writers conditioned each ticket in the margin with the amount of the wager and the number of spots chosen. Ink and brushes were eventually replaced by crayons, as shown in Figure 1.5. This figure shows two keno tickets written in 1963: a ticket from the Sahara Casino (now SLS) in which 8 spots have been marked with ink and 50¢ wagered, and a 50¢ 4-spot ticket from the California Club, marked with crayon.

FIGURE 1.5: Keno tickets from 1963: Sahara and California Club.

With the new drawing frequency introduced by Joseph Lyden, it became customary that prizes from winning tickets must be claimed before the next drawing starts. Casinos would provide bet slips with the winning numbers from a given drawing punched out, such as the one in Figure 1.6, to aid in identifying winning tickets. By placing this punched-out blank over a keno ticket, the markings on the caught numbers would show through the holes for easy counting.

Recently, this restriction has been relaxed at some casinos for tickets that play the same numbers for a sequence of successive drawings, with a sensible new rule in place that allows players to wait until all of their paid-for drawings are complete before cashing their winning tickets. The D Casino in downtown Las Vegas has gone a step beyond this relaxation, and advertises "The Most Liberal Late Pay Policy Downtown". Keno winners at the D have a week to claim their winnings if they've bet 1–9 consecutive games, and 360 days if they've paid for 10–999 straight games.

At some casinos, and in many state lotteries that include keno among their game offerings, advancing technology has seen crayons replaced by pencils and optical scanners introduced to translate player choices into valid keno tickets. Figure 1.7 shows an optical bet slip, from the Foxwoods Casino in

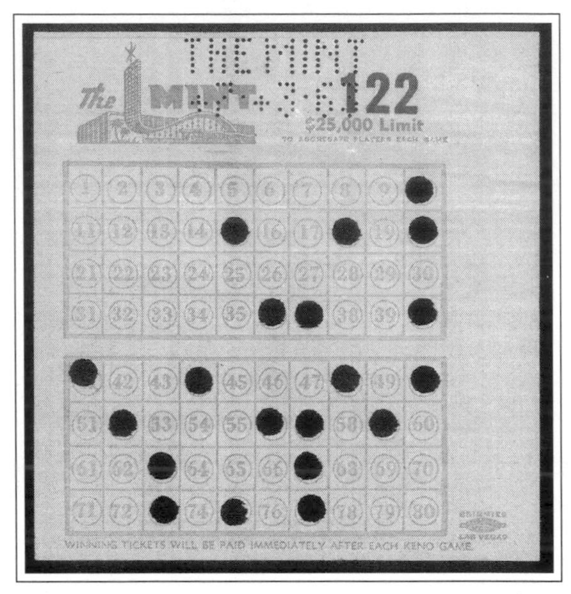


FIGURE 1.6: Mint Casino keno ticket with winning numbers punched out.

Mashantucket, Connecticut, which is filled out by the player and scanned.

Casinos originally chose their numbers by drawing numbered wooden balls by hand from a wooden container called a *goose*. This was later replaced by ping-pong balls shuffled by and drawn from an air blower, a method still in occasional use so that players may see the winning numbers being drawn and be confident in their randomness. Many casinos and lotteries now use a computerized random number generator to select the winning numbers. An additional keno innovation provided by the rise of computers is the Quick Pick gameplay option, which is visible on the bet slip in Figure 1.7. Quick Pick eliminates the need for players to pick their numbers; a player using

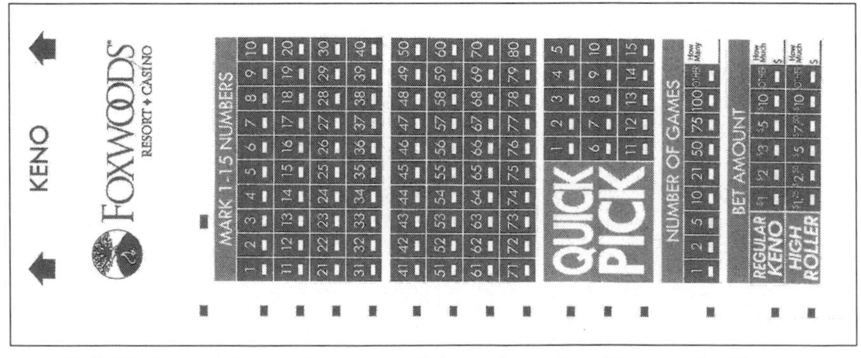


FIGURE 1.7: Optically-scanned keno bet slip, Foxwoods Casino.

this option merely specifies how many numbers she or she wishes to play. A computer then generates that many randomly-selected numbers from 1–80 and dispenses a ticket.

With keno's evolution into its modern form, available wagers expanded from pák kòp piú's Pick 10 option to games that invited players to choose from 1 to 15 numbers, as well as more complex games involving other combinations of numbers.

---

## 1.2 History of Lotteries

Gambling artifacts may be found among the relics of almost every civilization, and lotteries are part of that long history. The word "lottery" is derived from "lot", itself a derivative of the Teutonic word "hleut", an object used to resolve disputes or divide property under divine guidance, as expressed in the outcome of a random process [35]. The objects triggering these random processes include coins and dice as well as more involved devices used to choose a winner from among competitors.

### In the Beginning

The casting of lots to solicit the opinions of the gods as an aid to earthly decision-making eventually gave rise to the use of those devices for more recreational purposes. The Romans appear to have been the first to stage lotteries for fun; the emperor Augustus launched the forerunner of many government-run lotteries when he conducted a public works lottery with the net profit put toward funding civic improvements [28].

The first public lottery in more modern times was held in the Netherlands in1434, to raise money for fortifications for the Dutch town of Sluis. A public lottery was held in Augsburg, Germany, in 1470; the first state lottery in Germany followed at Osnabruck in 1521 [31].

Italian merchants during the Middle Ages used a form of lottery called the "Urn of Fortune" to dispose of unsold stock. The names of slow-selling items were listed on small pieces of paper and placed in an urn; customers were invited to pay some small amount intermediate among the prices of all the goods on offer and draw their prize from the urn [28]. In addition to the possibility of winning a prize worth more than the cost of entry, there was some entertainment value to patrons in the anticipation of a big win—much like we find in 21st-century lotteries.

The first lottery where all of the prizes were cash rather than a mixture of cash and merchandise is generally acknowledged to be La Lotto di Firenze, which began in Florence, Italy around 1530. Mathematically, these lotteries were quite simple, with winning tickets simply chosen in a random drawing from among all tickets purchased. The chance of winning a prize in such a lottery depends on the number of tickets sold; such a lottery is now often called a *raffle* (page 223). Some raffle drawings were conducted via separate drawings of tickets from one container and corresponding prizes, including many blank slips, drawn from a second container. Drawing continued until all tickets were matched with a prize. These drawings could take many days, as long as a month or more in some cities, as every ticket had its moment in the spotlight [29].

This idea spread to other Italian cities, and in 1576 was transformed in Genoa from a raffle drawing to a means for selecting 5 replacements for retiring members of the ruling colleges from 120 candidates [137]. This provided obvious opportunities for side bets on the outcome of the drawing. Over time, the idea of drawing numbers as the basis of a lottery spread throughout Italy, and following Italian unification in 1859–61, this evolved into the present-day Italian national lottery, or Lo Giuoco del Lotto [137]. This lottery is examined in detail beginning on page 234.

## Lotteries In America

In colonial America, lotteries were a common means of raising funds for public works and other construction projects. The charter of the Virginia Company in 1618 included the power to operate a lottery to fund the settlement of North America; the lottery itself was, however, held in England. A 1720 lottery offering a new brick house as the prize is cited as the first to be operated in the colonies [5]. The details of a 1760 lottery in Newbury, Massachusetts are shown in Figure 1.8 [18].

This advertisement clearly lays out everything needed to make sense of the lottery. Ten thousand dollars worth of tickets were to be sold for the purpose of raising $1000 in funds once the prizes were paid out; consequently, 10% of invested funds was kept by the lottery agents. This assumes, of course, that all of the tickets were sold. A sales shortfall would mean that less money was raised for the Newbury bridge.

After independence, the U.S. Congress legalized many state and city lotteries, including three, in 1812, 1820, and 1827, for the District of Columbia. The 1827 lottery included Thomas Jefferson's land so that his heirs could "rehabilitate the family fortunes", which were left in a poor state after Jefferson's death in 1826 [5]. Lotteries gained acceptance as a source of revenue for two reasons: citizens would not vote to tax themselves for civic needs, and government authorities could not guarantee the payment of interest owed on bonds that might be issued for the same purposes.

New Jersey lotteries of the era typically allocated 15% of ticket sales to fundraising, returning 85% to players. In the 1820s, New Jersey was home to a lottery that eschewed the raffle model in favor of a winning-number drawing, similar to modern lotteries. This number selection scheme was known as "Vannini's Patent Lotteries", after Joseph Vannini, the holder of an early

SCHEME of a LOTTERY

For raising a Sum of Money for the building and maintaining a Bridge over the River *Parker*, in the Town of Newbury, at the Place called Old Town Ferry (in pursuance of an Act of the general Court, passed in April 1760)

Wherein *Daniel Farnham, Caleb Cushing, Joseph Gerrish, William Atkins, Esq.*, and Mr. *Patrick Tracy*, Merchant, (or any Three of them) are appointed managers. The acting Managers are sworn to the faithful Performance of their Trust.

*Newbury*-Lottery Number Four, consists of 5000 Tickets, at Two Dollars each; 1655 of which are Benefit Tickets of the following Value.

| | | | | | | |
|---|---|---|---|---|---|---|
| 1 | of | 500 | Dollars, | is | 500 | Dollars. |
| 4 | of | 100 | | are | 400 | |
| 5 | of | 50 | | are | 250 | |
| 6 | of | 40 | | are | 240 | |
| 10 | of | 30 | | are | 300 | |
| 14 | of | 20 | | are | 280 | |
| 45 | of | 10 | | are | 450 | |
| 75 | of | 8 | | are | 600 | |
| 1495 | of | 4 | | are | 5980 | |
| 1655 Prizes, amounting to | | | | — | 9000 | Dollars. |
| 3345 Blanks. | | | | | | |
| 5000 Tickets at Two Dollars each | | | | | 10,000 | |
| To be paid in Prizes, | | | | | 9000 | |
| | | | | | 1000 | Dollars |

Remains to be applied for the Purpose aforesaid.

FIGURE 1.8: The Newbury Lottery, 1760 [18].

American patent on this way of choosing numbers [160]. The Queen's College Literature Lottery of 1824, organized to raise funds to endow a mathematics professorship at what is now Rutgers University, used tickets bearing three numbers in the range 1–35. Lottery patrons were not permitted to choose their own numbers, meaning that every ticket was different. Five numbers were drawn, and the winning tickets could be identified in minutes rather than days.

Despite the success of some lotteries in financing the early American republic, public opinion eventually turned against this form of gambling. By 1830, most states, with the notable exceptions of Maryland and Virginia, had banned lotteries [5]. The proximity of these states to the nation's capital and its considerable needs may account for the continued operation of government-approved lotteries there even as lotteries fell out of favor elsewhere.

## Louisiana and the End of State Lotteries

The last legal lottery in the USA in the 19th century was operated in Louisiana. Lotteries in Louisiana dated back before statehood, to 1810 and the Orleans Territory, where, as with the original British colonies, new settlements found lotteries a convenient means of raising needed money. As was the case in the eastern colonies, opposition to government-sanctioned gambling eventually won the day, and the state constitutions of 1845 and 1852 prohibited the buying and selling of lottery tickets [5].

This changed during the Civil War, which strained the economies of the southern states. Constitutional amendments in 1864 and 1866 provided for legal state lotteries and gambling houses. A provision of the lottery law was that the first $50,000 of revenues be given to the New Orleans Charity Hospital and the rest to the state's general fund. These games were cited as a necessary evil to keep money within the state; nearby states including Alabama, Georgia, and Kentucky had also established lotteries and enlisted agents to sell their tickets in other states [5].

The Louisiana Lottery filled a void caused by the banning of lotteries in most other states, and began its operations in 1868 [146]. Like numerous lotteries that preceded it both in and beyond America, Louisiana operated many of its drawings on the raffle model. The lottery pioneered what it called "approximation prizes": small prizes awarded to tickets whose numbers matched some of the digits of winning tickets. This idea would find use in a number of 20th-century lotteries. A drawing from March 1869 offered 600 prizes, drawn from 32,000 tickets sold; by 1878, a single drawing typically involved 100,000 tickets priced at $10 each, competing for 11,279 prizes. This number of tickets was cited in an 1889 pamphlet listing numerous objections to the lottery, which claimed that if 100,000 people each devoted 5–10 minutes to reading the lists of winning lottery numbers published in newspapers, that represented 694 days' time—nearly 2 years—that could have been put toward more useful pursuits [22]. Fractional tickets were available: a person could buy one-tenth of a full ticket for $1, and receive one-tenth of any prize drawn against that ticket. (By providing more ticket buyers than there were full tickets to be had, this might have pushed the lost-time number past the two-year mark.) By 1878, competition from other state lotteries had essentially been eliminated, and tickets for Louisiana drawings were distributed to buyers around the country, reaching every state by 1880.

Daily drawings, instituted in 1878, soon became a new target for lottery opponents. These games abandoned the raffle model for a method based on players choosing numbers from 1–75 or 1–78, while lottery agents chose 12 in that range. Elements of this lottery would later be incorporated into state-run lotteries such as Mega Millions and Powerball. The size of the selection was left to the player: a one-number ticket was called a *day number*, a ticket with two numbers was called a *saddle*, and with three a *gig*—the language of horse racing was borrowed for this game. A *capital saddle* prize was won when the player's two numbers were drawn among the first three winning numbers [5]. Winning numbers were selected with either a physical wheel spun to generate winning numbers or a bin filled with slips of paper bearing all of the numbers in use from which the winning numbers were drawn [149]. Tickets were sold for 25¢; the prize for matching one of the 12 chosen numbers was 20¢. It is certainly plausible that this prize was set to encourage further play for "only a nickel".

Legal battles over the lottery continued almost constantly after its sanction in the Louisiana constitution. 1879 saw the state legislature strip the lottery of its charter, an action which was nullified in the courts, and a new constitution adopted which legalized the lottery until 1895. This last action was a compromise: the lottery surrendered its monopoly rights, but gained the right to continue its operations—with a great head start over any possible competition, this was surely a tradeoff worth making. It was estimated that at the height of the lottery's popularity, over 90% of lottery proceeds were from ticket buyers outside Louisiana [5]. This distribution attracted the attention of federal officials. In Louisiana, there was tension between the lottery's state charter and the legislature's desire to ban it entirely. The United States Supreme Court ruled, in 1886 (*Louisiana Lottery Company v. Richoux*), that the state constitution overruled the legislature's right to strip the lottery of its charter.

Further suspicion was cast upon the lottery in 1890, when the March 11 drawing saw only 69% of tickets sold [5]. With 31% of the tickets held by the Louisiana Lottery Company, approximately one-third of the prizes were destined not to be awarded to paying customers. In that same year, the Anti-Lottery League of Louisiana was formed to oppose the renewal of the lottery's charter, which was slated to expire at the end of 1894. On the national level, the U.S. Congress moved in 1890 to outlaw the use of the U.S. Postal Service to transmit lottery tickets and newspapers bearing lottery advertisements, or to sell money orders for lottery play—at the time, it was estimated that 45% of all post office business in New Orleans was connected to the Louisiana Lottery [146]. While this did not lead, as lottery opponents hoped, to an amendment to the U.S. Constitution prohibiting lotteries, a second law, in 1895, barred lotteries from all forms of interstate commerce. This effectively spelled the end for the Louisiana Lottery, and state lotteries in America in general, by the end of the 1800s. The Louisiana Lottery attempted to continue operations by relocating to Puerto Cortez, Honduras, moving tickets in and out of America as personal baggage carried by traveling officials, and transmitting encoded results of drawings to New Orleans by telegraph and thence to other towns by courier. Various authorities eventually put a stop to these evasive actions; the Honduras experiment, and the Louisiana Lottery itself, ended by 1907 [5]. State-sanctioned lotteries did not return to Louisiana until 1991.

## The Numbers Game

The elimination of legal state lotteries did not diminish Americans' collective interest in gambling. The now-common daily state lottery drawings can trace their origins to an underground game called the *Numbers Game*, an illegal form of gambling formerly popular in large cities which filled the gap between the end of the Louisiana Lottery and the revival of state lotteries beginning in 1964 [135]. The Numbers Game is a very simple proposition: players bet on a number of their choice and win their bet if the number appearing in some legitimate and presumably unbiased source matches theirs. In its heyday, the Numbers Game attracted millions of dollars in bets and engaged a significant fraction of the adult population in some areas.

**Example 1.1.** A simple version of the Numbers Game called for gamblers to pick a three-digit number and paid 600 for 1 if their number was the winner. ■

The source of the winning number varied from city to city; it had to be both publicly available and above suspicion of influence by game operators. Some games used the last digits of the U.S. Treasury balance as published in a newspaper, and some winning numbers were based on payoffs at a specified racetrack. In New York City, a popular method called the Manhattan way derived its winning three-digit number from the win, place, and show payoffs at Aqueduct Racetrack in Queens. The six winning payoffs from the first three races were added together and the units digit of the number of dollars provided the first digit of the winning number, so if that total were $68.40, the first digit would be 8. Similar addition with results of the first five and first seven races produced the second and third digits [64].

By contrast, the Brooklyn number used the last three digits of the total daily handle at Aqueduct as the winning number, so if the track received $2,856,828 in wagers, the winning number was 828 [48].

**Example 1.2.** In New York City, one popular numbers game was the "You Pick 'Em Treasury Ticket" [135, p. 168–9]. Players chose a five-digit number and were paid off in accordance with the last five digits of the published U.S. Treasury balance each day. If a player matched all five digits, the payoff was 300 for 1; matching only the last four paid 30 for 1 and a match of the last three paid 3 for 1. ■

A different kind of numbers game, called the *policy game* (from the Italian *polizza*: "receipt" or "voucher") in some cities, used ideas seen in the Louisiana Lottery. Once again, the lottery called for gamblers to select numbers in the range 1–78. Saddles and gigs were joined by the *horse*—a four-number ticket. Winning numbers were either chosen from trustworthy sources, as above, or drawn using a *policy wheel*. This device was either a physical numbers wheel spun to generate winning numbers or a bin filled with slips of paper bearing all of the numbers in use from which the winning numbers were drawn [149].

A modern successor to the Numbers Game is the common state lottery game that invites players to choose their own two- to six-digit numbers. Their choices are then compared against numbers of the same length drawn by the lottery agency. Winning numbers for these games are drawn once or more per day, depending on the state lottery.

**Example 1.3.** In Pennsylvania, *Pick 2* is the simplest of the daily numbers games, asking players to select a two-digit number for drawings held twice a day. The simplest bet is a "straight" bet, where the gambler picks a two-digit number and wagers $1. If the player's number matches the number drawn by the state, the payoff is 50 for 1. ■

With the establishment of legal daily lotteries in many states, some illegal policy shops turned to the using the state's winning numbers as a trusted random source of their own winning numbers.

*Bolita* (Spanish for "little ball") is an illegal lottery akin to the Numbers Game played in Cuba and some Hispanic neighborhoods in the USA. In its original form, small balls numbered from 1–100 are placed in a bag, a game agent grabs hold of one ball through the bag, and the other 99 balls are poured out, leaving the agent holding the bag and the winning number. Other versions, like the Numbers Game, pull their winning numbers from accepted impartial sources, such as the handle at New York City horse races or the winning Florida lottery numbers. Until January 2003, Radio Martí, a shortwave radio station expressly established by the United States government to broadcast to Cuba, broadcast daily Florida lottery numbers to the island six days per week at listeners' request—possibly for use in bolita operations [45]. Upon concern that the USA should not be involved in encouraging illegal gambling in other countries, Radio Martí ceased transmitting lottery results, and bolita operators in Cuba turned to Spanish-language television stations for Florida's numbers.

*Nhit-lone* ("two digit") is a game similar to bolita that is played in Myanmar [136]. In this informal lottery, players choose a two-digit number. The winning number is derived from the Thai Stock Exchange: at a given time of day, the units digit of the main index and the final digit of the total value of shares traded are joined to form the prizewinner. Winning tickets are paid off at 80 for 1.

## State Lotteries Revived

State lotteries began to return beginning in 1964, in New Hampshire. Lotteries have since been introduced in 44 US states; only Alabama, Alaska, Hawaii, Mississippi, Nevada, and Utah are without a state lottery. Hawaii and Utah allow no legal gambling of any sort. In Nevada and Mississippi, the casino industry can be counted on to oppose lotteries. Alaska and Hawaii derive considerable revenue from tourism, and tourists are not thought to be a strong market for lottery tickets such as Powerball and its variants that can have a long time lag between ticket purchase and the drawing. Additionally since Alaska and Hawaii border no other states, there is no incentive to establish state lotteries to discourage citizens from crossing state lines to buy tickets and thus to take their gambling money out of state. Religious objections to the perceived immorality of gambling play a part in lotteries' absence from Alabama, Mississippi, and Utah [23].

In May 2017, the Mississippi state government established a task force to study lotteries in adjoining states, with an eye toward assessing how much money those states were raising each year through their lotteries. Curiously, the task force was not charged with making a recommendation one way or the other about establishing a state lottery in Mississippi. William Perkins, who edited the Mississippi Baptist Convention's weekly newspaper, raised his organization's objections to gambling using the following analogy [127]:

> It's almost like having an Olympic-size swimming pool and buying enough BBs to fill up the swimming pool, paint one of them red, put them in the pool and mix it up and then charge people $2 to reach in and try to find the red BB. That's how futile it is and that's how useless it is to depend on a lottery or any form of gambling to support your family.

While Perkins' conclusion that it's unwise to rely on a lottery to support one's family is correct, is this a mathematically apt comparison? An Olympic-sized swimming pool is 50 meters long, 25 meters wide, and at least 2 meters deep, and so has a minimum total volume of 2500 cubic meters. A standard spherical metal BB is 4.5 millimeters, or $4.5 \times 10^{-3}$ meters, in diameter, and so has a volume of

$$\frac{4}{3}\pi r^3 = \frac{4}{3}\pi \cdot (2.25 \times 10^{-3})^3 \approx 4.77 \times 10^{-8} \text{ cubic meters.}$$

Closely-packed identical spheres occupy approximately 74% of available volume [161], so it would take

$$\frac{.74 \cdot 2500}{4.77 \times 10^{-8}} \approx 3.88 \times 10^{10},$$

or nearly 39 billion, BBs to fill the pool.

By contrast, the most recent version of Powerball (page 277) only offers 292,201,338 different ticket combinations, which is less than 1% of the number of BBs that Perkins described, so the analogy falls short. It's about 133 times more likely that you'll win the Powerball jackpot than find that red BB.

# Chapter 2

# Mathematical Foundations

## 2.1 Elementary Probability

Informally, the *probability* of an event $A$ is an attempt to measure how likely that event is to occur, which is a number $P(A)$ between 0 and 1—or between 0 and 100%. Readers interested in a more formal axiomatic treatment of probability as it relates to games of chance should consult a text such as [14].

Simple probability is often a matter of little more than careful counting. For our work in probability, we will frequently be interested in the size of a set—that is, how many elements it has. For convenience, we introduce the following notation:

**Definition 2.1.** The expression $\#(A)$ denotes the number of elements in a set $A$.

This is most often used when $A$ is a finite set—while it is certainly possible to consider the size of an infinite set, such sets are not necessary for keno and lottery calculations, and are not considered in this book.

**Example 2.1.** If $A$ is a standard deck of playing cards, then $\#(A) = 52$. ■

**Example 2.2.** A Daily 3 lottery invites the player to choose his or her own 3-digit number. Since a player's number may start with the digit 0, the set $N$ of all possible choices has $\#(N) = 1000$, as it includes all three-digit numbers from 000 through 999. ■

The challenge here is that a set of interest in gambling mathematics can be very large. If we are interested in the set $K$ of all possible twenty-number keno draws, then $\#(K) = 3,535,316,142,212,174,320$, and we'd like to have a way to come up with that number without having to list all of the drawings and count them. Techniques for finding the size of such large sets will be discussed in Section 2.5.

We begin our study of probability with the careful definition of some important terms.

**Definition 2.2.** An *experiment* is a process whose outcome is determined by chance.

This may not seem like a useful definition. We illustrate the concept with several examples.

**Example 2.3.** Roll a standard six-sided die and record the number that results. ■

**Example 2.4.** Roll two standard six-sided dice (abbreviated as 2d6) and record the sum. ■

**Example 2.5.** Given a 10-spot keno ticket, draw 20 numbers in the range from 1–80 and count how many of those appear on the ticket. ■

**Example 2.6.** Roll 2d6 and record the larger of the two numbers rolled (or the number rolled, if both dice show the same number). ■

**Example 2.7.** Deal a five-card video poker hand and record the number of aces it contains. ■

An important trait of an experiment is that it leads to a definite outcome. While we will eventually concern ourselves with individual outcomes, we begin by looking at all of the possible results of an experiment.

**Definition 2.3.** The *sample space* $\mathbf{S}$ of an experiment is the set of all possible outcomes of the experiment.

**Example 2.8.** In Example 2.3, the sample space is $\mathbf{S} = \{1, 2, 3, 4, 5, 6\}$. The same sample space applies to the experiment described in Example 2.6. ■

**Example 2.9.** In Example 2.4, the sample space is $\mathbf{S} = \{2, 3, 4, \ldots, 12\}$. ■

It is important to note that the 11 elements of $\mathbf{S}$ in this example are not equally likely, as this will play an important part in our explorations of probability. Rolling a 7 is more likely than rolling any other sum on two dice; 2 and 12 are the least likely sums.

**Example 2.10.** In Example 2.5, the sample space is $\mathbf{S} = \{0, 1, 2 \ldots, 10\}$. These elements are also not equally likely. ■

When we're only interested in some of the possible outcomes of an experiment, we are looking at subsets of $\mathbf{S}$. These are called *events*.

**Definition 2.4.** An *event* $A$ is any subset of the sample space $\mathbf{S}$. An event is called *simple* if it contains only one element.

**Example 2.11.** In Example 2.5, one simple event would be "5 numbers on the ticket are caught". Because of the way that the sample space has been defined, a simple event covers one number of catches. There are, of course, many different 20-number keno draws that lead to 5 catches. ■

**Example 2.12.** Table 1.4 shows that a 10-spot keno ticket only wins money if 5 or more numbers are caught. The event "The ticket is a winner" in Example 2.5 can be described as a subset $W$ of $\mathbf{S}$ as $W = \{5, 6, 7, 8, 9, 10\}$. This is not a simple event, as it contains 6 elements. ■

**Definition 2.5.** Two events $A$ and $B$ are *disjoint* if they have no elements in common. In this case, we say that the event consisting of all outcomes common to $A$ and $B$ is the *empty set* $\varnothing$, which contains zero elements.

**Example 2.13.** In a 10-spot keno drawing, the two events $A$ = {Catch 6 numbers} and $B$ = {Catch 8 numbers} are disjoint. ■

A particularly important example of disjoint sets that speeds some probability calculations is the *complement* of a given event.

**Definition 2.6.** The *complement* of an event $A$, denoted $A'$, is the set of all elements of the sample space that do not belong to $A$.

**Example 2.14.** The New Jersey Lottery offers a \$1 "Fast Play" game called *Dollar Throwdown* which simulates 10 games of rock/paper/scissors (RPS) between the player and the lottery computer. RPS is played between two players, each of whom chooses "rock", "paper", or "scissors" and reveals this or her choice simultaneously. If the two players make the same choice, the result is a draw; if the choices differ, the winner is determined as follows:

- Rock breaks scissors.
- Paper covers rock.
- Scissors cut paper.

Considering just one of the 10 games on a ticket, the complement of the simple event {Win} is the event {Lose, Draw}. ■

The following result, called the *Complement Rule*, connects the probability of an event $A$ to the probability of its complement $A'$.

**Theorem 2.1.** ***(Complement Rule)****: For any event $A$,*

$$P(A') = 1 - P(A).$$

The Complement Rule frequently turns out to be useful in simplifying probability calculations.

**Example 2.15.** On a 10-spot keno ticket, the probability of matching *at least* one number can be computed directly by calculating the 10 individual probabilities $P(\text{Match } 1), \ldots, P(\text{Match } 10)$ and adding them together, or, more easily, by using the Complement Rule to recast the question as

$$P(\text{Match at least } 1) = 1 - P(\text{Match } 0),$$

and replacing 10 calculations by one. ■

To progress further, we need to develop procedures for assigning numerical probabilities to events.

**Definition 2.7.** Let **S** be a sample space in which all of the outcomes are equally likely, and suppose $A$ is an event within **S**. The *probability* of the event $A$, denoted $P(A)$, is

$$P(A) = \frac{\text{Number of elements in } A}{\text{Number of elements in } \mathbf{S}} = \frac{\#(A)}{\#(\mathbf{S})}.$$

Since the size of any event is necessarily less than or equal to the size of the sample space under consideration, the following theorem is a consequence of Definition 2.7.

**Theorem 2.2.** *For any event $A$, $0 \leqslant P(A) \leqslant 1$.*

There are several ways by which we might determine the value of $P(A)$. These methods vary in their mathematical complexity as well as in their level of precision. Each of them corresponds to a question we might ask or try to answer about a given probabilistic situation.

1. **Theoretical Probability**

   If we are asking the question *"What's supposed to happen?"* and relying on pure mathematical reasoning rather than on accumulated data, then we are computing the *theoretical probability* of an event.

   **Example 2.16.** For a single Pick 3 lottery ticket, there are 1000 possible outcomes, only one of which is a winner. The theoretical probability of winning is then $\frac{1}{1000}$. ■

   **Example 2.17.** If we roll 2d6, what is the probability of getting a sum of 7?

   An incorrect approach to this problem is to note that the sample space is $\mathbf{S} = \{2, 3, 4, 5, 6, 7, 8, 9, 10, 11, 12\}$, and since one of those 11 outcomes is 7, the probability must be $\frac{1}{11}$. This fails to take into account the fact that some rolls occur more frequently than others—for example, while there is only one way to roll a 2, there are 6 ways to roll a 7: 1-6, 2-5, 3-4, 4-3, 5-1, and 6-1. (It may be useful to think of the dice as being different colors, so that 3-4 is a different roll from 4-3, even though the numbers showing are the same.) Counting up all of the possibilities shows that there are $6 \cdot 6 = 36$ ways for two dice to land. Since six of those yield a sum of 7, the correct answer is $P(7) = \frac{6}{36} = \frac{1}{6}$. ■

2. **Experimental Probability**

   When our probability calculations are based on actual data drawn from repeated observations, the resulting value is the *experimental probability* of $A$. Here, we are answering the question *"What really did happen?"*

**Example 2.18.** Suppose that you toss a coin 100 times and that the result of this experiment is 48 heads and 52 tails. The experimental probability of heads in this experiment is $48/100 = .48$, and the experimental probability of tails is $52/100 = .52$. ■

This experimental probability is different from the theoretical probability of getting heads on a single toss, which is ½. This is not unusual.

Experimental probability can be a useful tool in its own right, as when processing a quantity of empirical data, and also as an approximation to theoretical probability when the numbers involved are too large for easy handling.

The connection between theoretical and experimental probability is described in a mathematical result called the *Law of Large Numbers*, or LLN for short.

**Theorem 2.3.** ***(Law of Large Numbers)*** *Suppose an event has theoretical probability $p$. If $x$ is the number of times that the event occurs in a sequence of $n$ trials, then as the number of trials $n$ increases, the experimental probability $x/n$ approaches $p$.*

Informally, the LLN states that, in the long run, things happen in an experiment the way that theory says that they do. What is meant by "in the long run" is not a fixed number of trials, but will vary depending on the experiment. For some experiments, $n = 500$ may be a large number, but for others—particularly if the probability of success or failure is small—it may take far more trials before the experimental probabilities get acceptably close to the theoretical probabilities.

---

## 2.2 Addition Rules

Our next challenge will be to extend our understanding of probability to *compound* events: events that can be broken down into several simple events. We can find the probability of these simple events using techniques from Section 2.1; this section allows us to combine those probabilities correctly to find probabilities of more complicated events.

**Definition 2.8.** Two events $A$ and $B$ are *mutually exclusive* if they have no elements in common—that is, if they cannot occur together.

**Example 2.19.** For a keno ticket with 6 numbers selected, the two events $A$ = "Match 5 numbers" and $B$ = "Match 6 numbers" are mutually exclusive. One or the other, or neither, can occur in a given single drawing, but not both. ■

**Example 2.20.** Most instant lottery tickets offer a variety of possible prizes, but any individual ticket can only win one prize. The two events $A =$ "Win a free ticket" and $B =$ "Win \$5" are mutually exclusive. ■

In computing probabilities, we may be in a situation where we know $P(A)$ and $P(B)$ and want to know the probability that either $A$ or $B$ occurs: $P(A \text{ or } B)$. The addition rules described next allow us to compute this new probability in terms of the known ones.

**Theorem 2.4.** ***(First Addition Rule)*** *If A and B are mutually exclusive events, then*

$$P(A \textit{ or } B) = P(A) + P(B).$$

If $A$ and $B$ are not mutually exclusive, a slightly more complicated formula can be used to calculate $P(A \text{ or } B)$.

**Theorem 2.5.** ***(Second Addition Rule)*** *If A and B are any two events, then*

$$P(A \textit{ or } B) = P(A) + P(B) - P(A \textit{ and } B).$$

*Proof.* By definition,

$$P(A \text{ or } B) = \frac{\#(A \text{ or } B)}{\#(\mathbf{S})}.$$

What we need to do is compute $\#(A \text{ or } B)$. Elements of $A$ or $B$ can be counted by adding together the number of elements of $A$ and of $B$, but if any elements belong to both, they have just been counted twice. In order that each element is only counted once, we must subtract out the number of elements that belong to both $A$ and $B$. This gives

$$\#(A \text{ or } B) = \#(A) + \#(B) - \#(A \text{ and } B).$$

Dividing by $\#(\mathbf{S})$ completes the proof. □

We can see that the First Addition Rule is a special case of the second, for if $A$ and $B$ are mutually exclusive, then they cannot occur together; hence $P(A \text{ and } B) = 0$.

**Example 2.21.** The probability of winning something on a Pick 5 keno ticket played against Table 1.4 involves the mutually exclusive probabilities of matching 3, 4, or 5 numbers, and may be written

$$P\text{Win}) = P(\text{Match } 3) + P(\text{Match } 4) + P(\text{Match } 5).$$

■

## Odds

In the world of gambling, probabilities are often encountered in terms of *odds.*

**Definition 2.9.** The *odds against* an event $A$ is the ratio $P(A') : P(A)$, or $P(A')/P(A)$.

This is often stated in a form like "$x$ to 1," as is the case in horse racing, for example. Most of the time when odds are quoted, they are odds against. We can also consider the odds in favor of, or *odds for*, an event.

**Definition 2.10.** The *odds for* an event $A$ is the ratio $P(A) : P(A')$—the reciprocal of the odds against $A$.

**Example 2.22.** When rolling 2d6, *snake eyes* is the name given to a roll of 1-1. Since there are 36 ways for the dice to land, and only one is a 1-1, the probability of snake eyes is 1/36. The odds against snake eyes would then be

$$\frac{35}{36} : \frac{1}{36},$$

or 35 to 1. ■

Contrary to common belief, a casino or lottery agency doesn't make its profit from wagers collected from losing bettors, but rather from paying off winners at less than true odds.

**Example 2.23.** A Pick 1 keno ticket pays off if the player's number is among the 20 numbers drawn. Since there are 80 numbers in play, the probability of winning is

$$p = \frac{20}{80} = \frac{1}{4}:$$

and the corresponding odds against winning are ¾ : ¼, or 3–1. However, since a winning ticket pays off at only 3 *for* 1 rather than 3 to 1, a net gain of twice the amount wagered, the payoff is made at less than true odds. To win an effective 2 to 1 payoff including the return of the original wager, it's necessary to beat odds of 3 to 1 against. ■

We can rearrange the formula for odds and derive the following result.

**Theorem 2.6.** *If the odds against an event $A$ are $x$ to 1, then* $P(A) = \dfrac{1}{x+1}$.

*Proof.* We have

$$\frac{1 - P(A)}{P(A)} = \frac{x}{1}.$$

Cross-multiplying gives

$$1 - P(A) = x \cdot P(A),$$

or

$$(x+1)\cdot P(A) = 1.$$

Dividing by $x+1$ gives

$$P(A) = \frac{1}{x+1},$$

completing the proof. □

---

## 2.3 Conditional Probability

**Definition 2.11.** Two events $A$ and $B$ are *independent* if the occurrence of one has no effect on the occurrence of the other one.

Two events that are mutually exclusive (Section 2.2) are explicitly *not* independent, since the occurrence of one eliminates the chance of the other occurring. Moreover, two events that are independent cannot be mutually exclusive.

**Example 2.24.** For the Maryland Lottery's Daily 3 drawing, the winning number is not drawn as a three-digit number, but as three separate digits drawn individually with ping-pong balls extracted from three air blowers, each containing balls numbered 0–9. We may conclude that the values of the different digits are independent. The probability of the third digit being 0 is not affected by any 0s that may or may not have occurred among the first two digits. ■

It is a fundamental principle of gambling mathematics that *successive trials of random experiments are independent.* This includes successive die rolls at craps, successive wheel spins at roulette, and successive weekly drawings of six Powerball numbers, but *not* successive hands in blackjack—for in blackjack, a card played in one hand is a card that cannot be played in the next hand. Since the composition of the deck has changed, we are not considering successive trials of the same random experiment.

This principle is not always well-understood by gamblers, and the inability or unwillingness to understand the doctrine of independent trials is sometimes called the *Gambler's Fallacy.* This fallacy is commonly committed by roulette players who have too strong a belief in the Law of Large Numbers, although it can crop up in any game where successive trials are independent.

**Example 2.25.** The Michigan Lottery maintains a Web page (https://www.michiganlottery.com/hot_cold) offering up lists of "hot" and "cold" numbers: numbers that have been drawn most frequently or least frequently in the recent history of its draw games. While some people use these lists as an aid to picking their numbers, in the belief that hot numbers are on a roll and more likely to be drawn again, there is no mathematical advantage to doing so. All possible lottery numbers are equally likely; this equiprobability is key to the success of lotteries. ■

Keno balls and lottery devices don't understand the laws of probability. They have no knowledge of the mathematics we humans have devised to describe their actions, and they certainly don't understand what the long-term distribution of results is supposed to be. For the same reason, in Example 2.25, it would be equally erroneous to bet on the "cold" numbers on the grounds that they're somehow "due".

If $A$ and $B$ are independent events, it is a simple matter to compute the probability that they occur together, with the use of a theorem called the *Multiplication Rule.*

**Theorem 2.7.** ***(Multiplication Rule)*** *If $A$ and $B$ are independent events, then*

$$P(A \text{ and } B) = P(A) \cdot P(B).$$

Informally, the Multiplication Rule states that we can find the probability that two successive independent events occur by multiplying the probability of the first by the probability of the second. The Multiplication Rule can be extended to any finite number of independent events: the probability of a sequence of $n$ independent events is simply the product of the $n$ probabilities of the individual events.

**Example 2.26.** In many state lotteries' Daily 4 drawings, the machines used to draw each digit are separate, and so the individual digits are independent of one another. A wager on a single number, such as 1729, has 1 chance in 10,000 of winning; this can be seen by looking at the probabilities of the four digits:

$$P(\text{Win}) = \left(\frac{1}{10}\right)^4 = \frac{1}{10^4} = \frac{1}{10,000}$$

—just what we calculate by thinking of the number 1729 as one number among 10,000 possibilities. ■

**Example 2.27.** Example 2.14 described New Jersey's Dollar Throwdown game, which simulates 10 games of rock/paper/scissors pitting the gambler against a computer. The probability of winning any one game of RPS is 1/3, since there are 3 equally likely outcomes: win, lose, and tie. If the 10 games are independent, the probability of winning all 10 games is

$$\left(\frac{1}{3}\right)^{10} = \frac{1}{59,049}.$$

If the game simply generated 10 consecutive independent matchups, this would be the probability of winning the $500 top prize.

In practice, the probability of a ticket with 10 wins is governed by the game's programming, and is fixed at 1/240,000 [108]. ■

If the events $A$ and $B$ are not independent, we will need to generalize Theorem 2.7 to handle the new situation. This generalization requires the idea of *conditional probability.* We begin with an example.

**Example 2.28.** If we draw one card from a standard deck, the probability that it is a king is $\frac{4}{52} = \frac{1}{13}$. If, however, we are told that the card is a face card, the probability that it's a king is $\frac{4}{12} = \frac{1}{3}$—that is, additional information has changed the probability of our event by allowing us to restrict the sample space. If we denote the events "The card is a king" by $K$ and "The card is a face card" by $F$, this last result is written $P(K|F) = \frac{1}{3}$ and read as "the (conditional) probability of $K$ given $F$ is $\frac{1}{3}$." ■

**Example 2.29.** In Example 2.26, the probability of 1729 being the winning number in a Daily 4 lottery drawing was found to be 1/10,000. If the first digit is drawn and found to be 1, the probability that 1729 will win has fallen to 1/1000. Of course, if the first digit is drawn and is not 1, the probability of 1729 winning has dropped all the way to 0. ■

The fundamental idea here is that more information can change probabilities. If we know that the event $A$ has occurred and we're interested in the event $B$, we are now not looking for $P(B)$, but $P(B \text{ and } A)$, because only the part of $B$ that overlaps with $A$ is possible. With that in mind, we have the following formula for conditional probability:

**Definition 2.12.** The *conditional probability* of $B$ given $A$ is

$$P(B \mid A) = \frac{P(B \text{ and } A)}{P(A)}.$$

This formula divides the probability of the intersection of the two events by the probability of the event that we know has already occurred. Note that if $A$ and $B$ are independent, we immediately have $P(B \mid A) = P(B)$, since then $P(B \text{ and } A) = P(A) \cdot P(B)$. This is one case where more information—in this case, the knowledge that $A$ has occurred—does not change the probability of $B$ occurring.

**Example 2.30.** When buying a lottery ticket for two consecutive Wednesday night drawings, the outcome on the first Wednesday has no effect on the results of the second Wednesday's drawing. Winning (or losing) the jackpot on the first drawing neither raises nor lowers the probability of winning the second drawing, since the two drawings are independent. ■

As with the addition rules, we can state a second, more general, version of the Multiplication Rule that applies to any two events—independent or not—and reduces to the first rule when the events are independent. This more general rule simply incorporates the conditional probability of $B$ given $A$, since we are looking for the probability that both occur.

**Theorem 2.8.** ***(General Multiplication Rule)*** *For any two events A and B, we have*

$$P(A \text{ and } B) = P(A) \cdot P(B|A).$$

*Proof.* This result follows from the fact that $P(A \text{ and } B) = P(B \text{ and } A)$ and from Definition 2.12. □

---

## 2.4 Random Variables and Expected Value

**Definition 2.13.** A *random variable* (RV for short) is an unknown quantity $X$ whose value is determined by a chance process.

This is another definition that, on its face, isn't terribly useful—indeed, this phrasing comes perilously close to using the words "random" and "variable" in the definition of random variable. Once again, a sequence of examples will illustrate this important idea far better than a formal definition.

**Example 2.31.** Roll 2d6 and let $X$ denote their sum. $X$ then takes on a value in the set $\{2, 3, 4, 5, 6, 7, 8, 9, 10, 11, 12\}$. ■

**Example 2.32.** In a five-card poker hand, let $X$ count the number of aces it contains. $X = 0$, 1, 2, 3, or 4. ■

**Example 2.33.** In a hand of blackjack, let $X$ denote the sum of the first two cards, counting the first ace as 11. Here, $X$ lies in the set $\{3, 4, 5, 6, \ldots, 19, 20, 21\}$. (A hand containing two aces would be counted here as 12, not 2 or 22.) ■

**Example 2.34.** Let $X$ be the number spun on a European roulette wheel. Then $X \in \{0, 1, 2, 3, \ldots, 35, 36\}$, a set with 37 elements. ■

**Definition 2.14.** A *probability distribution* for a random variable $X$ is a list of the possible values of $X$, together with their associated probabilities.

Probability distributions are most commonly presented in a table of values or as an algebraic formula, which is called a *probability distribution function* or *PDF*.

**Example 2.35.** Example 1.1 described a simple Numbers Game that called for gamblers to pick a 3-digit number and paid 600 for 1 if their number was the winner. Defining $X$ as the amount won on a \$1 wager gives the probability distribution in Table 2.1.

TABLE 2.1: Probability distribution for 3-digit Numbers Game

| $x$ | $P(X = x)$ |
|---|---|
| \$0 | .999 |
| \$600 | .001 |

Note that the two outcomes are not equally likely and that the cost of the ticket is not figured into the value of $X$. ■

**Example 2.36.** If $X$ denotes the number that appears when a fair d6 is rolled, then the probability distribution for $X$ is shown in Table 2.2.

TABLE 2.2: Probability distribution when a d6 is rolled

| $x$ | $P(X = x)$ |
|---|---|
| 1 | 1/6 |
| 2 | 1/6 |
| 3 | 1/6 |
| 4 | 1/6 |
| 5 | 1/6 |
| 6 | 1/6 |

Since the probabilities in the table are all the same, the simple events listed are equally likely. ■

**Example 2.37.** The Meskwaki Casino in Tama, Iowa offers a "Super 20 Special" keno game whose brochure advertises "18-Out-Of-21-Ways To Win," For a minimum bet of \$5, the player picks 20 numbers in the range 1–80 and wins unless 4, 5, or 6 of the numbers selected are caught—so indeed 18 out of the 21 possible numbers of catches are winners, although catching 2, 3, 7, or 8 numbers merely gets the players their money back. Let $X$ count the number of matches. A probability distribution for $X$ would be a list of the probabilities for catching 0 through 20 numbers; we will develop the formula for this probability in Chapter 3. ■

## Expected Value and House Advantage

The notion of *expected value* is fundamental to any discussion of random variables and is especially important when those random variables arise from a game of chance. The expected value of a random variable $X$ is, in some sense, an "average" value, or what we might expect in the long run if we were to sample many values of $X$.

The common notion of "average" corresponds to what statisticians call the *mean* of a set of numbers: add up all of the numbers and divide by how many numbers there are. For a random variable $X$, this approach requires some fine-tuning, as there is no guarantee that a small sample of values of $X$ will be representative of the range of possible values. Our interpretation of average will incorporate each possible value of $X$ together with its probability, computing what is in some sense a long-term average over a very large hypothetical sample.

**Definition 2.15.** Let $X$ be a random variable with a given probability distribution function $P(X = x)$. The *expected value* or *expectation* $E(X)$ of $X$ is computed by multiplying each possible value for $X$ by its corresponding probability and then adding the resulting products:

$$E(X) = \sum_x x \cdot P(X = x).$$

This expression may be interpreted as a standard mathematical mean drawn from an infinitely large random sample. If we were to draw such a sample, we would expect that the *proportion* of sample elements with the value $x$ would be $P(X = x)$; adding up over all values of $x$ gives this formula for $E(X)$.

We may abbreviate $E(X)$ to $E$ when the random variable is clearly understood. The notation $\mu = E(X)$, where $\mu$ is the Greek letter mu, is also common, particularly when the expected value appears as a term in another expression, as in Definition 2.23 (page 53).

**Example 2.38.** In 1693, Francis Roberts described a paradox arising from the expected return in two simple lotteries with six tickets apiece [133].

- Lottery A comprised three blanks and three prizes of 16 pence (d).
- Lottery B was composed of four blanks and two prizes of 2 shillings (24 d).

Tickets in each hypothetical lottery cost 1 shilling, or 12 pence. Deducting the cost of the ticket from the prize, a winner's profit in Lottery A was 4 pence and in Lottery B, 12 pence.

The expected value of a single ticket in Lottery A, in pence, was then

$$E = (4) \cdot \frac{3}{6} + (-12) \cdot \frac{3}{6} = -4\text{d}.$$

For Lottery B, the expectation was

$$E = (12) \cdot \frac{2}{6} + (-12) \cdot \frac{4}{6} = -4\text{d},$$

and so the paradox arises. The expected return on the two tickets is the same, even though the odds against winning are different: 1–1 for Lottery A and 2–1 for Lottery B.

We recognize that this paradox is easily resolved by noting that the prize in Lottery B is the larger, by an amount that exactly balances the lower odds of winning. ■

The notation used in Definition 2.15 does not indicate the limits of the indexing variable $x$, as is customary with sums; this is because those values may not be a simple list running from 1 to some $n$. When written this way, we should take the sum over *all* possible values of the random variable $X$, as in the next example.

**Example 2.39.** A bet on a single number at American roulette pays 35–1 if the chosen number is spun. The probability of winning is $\frac{1}{38}$, since there are 38 pockets on the wheel (numbered 1–36, 0, and 00). A random variable $X$ measuring the outcome of a \$1 bet can take on the two values 35 and –1, and the expected value of $X$ is then

$$E = (35) \cdot \frac{1}{38} + (-1) \cdot \frac{37}{38} = -\frac{1}{19} \approx -\$.0526,$$

or about –5.26¢. ■

A concept related to expected value is the *house advantage* associated with a game of chance.

**Definition 2.16.** The *house advantage (HA)* of a game with a wager of $N$ and payoffs given by the random variable $X$ is $-E(X)/N$.

If the expectation is negative, as it is in virtually every casino game, the HA will be positive. The house advantage of a game is frequently expressed as a percentage of the original wager, so the \$1 roulette wager above can be said to have a house advantage of 5.26%. The HA measures how much of the total amount wagered can be reliably expected to be retained by the casino or lottery agent, in the long run.

**Example 2.40.** In the Newbury lottery described on page 11, the lottery kept \$1000 from \$10,000 in ticket sales and paid out the rest in prizes. The house advantage would then be 10%: the ratio of the profit to the total sales. ■

**Definition 2.17.** If $X$ is a random variable measuring the payoffs from a game, we say that the game is *fair* if $E(X) = 0$.

**Example 2.41.** Suppose you gamble with a friend on the toss of a coin. If heads is tossed, you win \$1; if tails is tossed, you pay \$1. Since a fair coin can be expected to land heads and tails equally often, the expected value for this game is $E = (1)(½) + (-1)(½) = 0$. The game is fair. ■

If a game is fair, then in the long run, we expect to win exactly as much money as we lose, and thus, aside from any possible entertainment derived from playing, we expect no gain. This is often summarized in the following maxim:

*If a game is fair, don't bother to play.*
*If a game is unfair, make sure it's unfair in your favor.*

Failure to heed this maxim, of course, is responsible for the ongoing success of the gambling industry, for games which are unfair and favor the gambler are rare. Table 2.3 lists the house advantages for several popular wagers.

TABLE 2.3: House advantages for common wagers

| Wager | HA |
|---|---|
| Blackjack, single-deck, Las Vegas Strip rules | ~ 0.00% |
| Baccarat, Banker bet with 5% commission | 1.06% |
| Baccarat, Player bet | 1.23% |
| Craps, don't pass or don't come line | 1.34% |
| Craps, pass or come line | 1.41% |
| European roulette, all bets | 2.70% |
| Sports betting, one game with 10–11 payoff | 4.55% |
| Caveman Keno, Pick 2–10 games | ~ 5.00% |
| Game King video keno, Pick 6 game | 5.01% |
| American roulette, all bets except 5-number basket bet | 5.26% |
| Craps, field bet | 5.56% |
| American roulette, 5-number basket bet | 7.89% |
| Craps, hardway bet on 6 or 8 | 9.09% |
| Baccarat, Tie bet | 14.05% |
| Craps, Big Red/Any Seven bet | 16.67% |
| Pák kòp piú, with 5% commission and Table 1.1 as pay table | 24.04% |
| Keno, Pick 1 ticket paying 3 for 1 | 25.00% |
| Keno, Meskwaki Casino Super 20 Special | 34.60% |
| Caribbean Keno Pick 6 game | 37.92% |
| Michigan Lottery Daily 3 straight bet | 50.00% |
| Michigan Lottery Daily 3 box bet | 50.20% |
| Numbers Game: You Pick 'Em Treasury ticket (page 14) | 99.13% |

Some conclusions may be immediately drawn from the table:

- The best wagers are generally found on casino table games. Some gaming experts recommend avoiding any bet with a house edge over 2%; this eliminates all wagers in roulette as well as keno and lottery tickets. However, not every bet on a game such as craps or baccarat has the same HA.
- Video keno appears to be better for the player than live keno.
- Live keno and lotteries have some of the highest HAs of any legitimate casino game. Some of the reasons for these high HAs will be explored in Chapters 3 and 4.

In 1991, *Keno Newsletter* classified keno house advantages into three categories and simultaneously made some judgments about types of keno players [67]:

| | |
|---|---|
| 25% or less | Locals & Frequent keno players |
| 26–32% | Tourist games |
| 33% & more | Sucker games |

While the lottery bets shown in this table are exceptionally bad bets from the gambler's perspective, one fact that must be addressed here is that state lotteries are intended in part as a fundraising mechanism. Since its inception in 1972, the Michigan Lottery has contributed over \$16 billion to the state's public schools, and that number is far greater than it would be if players were getting a 95% return on their lottery dollars [96]. A 50% house advantage would be untenable in a casino, but is readily accepted in a state lottery when it is understood that a portion of the proceeds is supporting education.

In Example 2.39, a winning player's wager is returned with the payout; this is standard practice in most casino games, but is not the case with a keno bet or lottery ticket. Practical considerations dictate this difference: it makes sense for the casino or lottery agent to collect a wager when the ticket is purchased, and it may be some time before the bet is resolved. For these wagers, we have two options in accounting for the ticket price in expected value calculations:

- Use only the net payoffs in the calculations by reducing each payoff by the cost of a ticket, as in Example 2.38. This includes a negative payoff covering the case where the bet loses.
- Multiply the full payoffs by their corresponding probabilities, add up those products, and subtract the cost of the ticket at the end. When using this method, it is not necessary to compute the probability of losing.

**Example 2.42.** In Example 1.1, we considered a simple Numbers Game where players choose a 3-digit number and receive a 600 for 1 payoff if theirs is the winning number. The PDF for this game was shown in Table 2.1. Using the first method above, the possible net outcomes of a \$1 wager are \$599 and –\$1, and the expected value is then

$$E = (599) \cdot \frac{1}{1000} + (-1) \cdot \frac{999}{1000} = -\$0.40,$$

so the agents running the game hold a 40% house advantage. ■

If we used the second method of computing $E$ noted above, we would calculate

$$E = (600) \cdot \frac{1}{1000} - 1 = -\$0.40.$$

Since this approach, deducting the ticket price at the end, is computationally somewhat easier, we will use this method going forward.

**Example 2.43.** The Numbers Game wasn't always this bad a bet—sometimes it was much worse. In order to limit their exposure in case the winning number was especially popular, some Numbers Game operators offered lower payouts on the most popular numbers. One such game in New York City paid only 400 for 1 on the numbers 111, 222, and 100 [46]. A wager on one of these three numbers carried a house advantage of 60%. ■

**Example 2.44.** Some instant lottery games include a free ticket among the array of prizes. For example, consider a $1 instant ticket with the prize structure shown in Table 2.4.

TABLE 2.4: Hypothetical instant lottery ticket pay table

| Prize | Probability |
|---|---|
| $100 | .001 |
| $10 | .049 |
| $1 | .100 |
| Free ticket | .200 |
| Nothing | .650 |

How do we account for the free ticket prize when computing the expected value of this $1 ticket?

If the expected value of a ticket is $E$, then the value of a free ticket as a prize is $E+1$—the value of the ticket without subtracting its cost. Computing the expected value gives a linear equation in $E$:

$$E = (100) \cdot \frac{1}{1000} + (10) \cdot \frac{49}{1000} + (1) \cdot \frac{100}{1000} + (E+1) \cdot \frac{200}{1000} - 1$$

or

$$E = \frac{200E - 110}{1000}.$$

This equation has solution $E = -\$.1375$, which gives a house advantage of 13.75%. ■

An important principle of expected value is contained in the following theorem.

**Theorem 2.9.** *If $X_1, X_2, \ldots, X_n$ are random variables, then*

$$E(X_1 + X_2 + \cdots + X_n) = E(X_1) + E(X_2) + \cdots + E(X_n).$$

Another way to state this result is to say that *expectation is additive.*

**Example 2.45.** Theorem 2.9 can be used to show that buying more lottery tickets for a given drawing, while it may increase your probability of winning, does not change the house advantage. In 1970, it was proposed that the U.S.

government establish its own simple Numbers Game paying off 750 for 1 for the winning three-digit number. This payoff, better than the prevailing 600 for 1 payoff on offer in many cities, was suggested as an effort to eliminate the influence of organized crime [64]. Buying a single $1 ticket for this hypothetical game would give an expected value of

$$E = (750) \cdot \frac{1}{1000} - 1 = -\$0.25,$$

and so the house advantage is 25%.

If instead you purchase ten tickets, each one on a different number, the expectation of the combined wager is

$$E = (750) \cdot \frac{10}{1000} - 10 = -\$2.50.$$

This is simply 10 times the expected value of a single ticket, as Theorem 2.9 states. Dividing by the $10 risked gives the same HA, 25%, as when buying a single ticket. ■

### Delaware Sports Lottery

In the USA in 2017, betting on professional sports was only legal in Delaware, Montana, Nevada, and Oregon: the four states where sports betting was established prior to the 1992 passage of the Professional and Amateur Sports Protection Act, which banned sports wagering in states where it was not already legal. Delaware has incorporated professional sports into its Sports Lottery offerings. During the National Football League season, the state offers *parlay cards*, which challenge the bettor to pick the winner of 3–12 games against a point spread. Players may choose the number of games they pick, but only win if they correctly predict each game.

The *Early Bird* parlay card is played, as the name suggests, early in the week: on Monday and Tuesday during football season. The point spreads are all set to guarantee no ties by using half-points in each declared point spread, as, for example, when the Miami Dolphins are listed as a 4½-point favorite over the Jacksonville Jaguars. If Jacksonville's score in this game has 4½ points added to it for wagering purposes, there is no way that the decision can end in a tie. Players may choose their wager subject to a $2 minimum and a maximum determined by lottery management. The pay table is shown in Table 2.5.

**Example 2.46.** Under the assumption that the point spreads in each game are set to encourage equal action on both teams, as is the case in Nevada sports books, which of these ten wagers shown in Table 2.5 is the best for the player?

Equal action implies that the average player has a 50% chance of picking each game correctly against the spread. Accordingly, under the reasonable

TABLE 2.5: Delaware Early Bird parlay card pay table

| **Pick, $n$** | **Payoff, $A(n)$** |
|---|---|
| 3 of 3 | 6½ for 1 |
| 4 of 4 | 11 for 1 |
| 5 of 5 | 20 for 1 |
| 6 of 6 | 40 for 1 |
| 7 of 7 | 75 for 1 |
| 8 of 8 | 150 for 1 |
| 9 of 9 | 375 for 1 |
| 10 of 10 | 800 for 1 |
| 11 of 11 | 1400 for 1 |
| 12 of 12 | 2500 for 1 |

assumption that the games are independent, the probability of picking all $n$ games correctly is $\left(\frac{1}{2}\right)^n$. Assume the minimum \$2 wager for convenience. The expected value is then a function of $n$:

$$E(n) = 2A(n) \cdot \left(\frac{1}{2}\right)^n - 2,$$

where $A(n)$ is the payoff given in Table 2.5.

Computing the expectation for $n$ = 3–12 reveals that the smallest house advantage, 18.75%, is found when picking 3 games. See Figure 2.1.

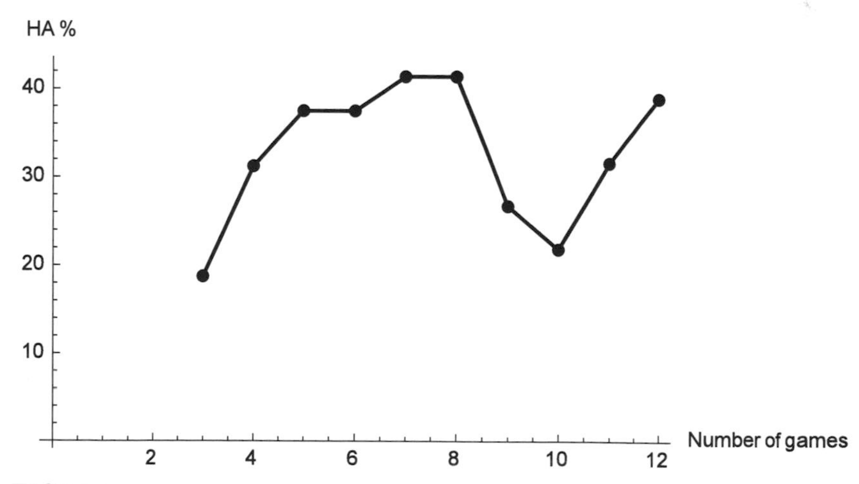


FIGURE 2.1: House advantage for Delaware Sports Lottery parlay cards.

The worst bet, with a HA of 41.41%, is found when picking either 7 or 8 games; since the prize doubles in moving from 7 to 8 games while the probability of winning is halved, the expectation remains thc same. ■

Sports betting is one of the few gambling games where skill plays a part in determining the outcome, and so if a gambler could consistently pick winners with probability greater than 50%, it would be possible to turn this game into one with a player edge. Let $p$ be the probability of picking a single game correctly. Under the assumption that picks in successive games are independent of one another, the expectation then becomes

$$E(n) = 2A(n) \cdot p^n - 2,$$

which is zero when

$$p = \sqrt[n]{\frac{1}{A(n)}}.$$

Using the tabulated values of $A(n)$ shows that the Pick 10 game has the lowest break-even probability: .5125. If you can pick winners at better than this rate, you can turn the Early Bird parlay into a positive-expectation wager.

**Example 2.47.** Delaware also offers a \$100,000 Parlay Card game which pays off \$100,000 for a \$5 bet if a player can pick 15 games out of 15 correctly against the point spread. This game has a house advantage of 38.96%, in line with Early Bird, and a break-even probability of .5167. ■

The 51.25%, or 51.67%, doesn't seem like an unreasonable barrier to overcome to turn Delaware's parlay cards in your favor. If you can flip a coin to predict winners at a 50% rate, it ought to be easy to study the games and increase your predicting acumen past the break-even probability.

That is no guarantee of success at the lottery game, though. Suppose that you could predict winners at a 60% rate. While that sounds good, your expected number of correct predictions is 60% of the number of games you select, and winning this parlay bet requires getting 100% correct. Even at 60% proficiency, your chance of winning on a Pick 10 parlay card is only

$$(.60)^{10} \approx \frac{1}{165}.$$

To have a 50% chance of winning at least once, you would need to bet $n$ tickets, where $n$ is the solution to

$$1 - \left[1 - (.60)^{10}\right]^n = \frac{1}{2},$$

which is

$$n = \frac{\ln .5}{\ln\left(1 - .60^{10}\right)} \approx 115,$$

rounding up to the nearest integer. At one ticket per week—which is not a limit in force in Delaware—that's over 6½ NFL seasons for a 50% chance at winning.

## 2.5 Combinatorics

*Combinatorics* is the branch of mathematics that studies counting techniques. In many applications of Definition 2.7, $P(A) = \#(A)/\#(S)$, the sheer number of elements comprising an event or the sample space is far too large to count them one by one. When computing probabilities, we seldom have need to consider each of these simple events individually; we are usually only interested in how many there are. Frequently in gambling mathematics, we find ourselves considering the number of ways in which several events can happen in sequence. If we know the number of ways that each individual event can happen, elementary combinatorics tells us that simple multiplication can be used to find the answer.

### Fundamental Counting Principle

**Theorem 2.10.** ***(Fundamental Counting Principle)*** *If there are $n$ independent tasks to be performed, such that task $T_1$ can be performed in $m_1$ ways, task $T_2$ can be performed in $m_2$ ways, and so on, then the number of ways in which all $n$ tasks can be performed successively is*

$$N = m_1 \cdot m_2 \cdot \ldots \cdot m_n.$$

That the Fundamental Counting Principle (FCP for short) is a reasonable result can be easily seen by testing out some examples with small numbers and listing all possibilities—for example, when rolling 2d6, one red and one green, each die is independent of the other and can land in any of six ways. By the FCP, there are $6 \cdot 6 = 36$ ways for the two dice to fall, and this may be confirmed by writing out all of the possibilities.

**Example 2.48.** In the You Pick 'Em Treasury Ticket described on page 14, the winning number has 5 digits. Since any digit from 0–9 may appear in any of the 5 positions, and since digits may be repeated, there are $10^5 = 100,000$ different possible winning numbers.

If repeated digits were not allowed in the winning number, there would be $10 \cdot 9 \cdot 8 \cdot 7 \cdot 6 = 30,240$ possible winning numbers. This would require a different impartial source for the winning number than the daily U.S. Treasury balance, which can repeat digits. ■

As an example of how the FCP is used, we can show that the number of possible events associated with a probability experiment is exponentially greater than the number of elements. Specifically, we have the following result.

**Theorem 2.11.** *If $\#(\boldsymbol{S}) = n$, then there are $2^n$ events that may be chosen from $\boldsymbol{S}$.*

Another way to state Theorem 2.11 is to say that a set with $n$ elements has $2^n$ subsets. For example, if we toss a coin, we have $\mathbf{S} = \{H, T\}$. Since we have $\#(\mathbf{S}) = 2$, we expect $2^2 = 4$ events. The complete list of events contained within $\mathbf{S}$ is $\varnothing, \{H\}, \{T\}, \{H, T\}$, which indeed numbers 4.

*Proof.* Let $A$ be an event contained in a sample space $\mathbf{S}$. For every element $x$ of the sample space, either $x$ belongs to $A$ or $x$ does not belong to $A$. That is, there are two possibilities for each element: "Yes, it is in $A$" or "No, it is not in $A$." Every arrangement of the $n$ possible yes and no answers corresponds to a different subset of $\mathbf{S}$, and thus to a different event. For example, choosing "no" for every element gives $\varnothing$, and at the other extreme, choosing "yes" each time gives $\mathbf{S}$ itself.

There are $n$ choices to be made, and two options for each of the $n$ choices. Therefore, by the Fundamental Counting Principle, there are

$$\underbrace{2 \cdot 2 \cdot \ldots \cdot 2}_{n \text{ terms}} = 2^n$$

possible events. □

A special case of the Fundamental Counting Principle arises when we consider the number of ways to arrange a set of $n$ elements, with no repetition allowed, in different orders. The first element may be chosen in $n$ ways, the second in $n - 1$, and so on, down to the last item, which may be chosen in only one way. The total number of orders for a set of $n$ elements is thus $N = n \cdot (n-1) \cdot (n-2) \cdot \ldots \cdot 3 \cdot 2 \cdot 1$. This number is given a special name, $n$ *factorial.*

**Definition 2.18.** If $n$ is a natural number, the *factorial* of $n$, denoted $n!$, is the product of all of the positive integers up to and including $n$:

$$n! = 1 \cdot 2 \cdot 3 \cdot \ldots \cdot (n-1) \cdot n.$$

$0! = 1$, by definition.

It is an immediate consequence of the definition that $n! = n \cdot (n-1)!$. Factorials get very big very fast. $4! = 24$, but then $5! = 120$ and $6! = 720$. $10!$ is greater than 3 million, and $52!$, which is the number of different ways to arrange a standard deck of cards, is approximately $8.066 \times 10^{67}$. In the 1970s and 1980s, the largest factorial that could be computed on a standard scientific calculator was $69! \approx 1.711 \times 10^{98}$, a limit well-known to many people whose interest in mathematics was generated in part by playing with early calculators.

## Where Order Matters: Permutations

**Definition 2.19.** A *permutation* of $r$ items from a set of $n$ items is a selection of $r$ items chosen so that the order matters.

For example, ABC is a different choice of three alphabet letters than CBA. It should be noted that "order" may appear in several forms. One way to determine whether or not order matters in making a selection is to ask if different elements of the selection are being treated differently once they are chosen.

**Example 2.49.** Most games of chance do not require that numbers, playing cards, or other game elements be chosen in order. One that did was *Mix & Match*, a Pennsylvania Lottery offering that ran from 2007–2010. Mix & Match called for players to choose 5 numbers in the range 1–19. The state chose 5 numbers twice weekly, and prizes were awarded to players whose tickets matched any of these numbers in the same order as the state's (the "match") payoff) or who matched 3 or more numbers without regard to order (the "mix") [126]. ■

**Example 2.50.** Another game in which the order of a selection mattered was *City Picks*, which was operated by the Wisconsin Lottery from March 2002 to February 2003. The object of City Picks was to arrange a list of nine Wisconsin cities in order and to match the order determined by the state in its nightly drawing [162]. The nine cities, chosen from throughout the state, were

| | | |
|---|---|---|
| Chippewa Falls | Dodgeville | Green Bay |
| Kenosha | Madison | Milwaukee |
| Superior | Two Rivers | Wisconsin Rapids |

Payoffs were based on how many of the player's ordered cities matched the order drawn by the state. ■

We are usually not as interested in a list of all of the permutations of a set as in how many permutations there are. The following theorem allows easy calculation of that number, which is denoted ${}_nP_r$.

**Theorem 2.12.** *The number of permutations of $r$ items chosen from a set of $n$ items is*

$${}_nP_r = \frac{n!}{(n-r)!}.$$

*Proof.* There are $n$ ways to select the first item. Once an item is chosen, it cannot be chosen again, so the second item may be chosen in $n-1$ ways. There are then $n-2$ items remaining for the third choice, and so on until there are $n-r+1$ numbers remaining from which to choose the $r$th and final term. By the Fundamental Counting Principle, we have

$${}_nP_r = n \cdot (n-1) \cdot \ldots \cdot (n-r+1).$$

Multiplying the right-hand expression by $1 = (n-r)!/(n-r)!$ gives

$$
\begin{aligned}
{}_nP_r &= n \cdot (n-1) \cdot \ldots \cdot (n-r+1) \cdot \frac{(n-r)!}{(n-r)!} \\
&= \frac{n \cdot \ldots \cdot (n-r+1) \cdot (n-r) \cdot \ldots \cdot 3 \cdot 2 \cdot 1}{(n-r)!} \\
&= \frac{n!}{(n-r)!}.
\end{aligned}
$$

□

**Example 2.51.** Order mattered for Mix & Match tickets. Since 5 numbers were being chosen from 19, the number of possible tickets was

$$ {}_{19}P_5 = \frac{19!}{14!} = 1,395,360. $$

■

**Example 2.52.** Since City Picks required players to arrange all 9 cities in order, the number of possible selections was

$$ {}_9P_9 = \frac{9!}{0!} = 9! = 362,880. $$

■

## Where Order Doesn't Matter: Combinations

Most of the time when gambling, we are not so concerned about the order of events, as when a hand of cards is dealt or a set of Powerball numbers is drawn. For counting these arrangements, we are interested in *combinations* rather than permutations.

**Definition 2.20.** A *combination* of $r$ items from a set of $n$ items is a subset of $r$ items chosen without regard to order. The number of such combinations is denoted ${}_nC_r$. This value is sometimes denoted $\binom{n}{r}$, which is read as "$n$ choose $r$."

Here, ABC and CBA are interchangeable combinations, as they are subsets of the alphabet consisting of the same three letters. The different order is not a concern here. If the elements of a selected subset are receiving the same treatment once selected, then the choice is a combination, not a permutation.

**Theorem 2.13.** *The number of combinations of $r$ items chosen from a set of $n$ items is*

$$ {}_nC_r = \frac{n!}{(n-r)! \cdot r!} = \frac{{}_nP_r}{r!}. $$

*Proof.* We begin with the formula for the number of permutations:

$$_nP_r = \frac{n!}{(n-r)!}.$$

Since we are looking for combinations, two permutations that differ only in the order of the elements are identical to us. Any combination of $r$ elements from a set of $n$ can be rearranged into $r!$ different orders, by the Fundamental Counting Principle. We then have

$$_nC_r = \frac{_nP_r}{r!} = \frac{n!}{(n-r)! \cdot r!},$$

as desired. □

**Example 2.53.** On page 11, we considered the Queen's College Literature Lottery, where tickets bore three numbers from 1–35. The order of the numbers was not considered either in printing the tickets or awarding the various prizes, so the number of possible tickets was $_{35}C_3 = 6545$. These tickets were sold for \$3.50 apiece.

Five numbers were drawn by lottery agents. Suppose that these numbers, in order, were 21, 1, 32, 27, and 35. While the tickets used combinations rather than permutations, the order of the drawn numbers mattered in identifying the winners. Winning tickets were determined as follows [160]:

- The first prize, \$5000, went to the ticket with the first 3 numbers: 1, 21, and 32.
- Second prize of \$2000 was awarded to the ticket bearing the last 3 drawn numbers: 27, 32, and 35.
- Smaller top-end prizes ranging from \$610 to \$100 went to tickets matching any other 3 of the 5 numbers.
- The next prize of \$50 was awarded to any ticket matching only the 4th and 5th numbers: 27 and 35 without any of the other winning numbers.
- Tickets matching only the 2nd and 3rd numbers, 1 and 32, with one non-matching number, won \$25.
- Any other ticket matching only two numbers then won \$9.
- Finally, a ticket matching one number won \$4.50.

The probability of winning \$100 or more by matching three numbers is simply

$$\frac{_5C_3}{_{35}C_3} = \frac{10}{6545} = \frac{1}{654.5}$$

■

The following theorem collects several simple facts about combinations.

**Theorem 2.14.** *For all* $n \geqslant 0$*:*

1. ${}_nC_0 = {}_nC_n = 1$ *and* ${}_nC_1 = n$.
2. *For all* $k$, $0 \leqslant k \leqslant n$, ${}_nC_k = {}_nC_{n-k}$.
3. $\sum_{r=0}^{n} {}_nC_r = 2^n$.

*Proof.* 1. Given a set of $n$ elements, there is only one way to select none of them—that is, there is only one way to do nothing, so ${}_nC_0 = 1$. Similarly, since the order does not matter, there is only one way to choose all of the items: ${}_nC_n = 1$.

If we are choosing only one item, we may select any element from among the $n$, and there are thus $n$ choices possible.

2. We note that every selection of $k$ items from a set of $n$ partitions the set into two disjoint subsets: one of size $k$ and the other of size $n-k$, and so choosing $k$ items to take is equivalent to choosing $n-k$ items to leave behind. The conclusion follows immediately.

Alternately, direct application of the formula for combinations gives the following:

$$ {}_nC_k = \frac{n!}{(n-k)! \cdot k!} = \frac{n!}{k! \cdot (n-k)!} = \frac{n!}{[n-(n-k)]! \cdot (n-k)!} = {}_nC_{n-k}. $$

3. If we think of a combination of $r$ items from a set $A$ with $\#(A) = n$ as choosing a subset of $A$ with $r$ elements, then the left side of this equation is simply the total number of subsets of $A$ of all sizes.

From Theorem 2.11, we know that the number of subsets of a set with $n$ elements is $2^n$. Since we have counted the set of all subsets of $A$ in two different ways, those two expressions must be the same, completing the proof.

□

## Way Tickets

*If you're not playing ways, you're not getting a kick out of keno.*—[Fitzgerald's Casino (Las Vegas) Keno Paybook]

As an application of the Fundamental Counting Principle and Theorems 2.13 and 2.14, we consider keno *way tickets*. Making a single $n$-number bet is only the beginning of keno excitement for a gambler. By making a way bet, it is possible to cover more combinations of numbers, frequently at a reduced price per wager, and increase the probability of drawing into a winning combination. All of this action can be accessed on a single keno bet slip. To make a way bet, the bettor selects several groups of numbers, as in Figure 2.2, where three groups of four numbers, indicated by circled blocks, have been chosen.

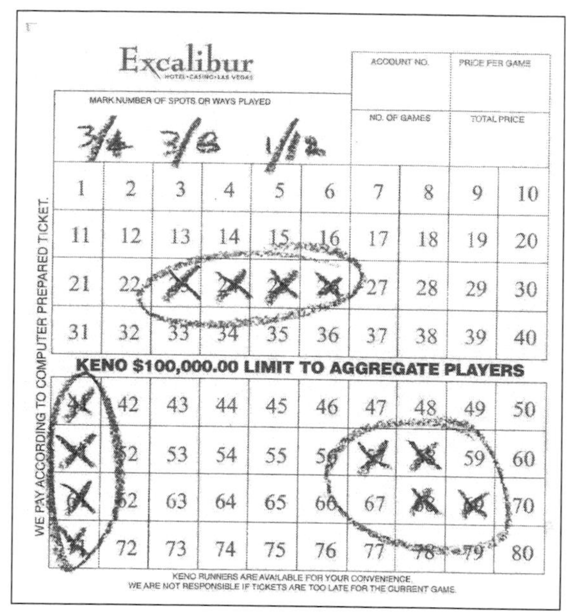


FIGURE 2.2: Keno bet slip with marked 3/4, 3/8, 1/12 way bet.

It should be noted that the numbers comprising a block need not be adjacent as shown here; by placing an X through the desired numbers and circling the entire set, any group of numbers may be combined into a block. The player may then specify how the blocks are to be combined. For the ticket shown in Figure 2.2, there are three 4-spot bets consisting of the three individual blocks, three 8-spot tickets, which arise from combining every pair of 4-spot blocks into a "virtual" 8-spot block, and a single 12-spot ticket encompassing all of the selected numbers. If all seven ways were then played, this ticket would be called a 3/4, 3/8, 1/12 seven-way keno ticket.

At the Excalibur Casino in Las Vegas, way tickets may be purchased at a discount from the standard $1 minimum: tickets with 3 or more ways can be purchased for 50¢ per way, 10 or more ways allow a price of 25¢ per way, and a ticket with 60 or more ways can be had for 10¢ a way. Payoffs, of course, are reduced in proportion to the amount wagered. A way ticket provides for the possibility of multiple wins, but this possibility is purchased with a greater cost for the ticket as a whole. This option is also more lucrative for the casino, as a player putting more money in play exposes a greater amount to the casino's inherent house advantage.

The D Casino is very player-friendly on way tickets, pricing them as low as 1¢ per way if 500 games are purchased. The target of 500 can be reached through a combination of ways and games, so a 10-way ticket purchased for 50 consecutive games would qualify for the lowest rate, and could be had for $5.

**Example 2.54.** The blocks on a way ticket need not be the same size. Consider the ticket shown in Figure 2.3, which has seven blocks: three 2-spots and four 4-spots covering 22 different numbers. How many wagers covering 10 spots or fewer are formed by the selected blocks?

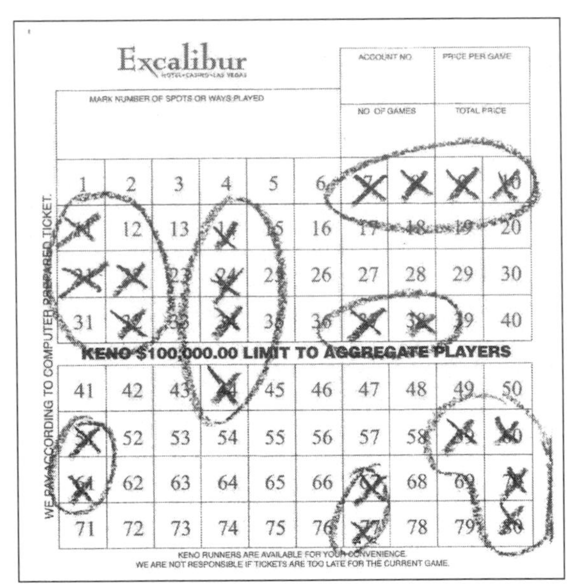


FIGURE 2.3: Keno bet slip with marked 3/2, 7/4, 13/6, 18/8, 22/10 way bet.

These blocks may be combined in the following ways to form virtual tickets with ten or fewer spots:

- The three 2-spots may be played individually.
- There are seven 4-spot tickets: the four 4-spots played individually and ${}_3C_2 = 3$ arising from all possible combinations of the three 2-spot blocks: {37,38,51,61}, {37,38,67,77}, and {51,61,67,77}.
- Each 4-spot block may be combined with each of the 2-spot blocks to form a separate 6-spot block; this gives $4 \cdot 3 = 12$ blocks. The three 2-spot blocks may be combined for one more 6-spot ticket, making 13 6-spot choices.
- There are 18 8-spot blocks:
  - ${}_4C_2 = 6$ from combining two 4-spot blocks.
  - ${}_3C_2 \cdot 4 = 12$ from combining any two 2-spot blocks with a 4-spot block.
- Finally, there are 22 10-spot blocks:

– Choose any two 4-spot blocks (${}_4C_2 = 6$ ways) and add in any of the three 2-spot blocks, for a total of 18.

– Use all three of the 2-spot blocks with any one 4-spot block, totaling four more.

This results in 63 different ways to assemble the seven blocks into keno tickets. A ticket such as this one, with multiple ways of different sizes, is called a *combination* ticket.

The pay table for the ticket in Figure 2.3, which is shown in Table 2.6, is taken from the Excalibur's $1 wager pay table.

TABLE 2.6: Pay table for a $1 keno wager, Excalibur Casino

| | Number of spots marked | | | | |
|---|---|---|---|---|---|
| **Catch** | **2** | **4** | **6** | **8** | **10** |
| 2 | $12 | $1 | | | |
| 3 | | $3 | $1 | | |
| 4 | | $120 | $4 | | |
| 5 | | | $88 | $9 | $4 |
| 6 | | | $1500 | $90 | $40 |
| 7 | | | | $1500 | $260 |
| 8 | | | | $20,000 | $2000 |
| 9 | | | | | $8000 |
| 10 | | | | | $50,000 |

Suppose now that the 20 numbers drawn against this ticket are the following, with the covered numbers in bold:

1, **9**, **11**, **14**, 16, 20, 23, **24**, 28, 40
**44**, 54, **61**, 68, 69, **70**, 71, 76, 78, 79

If you play the ticket in Figure 2.3 at the 10¢ per way minimum wager, representing a total investment of $6.30, find the total payoff.

The cash cow of this ticket is the block {**14**, **24**, 34, **44**}, from which three numbers were caught. Every winning way involves that block. The payoffs on this ticket are distributed as follows:

- Neither of the 2-spot blocks caught both numbers, and so there were no winners.

- {**14**, **24**, 34, **44**} pays 3 for 1 as a 4-spot ticket catching 3 numbers. No other block of 4 caught enough numbers to win money.

- The only 6-spot ways that caught at least 3 numbers combined {**14**, **24**, 34, **44**} with a 2-spot block.

  - {51, **61**} caught 1 number and so generated a 4-out-of-6 winner, paying 4 for 1.
  - {37, 38} and {67, 77} caught no numbers, but the corresponding 6-spot tickets {**14**, **24**, 34, 37, 38, **44**} and {**14**, **24**, 34, **44**, 67, 77} each paid off at 1 for 1 as a 3 of 6 catch.

- The 8-spot ways require at least 5 catches to pay off, and none of the 18 ways on this ticket reached that threshold.
- Winning 10-spot ways can be found by combining {**14**, **24**, 34, **44**} with any of the other three 4-spot blocks, each of which caught one of its numbers, and the block {51, **61**}, which caught one number. These combinations each make up a 10-spot ticket with 5 catches, which pays 4 for 1, and there are three of them.

The total payoff on this ticket is then \$.30 + \$.40 + \$.20 + \$1.20, or \$2.10. The ticket as a whole lost \$4.20. ■

Perhaps the ultimate way bet ticket is a *king* ticket. A single number played as part of a way bet is called a *king*, and a keno ticket consisting entirely of kings provides a multitude of betting combinations.

**Example 2.55.** A ticket with seven kings circled and all combinations activated has 127 possible winning ways. This number is $2^7 - 1$, and may be easily calculated without resorting to a list by referring to Theorem 2.11: There are $2^7$ ways to choose a subset of the seven kings, but since one of these, the empty set, does not correspond to a keno combination, we subtract 1 from this number.

At the Tahoe Biltmore Lodge in Crystal Bay, Nevada, this ticket could be played as the 7 Kings All Ways ticket, at 5¢ per way. The total price of the ticket was \$6.35; it was not necessary to track all 127 ways, because the pay table, shown in Table 2.7, was simply based on the number of catches.

TABLE 2.7: 7 Kings All Ways pay table, Tahoe Biltmore Lodge

| Catch | Payoff |
|---|---|
| 1 | \$0.15 |
| 2 | \$1.65 |
| 3 | \$6.85 |
| 4 | \$24.20 |
| 5 | \$116.75 |
| 6 | \$513.65 |
| 7 | \$2063.90 |

This pay table can be broken down to find the equivalent pay table for Catch 1 through Catch 7 events. Catching one number means winning on one

1-spot ticket, so we conclude that a Pick 1 ticket contained within 7 Kings All Ways paid off at 15¢ for 5¢, or 3 for 1. This is consistent with Table 1.4.

Similarly, catching 2 numbers corresponds to the following 18 wins:

- Two 1-spot wins. These account for 30¢ of the $1.65 payoff.
- One 2-spot win. Using Table 1.4's value of 12 for 1 on this win means that 60¢ of the win comes from this catch.
- Five Catch 2 wins on 3-spot tickets. These pay off at 1 for 1, covering a further 25¢.
- ${}_5C_2 = 10$ Catch 2 wins on 4-spot tickets. Paying these at 1 for 1 as well exactly completes the $1.65.

The house advantage on this wager is 29.40%.

■

While single numbers on a way ticket are called kings, two or more numbers taken together as a group are sometimes called *pawns*.

**Example 2.56.** During the 1960s, the Bonanza Club in North Las Vegas, Nevada (now Jerry's Nugget) offered a number of predetermined way tickets for the ease of patrons wishing to bet a way ticket more easily. One such ticket was the *Japanese King 12-Way*, which required picking 19 numbers. These numbers were divided into 10 kings and a group of 9 pawns, and the ticket was played as a 1-way 9-spot consisting of the pawns, and a 11-way 10-spot. Ten of the 10-spot ways were formed by combining the set of 9 pawns with each of the kings in turn, and the 11th collected all of the kings together. ■

Daily lotteries where players choose their own 2–6-digit numbers frequently offer the option of *boxing* a wager, which covers all permutations of the digits in a player's number on a single ticket. The payoffs on a boxed ticket are lower than for a straight ticket covering only one number and paying off only if the digits match in order.

**Example 2.57.** In Example 2.26, we considered a Daily 4 drawing based on the number 1729 and found that the probability of winning was 1 in 10,000. The probability of winning a box bet on the number 1729 is

$$\frac{{}_4P_4}{10,000} = \frac{24}{10,000},$$

since there are 24 distinct ways to arrange the four different digits in order.

If instead a number like 0330, with two pairs of digits, is chosen, some of the permutations are indistinguishable. The six distinct winning permutations are

$$0330 \quad 0303 \quad 0033 \quad 3300 \quad 3030 \quad 3003.$$

In this example, the pair of 3s and the pair of 0s can be interchanged without changing the four-digit number. The number 6 arises from dividing ${}_4P_4 = 24$ by $2 \cdot 2 = 4$ to account for the pair of duplicates. ■

Since the formulas for $_nP_r$ and $_nC_r$ involve factorials, which are known to grow very fast, it follows that these numbers also get very big very fast. Starting with a standard keno drum containing 80 balls, the number of possible subsets of a given size $r$ is quite large, even if order is not considered.

**Example 2.58.** The first, and for many years the most popular, keno ticket was the Pick 10 ticket, where players chose 10 numbers from 1–80. There are

$$_{80}C_{10} = 1,646,492,110,120,$$

or about 1.6 trillion, different possible Pick 10 tickets. ■

In Great Britain, a popular gambling pastime, akin to a lottery, is the football [soccer] pools. Bettors select subsets of 8 games from the list of the weekend's 49 designated games, attempting—if skill plays a part in their choices—to pick the games that will end in draws. Tied games are divided into two subsets: *score draws*, where each team scores at least once, and *scoreless draws*, where the final score is 0-0. The leading Pools manager is Littlewoods, which dates back to 1923 and is now part of The Football Pools [50]. Their Classic Pools game invites a player to choose 10–15 games from 49, and covers all possible combinations, or "lines," of 8 games from those selected. A 10-game ticket costs only £1, and thus covers $_{10}C_8 = 45$ possible combinations at a cost of less than 3 pence apiece. Players may select 11 matches for £2.75, covering $_{11}C_8 = 165$ lines, and 12 games, totaling $_{12}C_8 = 495$ lines, for £7.50 [39].

Points are awarded based on the outcomes of the eight games. A game ending in a score draw counts for 3 points, a scoreless draw or void match (one where the match is played on a different date than originally declared, for example) for 2, and a win for 1. Prize funds established from entry fees are divided among bettors whose lines score the top three point totals; if 8 games of the 49 end in score draws, the top prize pool is divided among any players with a line totaling 24 points. The top prize fund typically runs into millions of pounds.

**Example 2.59.** Assuming that exactly eight score draws occur, a player buying a 10-game ticket has probability

$$p = \frac{_{10}C_8}{_{49}C_8} = \frac{45}{450,978,066} \approx 9.978 \times 10^{-8}$$

of winning a share of the top prize pool. Here, the numerator counts the number of 8-game subsets that can be chosen from the player's 10 games, and the denominator counts the total number of 8-game subsets from the entire pool of 49 games. ■

Many Pools players simply pick their favorite numbers in the range from 1 to 49, thus making this at least as much a game of chance as a game of skill.

The Pools are so popular in Britain that in 1963, on a weekend when unusually bad weather led to the cancellation of most games, a "Pools Panel" was appointed to invent outcomes for games so as not to interfere with gambling [50].

## Combinatorics on Spreadsheets

Many calculations involving keno and lotteries are facilitated by using spreadsheet software such as Microsoft Excel. This section will introduce some Excel functions that are useful in computing probabilities and expected values; further applications will be shown in later sections.

- Factorials are calculated in Excel with the **FACT()** function, as the exclamation point is reserved for references to other sheets in a single workbook.
- Combinations and permutations use similar structures: ${}_nC_r$ is computed by using the command **COMBIN(n,r)**, and ${}_nP_r$ is found with the analogous **PERMUT(n,r)**.

**Example 2.60.** The spreadsheet shown in Figure 2.4 shows the results of some simple calculations, where columns A and B contain user-entered constants.

| | A | B | C | D | E |
|---|---|---|---|---|---|
| 1 | n | r | r! | nCr | nPr |
| 2 | 9 | 0 | 1 | 1 | 1 |
| 3 | | 1 | 1 | 9 | 9 |
| 4 | | 2 | 2 | 36 | 72 |
| 5 | | 3 | 6 | 84 | 504 |
| 6 | | 4 | 24 | 126 | 3024 |
| 7 | | 5 | 120 | 126 | 15120 |
| 8 | | 6 | 720 | 84 | 60480 |
| 9 | | 7 | 5040 | 36 | 181440 |
| 10 | | 8 | 40320 | 9 | 362880 |
| 11 | | 9 | 362880 | 1 | 362880 |

FIGURE 2.4: Spreadsheet calculations of combinatorial quantities.

- Column C computes the factorial of the corresponding entry in column B, as where cell C5 holds the command **=FACT(B5)**.
- Column D contains ${}_9C_r$, where the constant 9 is stored in cell A2 and $r$ is listed in column B, running from 0–9. This is computed, for example, by entering **=COMBIN(A2,B7)** in cell D7. Note that this column confirms part 2 of Theorem 2.14 in the case $n = 9$: ${}_9C_r = {}_9C_{9-r}$.

- Column E holds the values of ${}_9P_r$, where $r$, once again, is drawn from column B. Typing **=PERMUT($A$2,B2)** in cell E2 executes this command. By using dollar signs on the row and column labels of cell A2, copying and pasting this command into cells E3–E11 keeps the location of the first variable fixed while changing the second variable to run from B2 through B11 and fill out the column.

■

---

## 2.6 Binomial Distribution

We begin by considering an example:

> *Suppose you buy one Pennsylvania Daily 2 lottery ticket every day for 150 straight days. What is the probability that you will win 5 times?*

Solving this problem is facilitated by introducing the concept of a *binomial experiment.*

**Definition 2.21.** A *binomial* experiment has the following four characteristics:

1. The experiment consists of a fixed number of successive identical trials, denoted by $n$.
2. Each trial has exactly two outcomes, denoted *success* and *failure.*

   In practice, it is often possible to amalgamate multiple outcomes into a single category to get down to two. For example, in the Daily 2 lottery question above, with 100 different possible numbers, we can collect all 99 losing numbers into a single outcome—if we lose our bet, it matters little what the winning number was.
3. The probabilities of success and failure are constant from trial to trial. We denote the probability of success by $p$ and the probability of failure by $q$, where $q = 1 - p$.
4. The trials are independent.

**Definition 2.22.** A random variable $X$ that counts the number of successes of a binomial experiment is called a *binomial* random variable. The values $n$ and $p$ are called the *parameters* of $X$.

The experiment described in the example above meets the four listed criteria and is therefore a binomial experiment. If we change the experiment to "Start buying one lottery ticket per day, and let the random variable $X$ be the number of tickets required to win exactly 5 times," then the new experiment is not binomial. Since the number of trials is no longer fixed at the outset, criterion 1 is no longer true.

If $X$ is a binomial random variable with parameters $n$ and $p$, the formula for $P(X = r)$ can be derived through the following three-step process:

1. Select which $r$ of the $n$ trials are to be successes. This can be done in ${}_nC_r$ ways, as the order in which we select the successes does not matter.

   If we think of the trials as a row of $n$ boxes, each to be designated "success" or "failure," what we're doing here is determining which $r$ of the $n$ boxes are successes.

2. Compute the probability of these $r$ trials resulting in successes. Since the trials are independent, this probability is $p^r$.

3. We must now ensure that there are *only* $p$ successes. This is done by assigning the outcome "failure" to the remaining $n - r$ trials. The probability of this many failures is $(1 - p)^{n-r} = q^{n-r}$.

Multiplying these three factors together gives the following result, called the *binomial formula*:

**Theorem 2.15.** *If $X$ is a binomial random variable with parameters $n$ and $p$, then*

$$P(X = r) = {}_nC_r \cdot p^r \cdot q^{n-r} = {}_nC_r \cdot p^r \cdot (1 - p)^{n-r}.$$

This probability may be calculated on a spreadsheet with the command **BINOM.DIST(r,n,p,A)**, where the first three arguments are the numeric quantities $r, n,$ and $p$ noted in the theorem, and $A$ is a logical statement that evaluates to **true** or **false**. If $A$ is true, then Excel calculates $P(X \leqslant x)$; if $A$ is false, then this command returns $P(X = x)$. Simply typing "true" or "false" in the fourth variable slot will suffice.

We can now revisit the question that started this section with this new insight.

**Example 2.61.** Since the pool of two-digit numbers for Pennsylvania's Daily 2 game includes numbers with a leading zero such as 00 or 09, the probability of winning on a \$1 straight bet is $\frac{1}{100}$. We have $n = 150$ and $p = \frac{1}{100}$. The probability of winning on $r$ tickets is then

$$P(r) = {}_{150}C_r \cdot \left(\frac{1}{100}\right)^r \cdot \left(\frac{99}{100}\right)^{150-r}.$$

If $r = 5$, then $P(5) \approx .0138$, so the answer to the original question is "Slightly less than 1.4%". ■

If a random variable is binomial, computing its expected value is simple.

**Theorem 2.16.** *If $X$ is a binomial random variable with parameters $n$ and $p$, then $E(X) = np$.*

Put simply, the average number of successes is the number of trials multiplied by the probability of success on a single trial.

*Proof.*

$$\begin{aligned}
E(X) &= \sum_{x=0}^{n} x \cdot P(X = x) \\
&= \sum_{x=0}^{n} x \cdot {}_nC_x \cdot p^x \cdot q^{n-x} \\
&= \sum_{x=0}^{n} x \cdot \frac{n!}{(n-x)! \cdot x!} \cdot p^x \cdot q^{n-x}.
\end{aligned}$$

Since the $x = 0$ term is equal to zero, we can drop that term from the sum and renumber starting at 1:

$$\begin{aligned}
E(X) &= \sum_{x=1}^{n} x \cdot \frac{n!}{(n-x)! \cdot x!} \cdot p^x \cdot q^{n-x} \\
&= \sum_{x=1}^{n} \frac{n!}{(n-x)! \cdot (x-1)!} \cdot p^x \cdot q^{n-x} \\
&= np \cdot \sum_{x=1}^{n} \frac{(n-1)!}{(n-x)! \cdot (x-1)!} \cdot p^{x-1} \cdot q^{n-x} \\
&= np \cdot \sum_{x=1}^{n} \frac{(n-1)!}{[(n-1)-(x-1)]! \cdot (x-1)!} \cdot p^{x-1} \cdot q^{n-1-x+1}.
\end{aligned}$$

If we substitute $y = x - 1$ in this last sum, we have

$$\begin{aligned}
E(X) &= np \cdot \sum_{y=0}^{n-1} \frac{(n-1)!}{[(n-1)-y]! \cdot y!} \cdot p^y \cdot q^{n-1-y} \\
&= np \cdot \sum_{y=0}^{n-1} P(Y = y), \text{where } Y \text{ is a binomial random variable.}
\end{aligned}$$

This sum is the sum of all of the values in the probability distribution of a binomial random variable $Y$ with parameters $n - 1$ and $p$, and so adds up to 1, completing the proof. □

**Example 2.62.** In Example 2.61, the expected number of wins in 150 days of ticket-buying is

$$np = 150 \cdot \frac{1}{100} = 1.5.$$

Another way to reach this conclusion is to imagine 150 tickets purchased for the same drawing: one on every even number and two on every odd number. Half of the time, the winning number will be even, and there will be 1 winner. The other half of the time, the winning number will be odd, and thus there will be 2 winners. Averaging these two results gives an expected value of 1.5 winning tickets. ■

It follows from this result that your average winnings after making 150 of these bets would be

$$(1.5) \cdot 50 - 150 = -\$75.$$

On the average, you should expect to lose half of the money you wagered.

Aside from the general rule that "gambling games always favor the house," it might be reasonable to ask if there is any way to identify particularly bad bets such as this one before risking money. One tool that may be useful is the *standard deviation* of a random variable, which is denoted by the Greek letter sigma: $\sigma$.

**Definition 2.23.** The *standard deviation* $\sigma$ of a random variable $X$ with mean $\mu$ is

$$\sigma = \sqrt{\sum [x^2 \cdot P(X = x)] - \mu^2}$$

where the sum is taken over all possible values of $X$.

Informally, $\sigma$ is a measure of how far a typical value of $X$ lies from the mean. Computing $\sigma$ using the formula above is an arithmetically intense process:

- Compute the mean of the random variable.
- For each value of $x$ that $X$ can attain, multiply $x^2$ by $P(X = x)$ and add up the products.
- Subtract the square of the mean from this sum.
- Take the square root of the difference. Notice that the use of the square root guarantees that $\sigma \geqslant 0$.

Fortunately, it is seldom necessary to perform these calculations by hand, as calculators and computer software will readily compute $\sigma$. In the special case where the random variable is binomial, we have the following simple result:

**Theorem 2.17.** *If $X$ is a binomial random variable with parameters $n$ and $p$, then the standard deviation of $X$ is given by*

$$\sigma = \sqrt{np(1-p)} = \sqrt{npq}.$$

While the expected value of a random variable tells us about where a "typical" value of $X$ lies, the standard deviation gives us information about the "spread" of the values of $X$. The nature of the random variable $X$ allows us to use the mean and standard deviation to derive useful information about the distribution of the data set. For example, in considering a suitably large collection of 20-number keno draws, the distribution of sums is approximately bell-shaped (many values near the mean, and fewer values as we move away from the mean of 810 in either direction) and symmetrically distributed about the mean. See Figure 2.5, which shows a plot of the sums of 4 million simulated keno drawings, for an example.

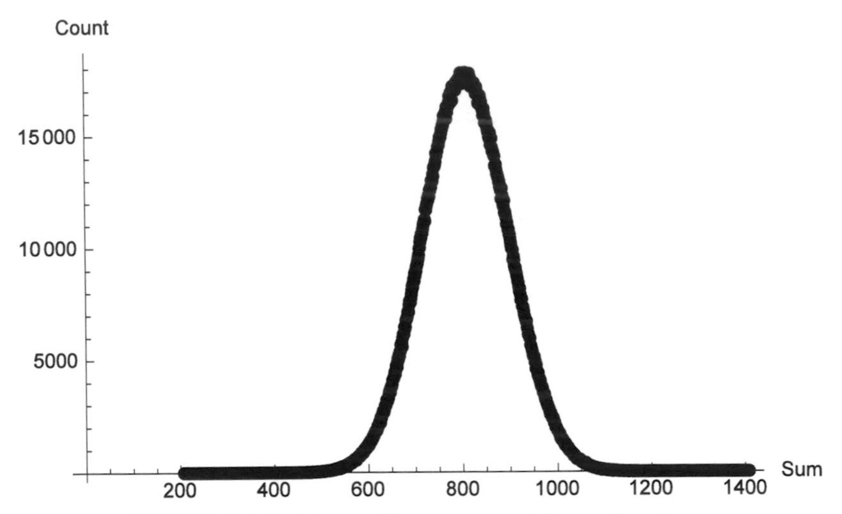

FIGURE 2.5: Four million sums of keno drawings.

Data sets that are distributed this way are called *normal*. In this circumstance, a result called the *Empirical Rule* is a good description of the data.

**Theorem 2.18.** ***(Empirical Rule)*** *If the values of many samples of a random variable with mean $\mu$ and standard deviation $\sigma$ are bell-shaped and symmetric, then we have the following result:*

1. *Approximately 68% of the data points lie within 1 standard deviation of the mean: between $\mu-\sigma$ and $\mu+\sigma$. (This is often stated as "about 2/3.")*

2. *Approximately 95% of the data points lie within 2 standard deviations of the mean: between $\mu - 2\sigma$ and $\mu + 2\sigma$.*

3. *Approximately 99.7% of the data points lie within three standard deviations of the mean: between $\mu - 3\sigma$ and $\mu + 3\sigma$. (This is often stated as "almost all.")*

For many practical purposes, an experimental result is deemed to be statistically significant, meaning that the result is likely to be the result of something other than random variation, if it is at least 2 standard deviations (SDs) away from the mean. This means that its probability—under the assumption that there is no unusual effect and all of the deviation from the mean is due to random chance—is less than 5%. Since the random variable is symmetric, this 5% is evenly distributed between results greater than 2 SDs above $\mu$ and results less than 2 SDs below $\mu$. This 2 SD standard is a convention agreed upon by the statistics community; it does not fall out of any equation as a rigorously derived standard.

By using 95% as a minimum, what we are saying is that 19 out of 20 times that we identify a result as due to something other than random variation, we will be correct, and this level of confidence is acceptable in many lines of inquiry. Some fields may have more exacting standards: in experimental particle physics, for example, the standard for confirming a discovery is "5 sigma," or at least 5 standard deviations away from the expected value, corresponding to $P\,(\text{Chance event}) < 1$ in 3.5 million.

**Example 2.63.** For the four million keno sums shown in Figure 2.5, the standard deviation of the sample data can be computed directly, and is found to be approximately 90. Using this number as an approximation to the population SD $\sigma$, we find that the Empirical Rule gives the following conclusions:

- Approximately 68% of all sums should lie between 720 and 900.
- Approximately 95% of all sums should lie between 630 and 990.
- Approximately 99.7% of all sums should lie between 540 and 1080.

The actual data show 68.3% within 1 SD of 810, 95.6% within 2 SDs, and 99.8% within 3 SDs, and so the Empirical Rule is confirmed. ■

Returning to the Pennsylvania Daily 2 lottery, repeated trials purchasing 150 tickets would yield a mean of 1.5 wins and a standard deviation of 1.2186 wins. The Empirical Rule tells us that 68% of the time, we would expect between .2814 and 2.7186 wins. Since the number of winning tickets must be an integer, we round this result inward (up or down, as needed, to the nearest integer within the interval) to [1,2], and 95% of the time, the number of wins will fall between −.9372 and 3.9372, which rounds inward to [0,3]. Scoring more than 5 wins in 150 tickets (the 3-sigma level) is highly improbable, and might be cause for an investigation into lottery operations.

---

## 2.7 Exercises

Solutions begin on page 303.

**2.1.** In the Chinese lottery (page 3), another method of resolving a 2-spot ticket divided the 12 numbers into 3 groups of 4. Each possible pair from one group was then combined with the 8 numbers in the other 2 groups to form a single 10-spot ticket. How many 10-spot tickets are determined by this approach?

**2.2.** An alternate approach to the Chinese lottery's 13-spot ticket breaks it into one group of 6 numbers and one group of 7 numbers. All possible subsets of 3 from the group of 6 are combined with the group of 7, and all possible subsets of 4 from the group of 7 are combined with the group of 6, to make 10-spot tickets. How many tickets result from this division?

**2.3.** A 14-spot Chinese lottery ticket could be configured by repeating each number three times in a 4-row array. How many 10-spot tickets does this array define? (Remember that an 11-spot ticket is, by default, rendered as a collection of 10-spot tickets.)

**2.4.** Mathematician Leonhard Euler wrote a series of papers on various mathematical aspects of the Genoese Lottery described on page 10. In addition to working with the probabilities of winning and the house advantage, Euler also considered several generalizations of the lottery which involved drawing fewer than 5 numbers in the range 1–90. In a game where only two numbers are drawn, find the probability that they are consecutive numbers. For the purposes of this problem, 90 and 1 should not be considered consecutive.

**2.5.** Some other questions considered by Euler: If 4 numbers are drawn in the range 1–90, find the probability of the following sequences.

a. Four consecutive numbers. The order in which the numbers are drawn does not matter. As in Exercise 2.4, "around the corner" sequences such as {88, 89, 90, 1} are not counted as 4 consecutive numbers.

b. Two sequences of two consecutive numbers each that do not form a sequence of 4 consecutive numbers.

c. More generally, if the range of numbers is $1$–$n$, find the probability that 4 drawn numbers are in sequence.

**2.6.** Bolita (page 15) pay tables vary among operators; one game with one ball drawn from 100 paid 80 for 1 on a winning wager [55]. Find the house advantage of this bet.

**2.7.** In the English-speaking Central American country of Belize, the name bolita has been Anglicized to *boledo*, but the game is the same as found in Cuba. Boledo asks players to pick a two-digit number from 00–99, and the payoff on a winning number is 70 for 1. Find the HA.

**2.8.** The Carousel Casino in downtown Las Vegas (later Mermaids and now part of the 18 Fremont development) offered a variety of 10-spot way tickets that could cover as many as all 80 numbers on a keno ticket. Players marked their tickets in groups of 5 numbers, and all possible combinations of 2 groups were played as 10-spot tickets. If a player marked 65 numbers in 13 groups of 5, how many way tickets were in play?

**2.9.** At Binion's Horseshoe, players could mark a 9-spot keno way ticket by selecting numbers in 3–20 groups of 3, which would then be combined in all possible combinations to reach 9 spots per way. A 54-spot ticket with 18 groups of 3 was offered at $285.60. How many 9-spot tickets did this single bet slip represent?

**2.10.** Binion's Horseshoe offered another way ticket where players picked 12 spots divided into four groups of 3, which were then combined into all possible 9-spot and 6-spot tickets. How many tickets in total did these 12 numbers comprise?

**2.11.** The Royal Casino in Las Vegas (now the nongaming Royal Resort) offered an 11-way 10-spot ticket that used all 80 numbers on a keno bet slip. Devise a way to achieve this ticket.

**2.12.** For the Queen's College Literature Lottery (Example 2.53), find the probability of winning the following amounts:

a. $9.

b. $4.50.

c. Any prize.

**2.13.** A later version of the Queen's College Literature Lottery sold 19,600 tickets each bearing three numbers from 1–$n$ [113]. Find $n$.

**2.14.** The *King of the Mountain* ticket offered at the Golden Nugget Casino in downtown Las Vegas called for players to choose 8 kings. If all possible tickets with 5 or more spots were formed from this collection of kings, find the cost of a ticket if it was priced at 50¢ per way.

**2.15.** *Mega Millions* is a lottery game played across America, in every state with a state lottery plus the District of Columbia and the Virgin Islands. Players choose 5 numbers in the range 1–70 and a 6th number from a separate pool from 1–25, which is shown in the drawing as a gold ball. Twice weekly, lottery officials select 6 winning numbers from the same sets, and payoffs are made based on the number of matches.

a. In how many ways may the player choose his or her set of 5 numbers?

b. How many different tickets are possible?

c. For a player to match exactly 3 of his or her 5 numbers, it's necessary for 3 of the winning numbers to be from the player's set of 5 and the other two to be from the 65 numbers not selected by the player. Using the Fundamental Counting Principle, find the number of different ways to match exactly 3 of 5 numbers.

d. Find the probability of matching 3 of 5 numbers, plus the gold ball.

**2.16.** In November 2017, *Lotto America*^SM^ was launched in 13 states. Similar to Mega Millions, Lotto America called for players to choose 5 white ball numbers from 1–52 and a Star Ball^SM^ number between 1 and 10. Tickets cost $1, half the price of a Mega Millions ticket, and drawings are held twice weekly.

a. In how many ways may the player choose his or her set of 5 numbers?

b. The top Lotto America prize, for matching all 5 white ball numbers and the Star Ball number, starts at $2 million and grows with each drawing without a winner until someone matches all of their numbers. Find the probability of winning the jackpot.

c. Find the probability that a Lotto America ticket matches no numbers.

# Chapter 3

# Keno

## 3.1 Standard Keno Wagers

*Keno is a guessing game.*—[Whiskey Pete's Casino Keno Paybook]

The most basic keno wager involves selecting some numbers and waiting to see how many are caught when the casino picks its 20 numbers, although the variety of games offered at most casinos on the same drawing of 20 numbers makes keno a game that can lead to mathematics far beyond basic counting. In this section, we shall examine some mathematical questions and game variations that arise from these simplest bets.

### Pick 1: Elementary Probability Suffices

The payoff for choosing and catching a single number on a keno ticket is largely standard across casinos, and the mathematics used to evaluate that wager is quite simple. A live keno game usually pays 3 for 1 when one number is caught; some lottery-run keno games may pay 2 for 1. A fair game would pay 4 for 1.

The probability of winning on a 1-spot ticket is ¼. This can be seen without recourse to the combinatorial formulas in Chapter 2. Suppose that you purchase 80 1-spot tickets, marking a different number on each one. Twenty of those tickets will be winners, and 60 will lose. This makes the probability of any one ticket winning 20/80, or ¼.

The expected return on a \$1 1-spot casino ticket is then

$$E = (3) \cdot \frac{1}{4} - 1 = -\frac{1}{4} = -\$0.25,$$

and the house advantage is 25% [135]. A Pick 1 ticket for a state lottery's keno game paying 2 for 1 on a single number is carrying a HA of 50%; we simply replace 3 by 2 in the equation above for $E$.

From time to time, better payoffs have been available.

- In the 1950s, Dick Graves' Nugget Casino in Sparks, Nevada (now the Nugget) paid 3.20 for 1 for catching the one marked number on a Pick 1 ticket. This ticket carried a lower HA of 20% [135].

- At the Luxor Casino in Las Vegas, a promotion called Keno On The Go offered players a minimum 10% bonus on their keno winnings, provided they played their ticket for 31 or more consecutive games. This 10% bonus applied to a Pick 1 ticket changed the payoff from 3 for 1 to 3.30 for 1 and cut the HA to 17.5%.

- At the D Casino, a special 40¢ keno ticket pays $1.40 if the chosen number is caught, rather than the $1.20 payoff that would result if 3 for 1 odds were offered. This is equivalent to paying $3.50 for a winning $1 wager, and the house edge on this ticket is 12.5%.

- The 75¢ Pick 1 ticket at the Reserve Casino in Henderson, Nevada (now the Fiesta Henderson) paid $2.75 on a winning 75¢ wager, a payoff of 3.67 for 1. The regular $1 Pick 1 rate at the Reserve at that time paid the standard $3, so carefully reading the keno rules at the casino, which was open as the Reserve from 1998–2001, could have paid off. The tickets may have seemed equivalent to a casual observer, since both the price and the payout have decreased by 25¢, but the HA on this ticket was a low 8.33%.

In the USA, the state of Nebraska is unusual in allowing municipal keno games administered under the auspices of local governments rather than state lottery authorities. These games were permitted by law prior to the founding of the statewide lottery in 1993, and are used to raise money for public works projects in their communities. As a result, widely varying game options and pay tables can be found from town to town, and keno accounts for nearly 70% of local lottery gameplay [104]. Some Nebraska keno outlets offer variations on even the simple Pick 1 ticket.

**Example 3.1.** In Bellevue, Nebraska and several other eastern Nebraska cities, a game called *Spotlight Keno* is on offer. Spotlight Keno is an optional feature that can be on or off for a given ticket, at the player's discretion. Fifty-six of the 80 keno numbers are colored red, 16 are white, and 8 blue. Payoffs for any ticket are then based on the color mix of the numbers that are caught: all-red or mixed combinations pay the lowest, followed by all-white and then all-blue catches.

The Pick 1 Spotlight Keno pay table used in Nebraska City, Nebraska is shown in Table 3.1. Two rates are available: the second column includes payoffs

TABLE 3.1: Spotlight Keno Pick 1 pay table

| Color | $1.25 ticket | 3 tickets for $1 |
|---|---|---|
| Red | $3 | $0.75 |
| White | $4 | $1 |
| Blue | $10 | $3 |

on a single \$1.25 ticket, while the third contains the payoffs for a special game offering three tickets for \$1.

How does this game compare with the standard Pick 1 game paying 3 for 1 and its 25% HA? We begin by equalizing all payoffs to their "for 1" equivalents by dividing by the cost of the ticket. The resulting payoffs are shown in Table 3.2.

TABLE 3.2: Spotlight Keno Pick 1 pay table, all payoffs "for 1"

| Color | \$1.25 ticket | 3 tickets for \$1 |
|---|---|---|
| Red | 2.4 | 2.25 |
| White | 3.2 | 3 |
| Blue | 8 | 9 |

Neither game is clearly superior to the other at all three payoff levels. Both games pay off less than 3 for 1 for catching a single red number; the question remaining is how far the increased white and blue payoffs go toward making up for this decrease.

Catching one marked number has probability ¼, which needs to be divided among the three events corresponding to the possible colors in proportion to their likelihood. Any one number in any one drawing has probability .7 of being red, .2 white, and .1 blue. Using $p$ for the ticket price and $x, y$, and $z$ for the red, white, and blue payoff amounts yields the following formula for the expected value of a single ticket:

$$E(x, y, z, p) = \left[\left(\frac{7}{10} \cdot x + \frac{2}{10} \cdot y + \frac{1}{10} \cdot z\right) \cdot \frac{1}{4}\right] - p.$$

Evaluating $E$ for the two games listed gives a HA of 22.00% for the \$1.25 ticket and 23.13% for the $33\frac{1}{3}$¢ ticket. Both represent a small, though perhaps not significant, improvement over the fixed 3 for 1 payoff. ■

## Pick 2: General Formula

When computing keno probabilities involving tickets with more than one marked number, it is necessary to consider every number drawn by the casino. For example, if you pick 2 numbers and want to find the probability of catching 1, you must account for the one number caught and also need to count the number of ways that the casino can draw the other 19 numbers *without* drawing your second number. In short, if $X$ is the number of catches, you want $P(X = 1)$ rather than $P(X \geqslant 1)$.

In general, suppose that you choose $n$ numbers and want to find the probability of catching exactly $k$, where $k \leqslant n$. The number of ways to catch $k$ of your $n$ numbers is ${}_nC_k$, and the number of ways to choose the remaining $20 - k$ numbers from the $80 - n$ that you did not pick is ${}_{80-n}C_{20-k}$. These two numbers are multiplied together to give the total number of tickets catching exactly $k$ of your numbers. The total number of ways for the casino to draw its numbers is ${}_{80}C_{20}$, the number of possible subsets of 20 keno balls from a set of 80, and the conditional probability of matching $k$ numbers from a set of $n$ chosen numbers is

$$\boxed{P(k \mid n) = \frac{{}_nC_k \cdot {}_{80-n}C_{20-k}}{{}_{80}C_{20}}.}$$

When $n$ is clear from the context of the game or the problem being solved, we may omit it and just write $P(k)$ for this conditional probability.

**Example 3.2.** In Example 2.37, we considered the Meskwaki Casino's "Super 20 Special" keno game whose brochure advertises "18-Out-Of-21-Ways To Win". Applying the formula above to this game, we have the following formula for $P(X = x)$:

$$P(X = x) = P(x \mid 20) = \frac{{}_{20}C_x \cdot {}_{60}C_{20-x}}{{}_{80}C_{20}}.$$

What can we conclude from this probability function?

- The probability of losing \$5 by matching 4, 5, or 6 numbers is

  $$P(X = 4) + P(X = 5) + P(X = 6) = .6285,$$

  which conforms to our notion that keno is not a very player-friendly game.

- Including the four break-even payoffs, the probability of not profiting from this bet is .9662, which definitely makes this look more like a casino game.

■

For a Pick 2 ticket, there are three possible outcomes, and typically the only payoff results from catching both numbers. We have the following probability distribution:

| $k$ | $P(k \mid 2)$ |
|---|---|
| 0 | $\frac{{}_2C_0 \cdot {}_{78}C_{20}}{{}_{80}C_{20}} = \frac{177}{316} \approx .5601$ |
| 1 | $\frac{{}_2C_1 \cdot {}_{78}C_{19}}{{}_{80}C_{20}} = \frac{120}{316} \approx .3797$ |
| 2 | $\frac{{}_2C_2 \cdot {}_{78}C_{18}}{{}_{80}C_{20}} = \frac{19}{316} \approx .0601$ |

The expected return then depends on the payoff. Suppose that this wager pays off at $X$ for 1 when both numbers are caught. It follows that the expected value of a $1 ticket is

$$E(X) = \frac{19X}{316} - 1.$$

A common value for $X$ is 12, as seen at the Orleans and El Cortez Casinos in Las Vegas. The expectation in this case is –27.85¢ per dollar wagered, and the house advantage is 27.85%.

**Example 3.3.** The function $P(k \mid n)$ and the associated expectation are easily computed in a spreadsheet. Figure 3.1 illustrates the probabilities and payoffs for $n = 2$.

| | A | B | C | D | E | F | G |
|---|---|---|---|---|---|---|---|
| 1 | Pick | Catch | Probability | Payoff | | C(80,20) = | 3.53532E+18 |
| 2 | 2 | 0 | 0.560126582 | 0 | | Ticket price: | 1 |
| 3 | | 1 | 0.379746835 | 0 | | | |
| 4 | | 2 | 0.060126582 | 12 | | | |
| 5 | | | | | | | |
| 6 | | | Expectation: | -0.27848 | | | |
| 7 | | | HA: | 27.85% | | | |
| 8 | | | | | | | |

FIGURE 3.1: Spreadsheet calculation of Pick 2 probabilities.

To calculate the probabilities in Excel, do the following:

1. Enter the indicated labels in row 1 and cell F2, and the ticket price of $1 in cell G2.

2. Since we'll be using the number ${}_{80}C_{20}$ often, we shall store it as a constant in cell G1. Enter

   ```
   =COMBIN(80,20)
   ```

   in G1.

3. Enter 2 (the value of $n$) in cell A2 and the numbers 0, 1, and 2 (values of $k$) in cells B2 through B4.

4. Compute the probabilities by entering

   ```
   =COMBIN($A$2,B2)*COMBIN(80-$A$2,20-B2)/$G$1
   ```

   in cell C2. The dollar signs are used since we want the values in A2 and G1 to be used in subsequent calculations.

5. Copy the formula in cell C2 into cells C3 and C4. This replaces cell B2 in the probability formula by cells B3 and B4, updating the values of $k$.

6. After entering the payoffs 0, 0, and 12 ($X$) in cells D2-D4, calculate the expected value and house advantage by entering

   **=SUMPRODUCT(C2:C4,D2:D4)-G2**

   in cell D6 and

   **=–D6/G2**

   in cell D7. **SUMPRODUCT** takes two equal-length columns or rows of numbers, multiplies the corresponding elements, and adds the resulting products. Format cell D7 as a percentage.

■

Other payoffs besides 12 for 1 exist, and they are frequently offered on games where the wager differs from $1. These differing ticket prices may make comparing tickets a challenge; the technique of determining $X$ by dividing the payoff by the ticket price and then using the formula on page 63 for $E(X)$ simplifies this comparison. Another option is to use the spreadsheet constructed above, changing the payoff in cell D4 and the ticket price in cell G2.

- The Kansas Lottery lists a keno game among its offerings that pays only 9 for 1 on a Pick 2 ticket catching both numbers. This payoff yields an HA of 45.89%; Kansas keeps nearly half of the money wagered on this ticket.

- In Holdrege, Nebraska, Gutterz Fun Center pays 10 for 1 on Pick 2 tickets [49]. The resulting HA is 39.87%. The same 10 for 1 payoff is found in the Penny Keno game option at Big Red Keno in Eagle, Nebraska, where a 1¢ 2-spot ticket pays off 10¢ when catching both numbers. (Regular keno in Eagle pays off at 12 for 1, with a $1 minimum bet.) This high HA is evidence that some keno outlets in Nebraska run the game very much like a state lottery.

- Pick 2 Keno as offered by the Ohio Lottery pays 11 for 1 on 2 catches. The house edge is 33.86%.

- Jerry's Nugget in North Las Vegas, Nevada offers a 70¢ Pick 2 ticket that pays $8.50 if both numbers are caught. This ticket has $X = 12\frac{1}{7}$ and a corresponding HA of 26.99%.

- The Keno Manager's Rate at Arizona Charlie's Decatur in Las Vegas offers a Pick 2 ticket that pays off $15.50 for a successful $1.25 wager. We have $X = 12.4$, and the HA is 25.44%.

- At the California Casino in Las Vegas, the Special Island Rate for a 2-spot ticket pays $10 on an 80¢ wager, for an effective payoff of 12.5 for 1. This cuts the house edge to 24.84%.

- The D Casino claims to offer "Simply The Best Special Rate Ever!", on 40¢ keno tickets. The Pick 2 game pays $5.40, which means that $X = 13.5$. The house advantage on this ticket is 18.83%.

- In contrast with Gutterz Fun Center, La Vista Keno in La Vista, Nebraska tops the D's "Best Rate Ever" by paying 14 for 1 on a 2-spot ticket [74]. The house advantage on this ticket is 15.82%—which is less than the HA of the worst standard bet (Any Seven, whose HA is 16.67%) on a craps table.

  Shopping around from city to city in Nebraska may turn up better keno payoffs in general, but this particular bit of knowledge is not guaranteed to be of much practical use, since Holdrege and La Vista are about 200 miles apart.

- Casinos frequently offer better playing conditions for games such as blackjack at tables with higher minimum bets, such as paying 3–2 on naturals at a $25 minimum table rather than the 6–5 payout on naturals at a $5 minimum table, or allowing players to double down on any two cards rather than restricting doubles to hands of 10 and 11. Two cities in Nebraska have offered local Pick 2 games with comparatively low house edges that require a $10 minimum wager.

  - A promotional Pick 2 rate briefly offered by Big Red Keno in Norfolk paid $145 for 2 catches, an effective payoff rate of 14.5 for 1, so the HA is 12.82%.
  - In Waverly, the keno game at the Trackside Bar & Grill pays out $150 on a winning $10 Pick 2 ticket (15 for 1) [157], so the HA—the best one we've seen so far—is only 9.81%.

  In both games, the payoff on a winning ticket was high, but the risk required for a chance at the big payoff was more substantial than on most keno tickets.

The relative risk among all of these pay tables can be quantified by a look at their standard deviations. The standard deviation $\sigma$ of a wager depends on the payoff $X$ for catching both numbers, the ticket price $p$, and the expected value $E$ of a single ticket. The probability of winning is fixed at $\frac{19}{316}$, and so we have

$$\sigma = \sqrt{\left[X^2 \cdot \frac{19}{316} + (-p)^2 \cdot \frac{297}{316}\right] - E^2}.$$

For the casino pay tables listed above, Table 3.3 shows the standard deviations.

TABLE 3.3: Standard deviation for Pick 2 keno wagers

| Location | HA | $\sigma$ | $\sigma$/Wager |
|---|---|---|---|
| Gutterz | 39.87% | 2.6064 | 2.6064 |
| Orleans/El Cortez | 27.85% | 3.0855 | 3.0855 |
| Jerry's Nugget | 26.99% | 2.1838 | 3.1197 |
| Arizona Charlie's | 25.44% | 3.9765 | 3.1812 |
| California | 24.84% | 2.5641 | 3.2051 |
| The D | 18.83% | 1.3777 | 3.4442 |
| La Vista | 15.82% | 3.5637 | 3.5637 |
| Norfolk | 12.82% | 3.6831 | 3.6831 |
| Trackside | 9.81% | 38.0246 | 3.8025 |

The standard deviation is clearly affected by the ticket price; this is seen in the Trackside row of Table 3.3, where tickets cost \$10. To eliminate this effect, we divide $\sigma$ by the cost of a ticket, which results in the values in the last column of the table. The relationship between the house advantage and $\sigma$/wager is almost perfectly linear: as the HA decreases, the standard deviation per unit wagered increases, indicating a more volatile wager.

The expected value on a single ticket does not tell the full story of how these bets perform over time. For a sequence of $n$ 2-spot keno wagers, the number of wins is a binomial random variable with success probability $p = \frac{19}{316}$. If $np > 30$, which corresponds to $n \geqslant 499$, the distribution of the results is approximately normal. The mean number of wins of that normal distribution is $np \approx 30.06$, and the standard deviation of the number of wins is $np(1-p) \approx 28.26$. Both the mean and standard deviation are independent of the bet size or expectation.

Consider a sequence of 500 wagers on each one of these 2-spot games. What is the probability that the bettor will be ahead after 500 bets—that is, what is the probability that the sum of the 500 results will be positive?

Let $X$ denote the total net winnings after 500 games. $X$ is the sum of 500 independent random variables $X_1, X_2, \ldots, X_{500}$, each of which represents the outcome of a single Pick 2 keno drawing. The *mean* of $X$, denoted $\bar{X}$, can be described using a result known as the Central Limit Theorem.

**Theorem 3.1.** ***(Central Limit Theorem)*** *Let $Y$ be a random variable with mean $\mu$ and standard deviation $\sigma$. If $n$ independent samples are taken, the mean of the sample is approximately normally distributed with mean $\mu$ and standard deviation $\sigma/\sqrt{n}$.*

Note that this theorem asserts that the mean is approximately normal, regardless of the underlying distribution of the variable $Y$. As the sample size increases, the standard deviation of the mean decreases, indicating that the sample mean gets closer to its theoretical value when the sample size is large.

The probability $P(X \geqslant 0)$ that we seek is equal to the probability $P(\bar{X} \geqslant 0)$. This can be computed in Excel with the command

**1 - NORM.DIST(0,mean,SD,true)**,

entering the addresses of the cells containing the mean and SD in the appropriate places in the formula. The "true" command directs Excel to compute the cumulative probability $P(\bar{X} \leqslant 0)$; we use the Complement Rule to find the desired probability.

The results of these calculations for the nine wagers above are shown in Table 3.4. For the four \$1 tickets, the chance of being ahead after 500 games

TABLE 3.4: Probability of a net win after 500 Pick 2 wagers

| **Location** | **Wager** | **HA** | $P(X > 0)$ |
|---|---|---|---|
| The D | \$0.40 | 18.83% | 11.08% |
| Jerry's Nugget | \$0.70 | 26.99% | 2.65% |
| California | \$0.80 | 24.84% | 4.15% |
| Norfolk | \$1.00 | 12.82% | 21.83% |
| La Vista | \$1.00 | 15.82% | 16.04% |
| Orleans/El Cortez | \$1.00 | 27.85% | 2.18% |
| Gutterz | \$1.00 | 39.87% | 0.03% |
| Arizona Charlie's | \$1.25 | 25.44% | 3.69% |
| Trackside | \$10.00 | 9.81% | 28.20% |

decreases as the HA increases, as one might reasonably expect. The high ticket price required for the Pick 2 bet at the Trackside Bar & Grill is accompanied by the highest probability of making a profit after 500 wagers, 28.20%.

**Example 3.4.** Paying 2-spot tickets only on two catches is not a universal practice. The Foxwoods Casino uses a modified 2-spot pay table with a \$2 minimum bet. This ticket pays \$2—so the gambler breaks even—if one of the two numbers is caught while lowering the payoff for catching 2 of 2 to \$10, effectively a 5 for 1 payoff. The expected value of a \$2 ticket is

$$E = (2) \cdot \frac{120}{316} + (10) \cdot \frac{19}{316} - 2 = -\$\frac{202}{316} \approx -\$.6392.$$

Dividing by the \$2 ticket price gives a house edge of 31.96%. ■

While it may seem like a more player-friendly wager because of the additional payoff for a single catch, this ticket is actually slightly worse for the gambler than the common ticket paying 12 for 1 for catching both chosen numbers. The chance of being ahead after 500 games is .00015%.

## Pick 3: Expanded Pay Table

At the Pick 3 level, keno pay tables typically start expanding, to include payoffs for catching either 2 or 3 of the 3 numbers chosen by the player. The probability of losing on this ticket is

$$P(0\,|\,3) + P(1\,|\,3) = \frac{3481}{4108} \approx .8474.$$

We can generalize our work on Pick 2 tickets by writing the expectation as a function of two variables. If the payoff for catching 2 of 3 is $X$ for 1 and that for catching all 3 is $Y$ for 1, the expected value of a $1 ticket is

$$E(X,Y) = (X) \cdot \frac{{}_3C_2 \cdot {}_{77}C_{18}}{{}_{80}C_{20}} + (Y) \cdot \frac{{}_3C_3 \cdot {}_{77}C_{17}}{{}_{80}C_{20}} - 1,$$

or

$$E(X,Y) = \frac{285}{2054}X + \frac{57}{4108}Y - 1.$$

The expected value increases, and the house advantage decreases, as $X$ and $Y$ increase, but there's a limit to how high a casino can set these values and still make money on keno. Table 3.5 gives the current or historical values of $X$ and $Y$, and the corresponding house advantages, for several casinos. All payoffs are converted to the equivalent "for 1" payoff on a $1 wager.

TABLE 3.5: Pick 3 payoffs and HAs

| **Casino** | ***X*** | ***Y*** | **HA** |
|---|---|---|---|
| Harvey's: South Lake Tahoe, NV (1982) | 1 | 52 | 13.97% |
| California: Las Vegas, NV | 1 | 50 | 16.75% |
| The Reserve: Henderson, NV (1998) | 1 | 48 | 19.52% |
| Big Red Keno: Blair, NE | $\frac{1}{2}$ | 50 | 23.69% |
| Orleans: Las Vegas, NV | 1 | 45 | 23.69% |
| FireKeepers: Battle Creek, MI | 1 | 44 | 25.07% |
| Harrah's: Reno, NV (1969) | 1 | $43\frac{1}{3}$ | 26.00% |
| El Cortez: Las Vegas, NV | 1 | 43 | 26.46% |
| Meskwaki: Tama, IA | 1 | 42 | 27.85% |
| Harrah's: Reno, NV (2010) | 1 | 41 | 29.24% |
| Foxwoods: Mashantucket, CT | $2\frac{1}{2}$ | 25 | 30.62% |
| Carson Station: Carson City, NV | 1 | 40 | 30.62% |
| Texas Station: North Las Vegas, NV | 0 | 50 | 30.62% |
| Big Red Keno: Omaha, NE | $\frac{1}{4}$ | 46 | 32.70% |

**Example 3.5.** At the Opera House Casino in North Las Vegas, Nevada, a $1 Pick 3 ticket offering a free ticket for catching 2 numbers and a $43 payoff for catching all 3 was available. Substitution of the free ticket for the $1 cash payoff on a comparable ticket at the El Cortez raises the HA of this ticket from 26.46% to 30.72%. ■

**Example 3.6.** The regular $1 Pick 3 pay table at Arizona Charlie's Decatur pays $1 for catching 2 numbers and $45 for catching 3, matching the game on offer at the Orleans shown in Table 3.5. The casino offers a different Pick 3 game with slightly better payoffs to players willing to commit to 21 or more consecutive games in advance. The bonus rate on a $1 Pick 3 ticket retains a $1 payoff for catching 2 numbers, but pays a winner $48 if all 3 numbers are caught. The HA on the bonus rate ticket drops from 23.69% to 19.72%. ■

In addition to a slightly smaller HA, gamblers choosing this option at Arizona Charlie's Decatur are given a year from the time of their bet to collect their winnings. This can be handy when buying into hundreds of tickets at once, especially for travelers. Additionally, the casino offers lower prices to these long-term players: by committing to 100 or more games, the ticket price can be reduced to 5¢ with payoffs dropping in proportion, of course.

**Example 3.7.** Table 3.5 shows that Big Red Keno in Blair, Nebraska's regular Pick 3 game carries a 23.69% HA. Big Red Keno operates in over 240 locations in eastern Nebraska, and each city's outlets offer a monthly special game with an improved pay table [9]. In September 2017, Blair's special was a $1 Pick 3 ticket paying 50¢ on 2 catches and $55, up from $50, if all three numbers were caught. This change in the pay table lowers the HA to 16.75%. ■

In designing a keno pay table, a casino has two options: to set the payoffs and let them determine the HA, or to choose the HA first and adjust the pay schedule to deliver that edge. At La Vista Keno, several different Pick 3 pay tables all lead to the same house advantage: 16.75%.

- **Regular Keno**: A $1 ticket pays 50¢ on 2 catches and $55 for catching all 3 numbers—the same payoffs as the September 2017 special in Blair.
- **Nifty Fifty**: These 50¢ tickets pay 25¢ for 2 catches and $27.50 for catching 3. This ticket is simply the Regular Keno ticket with the price and all payoffs divided by 2, which leaves the HA unchanged.
- **$1.50 Special**: Tickets cost $1.50 and pay 50¢ on 2 catches, $85 for catching all 3.

However, **Pick 3 Quarter Madness** at La Vista breaks this pattern. Catching 2 of 3 numbers on this 25¢ ticket pays 25¢, but catching all 3 numbers pays only $12. The resultant HA is 19.52%; if the Catch 3 payoff was $12.50, the ticket would also have a 16.75% HA.

## Pick 4: Catch All Option

In Table 3.5, the entry for Texas Station illustrates the "Catch All" option offered on some keno tickets. This ticket has a high payoff when all numbers are caught, but extracts a price from players by eliminating the payoff on smaller catches. At Treasure Island (TI) on the Las Vegas Strip, these games begin at the Pick 4 level and are offered alongside other Pick 4 tickets that also pay off on fewer than 4 catches. Pay tables for TI's $2 standard and Catch All Pick 4 games are listed in Table 3.6.

TABLE 3.6: Treasure Island $2 Pick 4 game options

| **Standard** | | **Catch All** | |
|---|---|---|---|
| Catch | Payoff | Catch | Payoff |
| 2 | 2 | 2 | 0 |
| 3 | 4 | 3 | 0 |
| 4 | 320 | 4 | 450 |

Players must choose which Pick 4 game they wish to play when marking a ticket. Which is the better choice?

For a Catch All ticket picking $n$ numbers and paying $X$ for 1 when all are caught, the general expected return is simply

$$E = (X) \cdot \frac{{}_{80-n}C_{20-n}}{{}_{80}C_{20}} - 1$$

—since all $n$ numbers must be caught to win, there is only one factor in the numerator of the probability. $P(n \mid n) = {}_nC_n = 1$, and so this factor may be dropped.

For Treasure Island's Pick 4 game, we have

$$E = (450) \cdot \frac{{}_{76}C_{16}}{{}_{80}C_{20}} - 2.$$

The expectation is approximately −62.15¢, and the house advantage is half this, or 31.07%.

The expected value of a $2 wager on TI's standard Pick 4 ticket is

$$E = (2) \cdot \frac{{}_4C_2 \cdot {}_{76}C_{18}}{{}_{80}C_{20}} + (4) \cdot \frac{{}_4C_3 \cdot {}_{76}C_{17}}{{}_{80}C_{20}} + (320) \cdot \frac{{}_{76}C_{16}}{{}_{80}C_{20}} - 2 \approx -\$.4214.$$

The HA is 21.07%. Figure 3.2 shows a spreadsheet calculation of these house edges.

Everything in a casino comes at a price, and the price of the increased Catch 4 payoff on the Catch All ticket is a zero payoff for catching 2 or 3 numbers. This change in the pay table leads to a 10% higher house advantage, which is ultimately the cost borne by the player pursuing the higher payoff.

**Example 3.8.** As is often the case, looking around a bit in a casino's keno paybook may turn up better odds. At the Foxwoods Casino, a $2.50 Pick 4 ticket pays off $10 when catching 3 of 4 numbers and $400 when catching all numbers, for a house edge of 33.69%. This is the equivalent of a 160 for

| | A | B | C | D | E | F | G | H |
|---|---|---|---|---|---|---|---|---|
| 1 | Pick | Catch | Probability | Standard | Catch All | | C(80,20) = | 3.53532E+18 |
| 2 | 4 | 2 | 0.212635466 | 2 | 0 | | Ticket price: | 2 |
| 3 | | 3 | 0.043247891 | 4 | 0 | | | |
| 4 | | 4 | 0.003063392 | 320 | 450 | | | |
| 5 | | | | | | | | |
| 6 | | | Expectation: | -0.42145 | -0.62147 | | | |
| 7 | | | HA: | 21.07% | 31.07% | | | |
| 8 | | | | | | | | |

FIGURE 3.2: Spreadsheet calculation of Pick 4 expected values.

1 payoff when all 4 numbers are caught. A Foxwoods keno player wishing to forgo both the 3-number payoff and the option of choosing his or her own numbers can get a 250 for 1 payoff, a $250 payoff on a $1 bet, by playing *Four Corners Keno.* A Catch All game, Four Corners Keno pays off if the four numbers 1, 10, 71, and 80—the numbers at the corners of the ticket—are drawn.

Specifying the numbers in advance, rather than leaving them to player choice, has no effect on the probability of winning. The expected value of a $1 bet is

$$E = (250) \cdot \frac{{}_4C_4 \cdot {}_{76}C_{16}}{{}_{80}C_{20}} - 1 = -\$.2342$$

—giving a house edge that is 10% smaller than in the first game: 23.42%. Giving up choice is rewarded here. ■

## Pick 5: The Price Also Matters

When assessing keno games in search of the best one—however "best" might be defined—the ticket price can be just as important as the pay table.

A look at the vast array of slot machines on a typical casino floor, with denominations ranging from ½¢ to $5000 (in high-roller areas), might lead a casual observer to conclude that if you have a coin or bill of any denomination in your pocket, the casino will offer a way for you to leave it with them. Different denominations of slot machines, of course, have different return percentages, with the fraction of wagers returned as payouts typically increasing as the machine's denomination increases.

With the near-total replacement of coin-operated slot machines by ticket in/ticket out (TITO) machines which accept currency and pay out in tickets to be redeemed with the cashier or at an ATM-like machine, the list of places to drop a coin in a casino is admittedly shorter, but the El Cortez Casino in downtown Las Vegas has valiantly attempted to fill that void by offering keno games in a wide array of denominations. A gambler at the El Cortez will find a keno paybook offering standard games for as little as 40¢ as well as certain specialized games running as much as $5 per ticket. The Pick 5 games at the

El Cortez show a particularly wide range of playable wagers and pay tables, with denominations of 40¢, 65¢, 70¢, 85¢, 90¢, 95¢, $1.00, $1.10, and $1.25 available. The payoffs in these games are shown in Table 3.7.

TABLE 3.7: El Cortez Pick 5 keno payoffs ($)

| | Wager | | | | | | | | |
|---|---|---|---|---|---|---|---|---|---|
| **Catch** | **40¢** | **65¢** | **70¢** | **85¢** | **90¢** | **95¢** | **$1.00** | **$1.10** | **$1.25** |
| 3 | 0.40 | 0.65 | 1.20 | 0.50 | 0.50 | 0 | 1 | 0.60 | 1.25 |
| 4 | 3 | 3 | 15 | 15 | 15 | 5 | 25 | 10 | 22 |
| 5 | 400 | 600 | 340 | 600 | 650 | 1000 | 500 | 1000 | 1000 |

What does this table tell us?

- Investing more money on a ticket is no guarantee of higher payoffs if you hit a particular winning combination.
- A change in the ticket price of as little as 5¢ may dramatically change the payoff schedule.
- Some of these "wins" are merely break-even payoffs, and some are net losses when the cost of the ticket is subtracted.

One important conclusion is less clear (perhaps intentionally): which of these pay tables offers the lowest house advantage. A look at Figure 3.3 provides the answer, and perhaps reveals some surprising additional information.

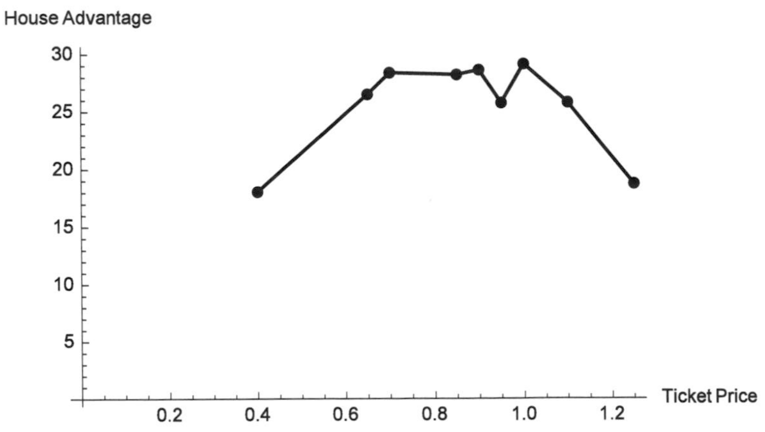


FIGURE 3.3: House advantage for El Cortez keno Pick 5 bets.

- The bets with the lowest HA are the extreme values: the 40¢ and $1.25 games. The 40¢ game barely beats out the $1.25 game when the percentages are computed: 18.04% to 18.73%.

- The 95¢ game is somewhat more player-friendly (or less player-unfriendly) than its neighbors at 90¢ and $1, to the tune of about 3% even as it eliminates the payoff for catching 3 of 5 numbers.
- Nonetheless, the house edge for all of these games exceeds 15%. Most of the games have a house edge over 25%.

**Example 3.9.** For its 75th anniversary in 2017, the El Cortez offered a special 75¢ Catch All keno rate. The 5-spot ticket paid $875 if all 5 numbers were caught, giving a house advantage of 24.76%. This is right in line with the house edge on the casino's other Pick 5 games. ■

Taking the $1 ticket as the standard, how does the $1 game at the El Cortez compare to the $1 game at Treasure Island?

The two games' pay tables are shown in Table 3.8.

TABLE 3.8: El Cortez and Treasure Island $1 Pick 5 payoffs

| Catch | El Cortez | Treasure Island |
|---|---|---|
| 3 | 1 | 1 |
| 4 | 25 | 8 |
| 5 | 500 | 900 |

The El Cortez ticket carries a house advantage of 29.13%; at Treasure Island, the HA on a $1 ticket is 23.89%. The better game lives at Treasure Island.

This result stands as a counterexample to the general rule that gambling conditions are typically more favorable to players in downtown Las Vegas than on the Las Vegas Strip. There are occasionally better wagers on the Strip, but they require some careful calculation to find.

In the USA, gambling winnings are taxable as ordinary income. Gambling losses may be used to offset winnings, as when $1100 in documented losses may be applied to $1150 in winnings to reduce a player's taxable gambling income to $50, though losses are not otherwise deductible from other income. Casinos are required to notify the Internal Revenue Service of any player's net keno winnings (money won minus the cost of the ticket) that exceed $1500; net wins that fall short of $1500 are left to the gambler's honor to report.

Players Keno in Papillion, Nebraska offers a "No Tax" 5-spot game designed so that player winning the top prize is not subject to mandatory IRS reporting. The tickets may be purchased for a minimum wager of $1.50 and are paid off according to Table 3.9.

Catching all 5 numbers results in a net payoff of $1498.50, just under the IRS reporting limit. The winning gambler should declare this payoff as income when filing a tax return.

For a look at how Pick 5 games can be an important part of video keno strategy, see Section 3.4.

TABLE 3.9: Players Keno: No Tax Pick 5 pay table ($1.50 ticket)

| Catch | Payoff |
|---|---|
| 3 | $1.50 |
| 4 | $8.00 |
| 5 | $1500.00 |

## Pick 6: Casinos Versus Lotteries

A number of state and provincial lotteries offer keno games as an adjunct to other lottery ticket options. Typical lottery-run keno games use computer-generated numbers, and can be drawn multiple times per day. In Michigan, the Club Keno game is played in bars throughout the state linked through a computer network, and draws new numbers every 5 minutes. An exception is Washington, where state law limits all lottery games, including Daily Keno, to one drawing per day. These games tend to offer fewer options than casino keno; game choices are usually restricted to standard Pick 1 through Pick 10 games, and options such as way tickets, Catch All games, and other unusual tickets are absent from the lottery game menu.

Lotteries are, of course, intended in part to raise funds for education and other public projects and as a result, house advantages on lottery tickets tend to run around 50%. Since this far outstrips even the worst bets in any legitimate casino, it would be reasonable to expect that lottery keno games would offer a higher HA than their casino counterparts. We can test this assumption by examining a collection of casino and lottery Pick 6 games.

Why Pick 6? Some gambling authorities recommend picking a "middle of the road" number of spots—4 to 8—for best results when playing keno [41]. These tickets are touted as having the best return rate within a given machine; the biggest payoffs are more likely than when picking more spots, and the payoffs are better than when picking fewer spots.

Pick 6 pay tables and house advantages for 11 casino games and 5 lottery games, are listed in Table 3.10. All casino pay tables are taken from standard games, and all payoffs are based on a $1 wager.

All of the casinos and lotteries agree that catching 3 numbers out of 6 should pay back the $1 wagered. After that, the pay tables go in different directions; the lotteries usually pay more for catching 4 of 6, but the casinos generally offer better payback on catches of 5 and 6 numbers. However, the house advantage for the five lottery games exceeds that for the eleven casino games by about 5–25%, depending on the games. Washington's Daily Keno game has a HA well out of the range of the other games; this is not totally surprising for a state lottery.

The Catch All Pick 6 options at some casinos offer HAs that are slightly higher. At the Angel of the Winds, the Catch All Pick 6 ticket pays $5250 for

TABLE 3.10: Pick 6 pay tables: $1 tickets.

| | Catch | | | | |
|---|---|---|---|---|---|
| **Casino** | **3** | **4** | **5** | **6** | **HA** |
| Angel of the Winds | 1 | 4 | 92 | 1500 | 27.78% |
| Arizona Charlie's Decatur | 1 | 3 | 110 | 1500 | 25.06% |
| Barton's 93 Club | 1 | 3 | 90 | 1800 | 27.38% |
| D | 1 | 4 | 90 | 1495 | 28.46% |
| El Cortez | 1 | 4 | 90 | 1500 | 28.39% |
| Excalibur | 1 | 4 | 88 | 1500 | 29.01% |
| FireKeepers | 1 | 2 | 75 | 2500 | 25.85% |
| Four Queens | 1 | 4 | 85 | 1500 | 29.94% |
| Golden Nugget | 1 | 4 | 85 | 1480 | 30.20% |
| Orleans | 1 | 4 | 80 | 2000 | 25.04% |
| Treasure Island | 1 | 4 | 80 | 2200 | 22.46% |
| | **Catch** | | | | |
| **Lottery** | **3** | **4** | **5** | **6** | **HA** |
| Caribbean Keno | 1 | 5 | 50 | 1500 | 37.92% |
| Club Keno (Michigan) | 1 | 7 | 57 | 1100 | 35.21% |
| Daily Keno (Washington) | 1 | 4 | 40 | 1000 | 50.32% |
| Keno (Delaware) | 1 | 5 | 50 | 1000 | 44.37% |
| Keno (Oregon) | 1 | 5 | 55 | 1600 | 35.09% |

1, yielding a HA of 32.28%. Treasure Island pays the same catch at $5500 for 1; the HA there is 29.06%.

Some casino Pick 6 games also offer opportunities for other creative wagering options. A 1993 promotional Pick 6 game at the Peppermill Casino in Reno, Nevada gave players one chance in four at an enhanced 6-catch payout with *Super Jackpot Keno.* The standard Pick 6 pay table for a $2 ticket at the Peppermill and the Super Jackpot Keno payoffs are shown in Table 3.11.

TABLE 3.11: Peppermill Casino's Pick 6 regular and Super Jackpot pay tables

| Catch | Standard | Super Jackpot |
|---|---|---|
| 3 | $2 | $2 |
| 4 | $6 | $6 |
| 5 | $180 | $180 |
| 6 | $3600 | $6300 |

Just prior to the drawing, there was a 25% chance that that game would be randomly designated a Super Jackpot game, and the prize for catching all 6 numbers raised [78]. The standard pay table carried a HA of 27.38%;

the Super Jackpot pay table's HA was a low 9.97%. Since the only prize that changed was the top prize, and that only 25% of the time, the effective top prize was $.25 \cdot 6300 + .75 \cdot 3600 = \$4275$, and the expected value of a single \$2 ticket was

$$E = (2) \cdot P(3 \mid 6) + (6) \cdot P(4 \mid 6) + (180) \cdot P(5 \mid 6) + (4275) \cdot P(6 \mid 6) - 2 \approx -\$0.46,$$

giving a composite house advantage of 23.03%.

At Caesars Tahoe (now the MontBleu) in Stateline, Nevada on the California border, keno patrons were offered their choice of three different Pick 6 pay tables. These are shown with their house advantages in Table 3.12.

TABLE 3.12: Caesars Tahoe Pick 6 pay tables

| Catch | Table A | Table B | Table C |
|---|---|---|---|
| 3 | \$1 | \$0 | \$0 |
| 4 | \$3 | \$3 | \$4 |
| 5 | \$100 | \$100 | \$125 |
| 6 | \$1480 | \$2500 | \$3000 |
| **Price** | \$1 | \$1 | \$1.25 |
| **HA** | 28.41% | 28.24% | 28.96% |

The differences among the house edges amount to less than 1%; all three HAs fall into the range determined by the other casino edges in Table 3.10. The only real difference among the three choices is the replacement of a break-even payoff for catching 3 numbers with a higher jackpot for catching all 6.

The Marina Casino, which sat on the site of the current MGM Grand Casino in Las Vegas, ran a keno game called *High Frequency Pay*, which offered more payoffs for a given game, but in smaller amounts. The Pick 6 game paid off on as few as 2 catches, following Table 3.13 for a \$1 ticket.

TABLE 3.13: Marina High Frequency Pay 6-spot pay table

| Catch | Payoff |
|---|---|
| 2 | \$.50 |
| 3 | \$1 |
| 4 | \$8 |
| 5 | \$35 |
| 6 | \$600 |

The top payoff has fallen considerably from the values in Table 3.10. Balancing this change by the partial refund on 2 catches offers only slight compensation, as the HA stands at 30.20%—equal to the Golden Nugget's game, which is the highest listed.

## Pick 7: Every Ticket a Winner

Many keno paybooks start to offer games, at the Pick 7 level and higher, that a player cannot lose because the ticket pays off no matter how many numbers are caught. The Sam's Town 7 Spot Special game, found at Sam's Town in Las Vegas, is an example of this different payoff option which is in some sense the opposite of the Catch All ticket:

> Here is a ticket that is unique for those of you who are always saying...**"I NEVER WIN!"**
>
> Mark any 7 numbers on your Keno ticket. Pay $1.00 and you win on...ALL CATCHES!!!

This new pay table, Table 3.14, pays off no matter how many of a player's seven numbers are drawn—even if that number is 0.

TABLE 3.14: Sam's Town 7 Spot Special pay table

| Catch | Payoff |
|---|---|
| 0 | $1.00 |
| 1 | $0.30 |
| 2 | $0.30 |
| 3 | $0.50 |
| 4 | $2.00 |
| 5 | $10.00 |
| 6 | $100.00 |
| 7 | $2500.00 |

Of course, a moment's careful perusal of the pay table reveals that three of the eight payoffs—on the three most likely outcomes—result in a net loss for the player after the $1 price of the ticket is factored into the game.

**Example 3.10.** Find the probability that this ticket results in a loss.

We have $P(\text{Loss}) = P(1\,|\,7) + P(2\,|\,7) + P(3\,|\,7)$, or

$$P(\text{Loss}) = \frac{({}_7C_1 \cdot {}_{73}C_6) + ({}_7C_2 \cdot {}_{73}C_5) + ({}_7C_3 \cdot {}_{73}C_4)}{{}_{80}C_{20}} \approx .8168.$$

■

The house edge on this wager is 27.34%, which is somewhat worse than the standard Sam's Town keno with 7 numbers selected. That game has the slightly less-unfavorable HA of 24.83%. But at least you're guaranteed to "win" something with the 7 Spot Special.

An alternate way to look at the 7 Spot Special ticket is to deduct the minimum payoff of 30¢ from the ticket price—since every ticket wins at least that amount—and from each payoff. The resulting pay table, for a 70¢ ticket, is Table 3.15.

TABLE 3.15: 7 Spot Special pay table reduced by 30¢ minimum win

| Catch | Payoff |
|---|---|
| 0 | $.70 |
| 1 | 0 |
| 2 | 0 |
| 3 | $0.20 |
| 4 | $1.70 |
| 5 | $9.70 |
| 6 | $99.70 |
| 7 | $2499.70 |

Regarded this way, the ticket has a considerably higher HA: 39.06%.

From the casino's perspective, the 7 Spot Special ticket has the advantage that gamblers receiving a very small payoff at the keno desk might be more motivated than big winners to roll their meager winnings into another ticket, thus giving the casino another shot at retaining that money. Many casino officials think of large sums of money won by a high roller as a "temporary loan" from the casino to the gambler, with the mindset being that the casino will eventually retrieve these funds as the player continues to gamble. The 7 Spot Special extends that way of thinking to low-rolling gamblers.

Seven, of course, is regarded by many people as a particularly lucky number. Some casinos recognize this superstition with special Pick 7 keno games, such as several Caesars casinos in Las Vegas—Bally's, Harrah's, and the Rio—which offer a bet called "Just 7777s". This is a Pick 7 game with pay table shown in Table 3.16.

TABLE 3.16: Just 7777s pay table

| Catch | Payoff |
|---|---|
| 4 | $0.70 |
| 5 | $7 |
| 6 | $77 |
| 7 | $7777 |

Continuing the focus on the lucky number 7, the ticket price for this game is 70¢. The emphasis on luck runs out when the game is played, though. The house edge for this game exceeds 50%, putting it firmly on a par with state lotteries.

### Keno in Israel

Israel's national lottery offers a keno variant in which the lottery agency picks 17 numbers from 1–70 instead of 20 from 1–80. As with some lottery-sponsored keno games in U.S. states, the game is drawn multiple times daily and the results broadcast to sites across the country [97].

Three tickets are offered, all of which are interpreted as Pick 7 way tickets:

- A standard 7-spot ticket.
- A Systematic 8-spot ticket which is played as an 8-way 7-spot ticket.
- A Systematic 9-spot ticket comprising ${}_9C_7 = 36$ 7-spot tickets.

The pay table for this game, which includes a 1 for 1 payoff for no catches, is shown in Table 3.17. It should be noted that although the payoffs in this table are stated "for 1", the bet range in Israel is 2–7 new Israeli shekels (NIS).

TABLE 3.17: Israel Lottery Pick 7 keno pay table [97]

| Catch | Payoff (for 1) |
|---|---|
| 0 | 1 |
| 3 | 2 |
| 4 | 5 |
| 5 | 7 |
| 6 | 40 |
| 7 | 500 |

**Example 3.11.** How do the probabilities of the various catches in this game compare to those of a standard 80-ball Pick 7 keno game?

The probability distribution function for Israel's game is

$$p_1(k\,|\,7) = \frac{{}_7C_k \cdot {}_{63}C_{17-k}}{{}_{70}C_{17}},$$

while a standard keno game has PDF

$$p_2(k\,|\,7) = \frac{{}_7C_k \cdot {}_{73}C_{20-k}}{{}_{80}C_{20}}.$$

The two functions turn out to be nearly identical. See Table 3.18.

■

While the probabilities of the different catches are nearly the same, the payoffs for catching 0 or 3 spots, which standard 7-spot tickets do not offer, give a small edge to the Israeli game. The HA of a standard Pick 7 game using the Orleans Casino pay table on page 6 is 28.14%, while the HA in Israel is 22.69%.

TABLE 3.18: Comparison of Israeli lottery and standard keno probabilities

| Catch ($k$) | $p_1(k \mid 7)$ | $p_2(k \mid 7)$ |
|---|---|---|
| 0 | .128584 | .121574 |
| 1 | .325563 | .315193 |
| 2 | .325563 | .326654 |
| 3 | .166104 | .174993 |
| 4 | .046509 | .052191 |
| 5 | .007113 | .008639 |
| 6 | .000547 | .000732 |
| 7 | $1.62232 \times 10^{-5}$ | $2.44026 \times 10^{-5}$ |

**Example 3.12.** The probability of catching 7 numbers on a 9-spot ticket is

$$p_1(7 \mid 9) = \frac{{}_9C_7 \cdot {}_{61}C_{10}}{{}_{70}C_{17}} \approx \frac{1}{2427}.$$

What is the payoff on this catch?

- One ticket catches all 7 numbers, and pays at 500 for 1.
- There are ${}_7C_6 \cdot {}_2C_1 = 14$ tickets that catch 6 of 7 numbers; each one pays 40 for 1.
- The remaining ${}_7C_5 = 21$ tickets catch 5 numbers each, qualifying for a 7 for 1 payoff.

The total prize is NIS 1207 for each shekel wagered. ■

## Pick 8: Balancing Act

For many years, the only keno bet available was a 10-spot ticket. As keno became established and betting options expanded, bettors came to prefer 8-spot tickets over all others [90]. This is possibly because players perceived an 8-spot ticket as striking a balance between a higher probability of winning (with a smaller number of spots marked) and a larger prize (by choosing and catching more numbers). Betting a $1 8-spot ticket against the Orleans pay table (page 6) gives a small chance of winning $50,000; this cannot be matched by a $1 ticket with fewer spots marked. At the same time, playing a $1 9-spot ticket to win $50,000 requires catching all 9 numbers, which is clearly less likely than catching 8 out of 8, and can easily be shown to be exactly 6 times less likely.

This preference was not, however, based on good mathematics. Figure 3.4 shows the probability $p(k)$, expressed as a percentage, of winning any amount on a Pick $k$ ticket played against the standard pay table, for $1 \leqslant k \leqslant 10$.

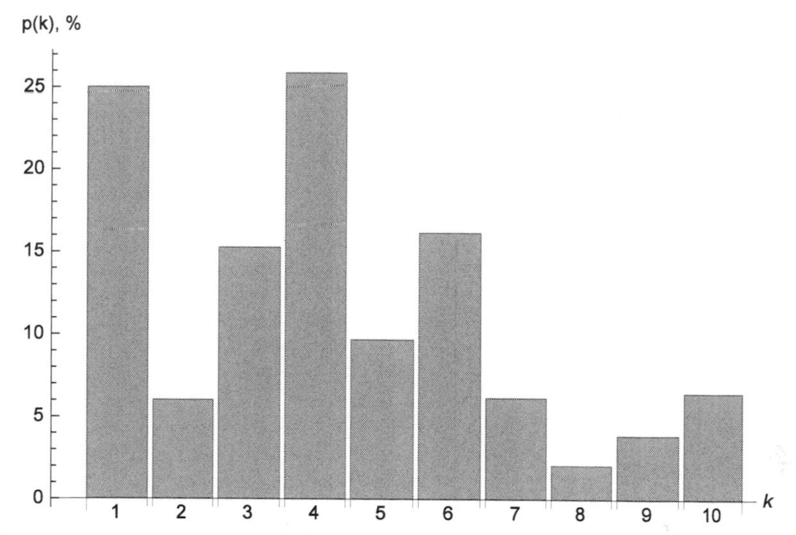


FIGURE 3.4: Probability $p(k)$ of winning keno ticket with $k$ spots marked.

A Pick 8 ticket turns out to have the *lowest* probability of winning anything: 2.08%. The best chance of holding a winning ticket comes on a pick-4 ticket, at 25.89%, but 82% of this probability is a simple break-even payoff for catching 2 numbers out of 4. The best ticket for making a profit is the lowly pick-1 ticket, where every win scores a profit. This, however, lacks a certain excitement, as the payoff is a mere 3 for 1.

Of course, pay tables can vary among casinos. At the Excalibur Casino, the Red Hot Keno Special's Pick 8 pay table starts when 4 numbers of 8 are caught, instead of 5 as in the pay table on page 6. The probability of catching 4 numbers is approximately .0815, which raises the probability of a win on an 8-spot ticket from 2.08% to 10.23%, and moves a Pick 8 ticket to the middle of the pack: 5th out of 10 game options.

## Aggregate Limit

> *The more you play your ticket for, the easier it is to win $50,000.00.*—[Sentence found in multiple keno paybooks]

An important feature of most keno schedules, although one that may not have much of an effect on actual gameplay, is the *aggregate limit*. This value represents the maximum amount that the casino will pay winners on any one drawing. For example, the aggregate limit at the Four Queens Casino in Las Vegas is $100,000. If the winnings due players on any one keno drawing exceed this amount, the payouts to winners are prorated in accordance with established rules. One casino practice is simply to pay each winning ticket the same percentage of the amount won, so if the amount due players at the Four Queens was $110,000, each ticket would receive $\frac{10}{11}$ of the amount won. This might turn some break-even payoffs into losing tickets. Some casinos pay off small wins in full and reserve prorating for larger prizes.

In some keno lounges in Nebraska, the aggregate limit applies to a day's total payouts from all drawings, possibly across multiple locations. Keno Casino imposes a $25,000 cumulative payout limit per day at all of its locations in Cass and Otoe counties, though only wins exceeding $1500 contribute toward reaching this limit.

The aggregate limit in keno plays a role similar to that of the maximum allowable bet on casino table games: it limits the casino's exposure in the event that a patron, or several patrons, hits a particularly lucky outcome. Rather than discouraging high-rolling play by capping the amount that gamblers are allowed to bet, an artificial limit on payouts (which, in practice, is seldom reached) allows the casino to keep its losses under control.

Nevada gaming regulations imposed a $25,000 aggregate limit on keno until 1979, when the limit was raised to $50,000, in part due to inflation. In 1989, the cap on aggregate limits was removed, and casinos were set free to determine the amount of risk they wished to assume in their keno lounges [11].

While it is unlikely that the aggregate limit will be reached in any keno drawing, a smart keno player will nonetheless take this into account when sizing his or her bets. If a particular game has a top prize of $50,000 for a $1 wager and a $100,000 aggregate limit, but can be bet for any multiple of $1 the player desires with payoffs below the aggregate limit increasing in proportion to the wager, it would be foolish to bet more than $2 at this game. If a $3 ticket hit the jackpot, the payout would be $100,000, not the expected $150,000.

This cap on overall payoffs has a slight effect on the house advantage when the payoff for a single game is subject to the limit. Consider the Pick 8 pay table in Table 3.19, drawn from the basic game at the Avi Casino in Laughlin, Nevada, the only Native American casino resort in that state, at a time when the casino's aggregate limit was $50,000.

TABLE 3.19: Avi Casino's Pick 8 pay table with $50,000 aggregate limit

| **Catch** | **Payoff (for 1)** |
|---|---|
| 4 | 1 |
| 5 | 8 |
| 6 | 80 |
| 7 | 1500 |
| 8 | 20,000 |

If a player wagers more than a dollar against this table, the payouts increase in proportion—except when that new prize would exceed the $50,000 aggregate limit. The HA of this ticket is 25.51% on a $1 wager; this increases to 26.96% if the bet is $3 and hits 29.86% on a $5 ticket. While the increase in the HA is small due to the low probability of catching all 8 numbers, a ticket like this should never be bet for more than $2.50.

**Example 3.13.** A 1963 keno ticket from the Sahara Casino offered some questionable advice regarding the aggregate limit. Under the heading "How To Win $25,000.00", the Pick 15 pay table in Table 3.20 is shown. This table is based on a $5 wager.

TABLE 3.20: Sahara Pick 15 pay table: 1963

| Catch | Payoff (for 5) |
|---|---|
| 7 | 40 |
| 8 | 140 |
| 9 | 820 |
| 10 | 3150 |
| 11 | 13,000 |
| 12 | 25,000 |
| 13 | 25,000 |
| 14 | 25,000 |
| 15 | 25,000 |

While the table repeats the aggregate limit payoff for catching 12 or more numbers, the payoffs are artificially capped at that level. Catching 13 numbers out of 15 is approximately 31 times less likely than catching 12, and so is deserving of a considerably higher payoff. Similar mismatches between probabilities and payoffs occur at the catch-14 and catch-15 levels. ■

If you were truly out to win $25,000 in 1963, you would have had a far better chance of doing so—and less of a chance of having your winnings capped by the aggregate limit—by playing $7 on the Sahara's Pick 8 game advertised on the same ticket, which uses Table 3.21 as its pay table. This game paid $25,000 for catching 8 out of 8 numbers, which is nearly 6½ times more likely than catching 12 or more numbers out of 15.

TABLE 3.21: Sahara Pick 8 pay table: 1963

| Catch | Payoff (for 7) |
|---|---|
| 4 | $10.50 |
| 5 | $100.80 |
| 6 | $756.00 |
| 7 | $5600.00 |
| 8 | $25,000.00 |

As the aggregate limit rose to $25,000, Harolds Club's paybook highlighted the easiest ways to hit that top prize. The 15¢ minimum wager at the time wasn't enough to win $25,000, but a Pick 8 game showed Table 3.22 as its pay table for the gambler willing to risk $1736.60—catching a mere 5 numbers would hit the maximum jackpot.

TABLE 3.22: Harolds Club Pick 8 pay table for a $1736.60 bet

| **Catch** | **Payoff** |
|---|---|
| 3 | $173.65 |
| 4 | $2691.65 |
| 5 | $25,000.00 |
| 6 | $25,000.00 |
| 7 | $25,000.00 |
| 8 | $25,000.00 |

An unusual feature of Table 3.22 is that it offers a payoff—albeit one just less than 10% of the cost of the ticket—for catching only 3 of 8 numbers. This was a common feature of Reno keno paybooks of that era. While this raises the probability of winning something from the 10.03% or 2.08% noted on page 81 for Pick 8 games to 31.71%, it masks the fact that the HA on this ticket is 55.23% due to the repeated appearance of $25,000 on the pay table. The average return on one of these tickets is –$959.07.

**Cover All**

It's possible to have all 80 numbers working for you on a keno ticket by playing a *Cover All* way ticket that divides all of the numbers into 20 blocks of 4 and offers the entire collection as a 190-way 8-spot ticket. $190 = {}_{20}C_2$, the number of ways to choose any two of these 4-spot blocks to make an 8-spot ticket. Typically, this is a preprogrammed option that eliminates the need for the player to mark up a bet slip. Standard arrangements of the blocks partition the ticket vertically (Figure 3.5) or in square blocks of 4 (Figure 3.6).

| 1 | 2 | 3 | 4 | 5 | 6 | 7 | 8 | 9 | 10 |
|---|---|---|---|---|---|---|---|---|---|
| 11 | 12 | 13 | 14 | 15 | 16 | 17 | 18 | 19 | 20 |
| 21 | 22 | 23 | 24 | 25 | 26 | 27 | 28 | 29 | 30 |
| 31 | 32 | 33 | 34 | 35 | 36 | 37 | 38 | 39 | 40 |
| 41 | 42 | 43 | 44 | 45 | 46 | 47 | 48 | 49 | 50 |
| 51 | 52 | 53 | 54 | 55 | 56 | 57 | 58 | 59 | 60 |
| 61 | 62 | 63 | 64 | 65 | 66 | 67 | 68 | 69 | 70 |
| 71 | 72 | 73 | 74 | 75 | 76 | 77 | 78 | 79 | 80 |

FIGURE 3.5: Cover All Keno bet slip with vertical blocks of 4 indicated.

| | | | | | | | | | |
|---|---|---|---|---|---|---|---|---|---|
| 1 | 2 | 3 | 4 | 5 | 6 | 7 | 8 | 9 | 10 |
| 11 | 12 | 13 | 14 | 15 | 16 | 17 | 18 | 19 | 20 |
| 21 | 22 | 23 | 24 | 25 | 26 | 27 | 28 | 29 | 30 |
| 31 | 32 | 33 | 34 | 35 | 36 | 37 | 38 | 39 | 40 |
| 41 | 42 | 43 | 44 | 45 | 46 | 47 | 48 | 49 | 50 |
| 51 | 52 | 53 | 54 | 55 | 56 | 57 | 58 | 59 | 60 |
| 61 | 62 | 63 | 64 | 65 | 66 | 67 | 68 | 69 | 70 |
| 71 | 72 | 73 | 74 | 75 | 76 | 77 | 78 | 79 | 80 |

FIGURE 3.6: Cover All Keno bet slip with square blocks of 4 indicated.

The Angel of the Winds Casino in Arlington, Washington invites players to use their imagination on this bet by creating their own set of 20 blocks of 4 that covers all 80 numbers. A fan of the video game Tetris could have great fun designing such a ticket.

Cover All tickets are frequently offered at a reduced rate—as low as 5¢ per way, at the D Casino—and often use the standard Pick 8 pay table, scaled down in proportion to this reduced bet. The D's Pick 8 pay table, for a 5¢ bet, is shown in Table 3.23. The equivalent payoffs for 1 are included in the 3rd column for ease of comparison.

TABLE 3.23: Pick 8 pay table for 5¢ bet at the D Casino

| **Catch** | **Payoff** | **For 1** |
|---|---|---|
| 5 | $0.45 | 9 |
| 6 | $4.00 | 80 |
| 7 | $74.75 | 1495 |
| 8 | $1250.00 | 25,000 |

The Angel of the Winds, by contrast, uses a separate pay table for Cover All bets. See Table 3.24, which shows the two pay tables.

TABLE 3.24: Pick 8 pay tables (for 1) at Angel of the Winds

| **Catch** | **Standard** | **Cover All** |
|---|---|---|
| 5 | 8 | 5 |
| 6 | 102 | 60 |
| 7 | 1500 | 2000 |
| 8 | 18,000 | 36,000 |

Since the better payoff on the two games varies with the number of catches, determining the better game requires computing the expectation per way. For the standard ticket with a 1-unit wager,

$$E = (8) \cdot P(5) + (102) \cdot P(6) + (1500) \cdot P(7) + (18,000) \cdot P(8) - 1,$$

where

$$P(k) = P(k \mid 8) = \frac{{}_8C_k \cdot {}_{72}C_{20-k}}{{}_{80}C_{20}}.$$

This gives $E \approx -.2933$ units. For the Cover All option,

$$E = (5) \cdot P(5) + (60) \cdot P(6) + (2000) \cdot P(7) + (36,000) \cdot P(8) - 1 \approx -.2891,$$

so the expectation is very slightly better when betting Cover All. However, the Cover All ticket gains its advantage by paying larger amounts for the less-likely events of catching 7 or 8 numbers. This should be taken into consideration: if you don't hit 7 or 8 spots on a ticket, you're not getting the full advantage of the pay table.

At the D, the house advantage on Cover All is 29.74%, which is marginally worse than the Angel of the Winds games.

## Pick 9: 10 Spots Reimagined

In the interest of diminishing the need for keno players and employees to handle lots of small change, some keno operations have issued special chips for the keno lounge. Four such chips are shown in Figure 3.7.

FIGURE 3.7: Keno chips.

From left to right, top to bottom:

- 50¢ chip from The Diggin's, a keno bar in Helena, Montana.

- Non-denominated chip from the El Cortez in Las Vegas.
- 70¢ chip from the Union Plaza Casino, now the Plaza, in Las Vegas.
- $1 chip from the Pioneer Casino in Reno.

The 70¢ Union Plaza chip shown here is certainly an unusual denomination. This originates from a time when a 9-spot keno ticket was the second most popular wagering option, trailing only the 10-spot ticket [91]. Such a ticket was interpreted as a 71-way 10-spot ticket, with each of the unmarked numbers functioning as a king, just as was done in the Chinese lottery. These tickets were priced at a penny per way, rounded from 71¢ down to 70¢ for everyone's convenience. The special 70¢ chips made placing and paying bets easier on players and casino employees alike.

These 70¢ tickets were played against Table 3.25.

TABLE 3.25: 9-spot (71-way 10-spot) pay table, 1¢ per way [90, p. 98]

| Catch | Payoff |
|---|---|
| 4 | $0.30 |
| 5 | $3.60 |
| 6 | $35.60 |
| 7 | $221.40 |
| 8 | $747.00 |
| 9 | $1476.00 |

**Example 3.14.** Consider the case where 7 of 9 numbers are caught, an event with probability

$$P(7\,|\,9) = \frac{{}_9C_7 \cdot {}_{71}C_{13}}{{}_{80}C_{20}} \approx 5.917 \times 10^{-4}.$$

Marking this ticket as a 71-way 10-spot ticket means that the ticket has caught 8 out of 10 spots, 13 times (once for each of the unmarked numbers that was drawn). With a wager of 1 cent per way, the payoff for catching 7 of 10 numbers translates to $221.40/13 \approx 17.03$, a payoff of 1703 for 1 on each 8-spot catch. ■

By contrast, a standard 10-spot ticket (see page 6) pays 1000 for 1 when catching 8 numbers, and a standard 9-spot ticket pays 300 for 1 on a 7-number catch, so there was an advantage in playing this 9-spot ticket as a 71-way 10-spot. The advantage was diminished by the low cost of entry, which leads to a low absolute payoff.

**Example 3.15.** In some Asian countries, 9 is considered to be a lucky number. In Omaha, Nebraska, June and July 2017 saw Big Red Keno offer "Lucky 9" as its monthly special game. A $1 ticket was played against Table 3.26. This

TABLE 3.26: Big Red Keno's Lucky 9 pay table

| Catch | Payoff |
|---|---|
| 7 | $99 |
| 8 | $9999 |
| 9 | $99,999 |

game was only lucky for Big Red Keno, however, as the chance of winning was only 1 in 1600 and the house advantage was 54.31%. The loss of payoffs for catching 4–6 numbers looks severe, but if the pay table was revised in keeping with the game's theme to pay $9 for catching 6 of 9 numbers and $.99 (almost breaking even) for catching 5 of 9, the HA would still be high: 45.94%. ■

## Pick 10: The First Choice

Pick 10 is, of course, the original keno wager, going back to the days of pák kòp piú. The original pák kòp piú pay table (Table 1.1, page 3) carried a 24.04% house advantage, assuming a 5% commission was charged. The house edge went up as the game evolved to pakapoo in New Zealand; the two pay tables in Table 1.2 (page 3) carry HAs of 39.35% for Game A and 62.67% for Game B—we see that eliminating small payoffs for matching fewer numbers is not balanced by higher payoffs for less-likely matches of more numbers, as was done in the pay table for Game B. This is a phenomenon that recurs in keno game design.

The HAs didn't stop rising there, though. In 1989, just after the cap on aggregate limits was removed in Nevada, Caesars Palace introduced a $2 "Winner Take All" Pick 10 ticket with a million-dollar prize [91]. This was a Catch All ticket, with no prizes for catching fewer than 10 numbers. The expectation on this ticket was

$$E = (1,000,000) \cdot P(10 \mid 10) - 2 = (1,000,000) \cdot \frac{_{70}C_{10}}{_{80}C_{20}} - 2 \approx -\$1.89,$$

and the house advantage was *94.39%*. This ticket's high HA was exposed in the gaming press, and was then soon discontinued [26].

**Example 3.16.** Caesars Palace's million-dollar ticket was superseded as the worst keno ticket ever by one introduced at Harrah's Casino in Las Vegas in 2010. The special "Stimulus" game, which the casino advertised as "Doing Our Part To 'Boost' Your Economy", was a 5¢ Catch All game that required buying in for at least 100 tickets; a minimum investment of $5. The game was offered in Pick 5 through Pick 10 versions; the Pick 10 game had a prize of $15,000, which, at 300,000 for 1, sounds good compared to the 5¢ ticket price [139].

The tempting payoff masked an astronomical house advantage. The expectation on a single ticket was

$$E = (15,000) \cdot P(10 \mid 10) - .05 \approx -\$.0483$$

—the HA was 96.63%. ■

This ticket returned in 2017. "Boost Your Economy" reappeared as the label for a sequence of five 5¢ Catch All tickets from Catch 6 through Catch 10. While the other tickets all had HAs exceeding 48%, none was nearly as bad for the player as the Pick 10.

## A Game Evolves

Both the Caesars Palace and Harrah's tickets were special games, separate from their standard 10-spot games. In a competitive gaming market, standard games with HAs near or exceeding even 40%, to say nothing of over 90%, would be untenable, as a rival game operator could easily offer more favorable odds while still making a good profit. We shall examine a number of Pick 10 pay tables from Nevada casinos to explore how the house edge has changed as the aggregate limit, and thus the pay table, has changed.

It should be noted at the outset that the probability of catching all 10 numbers on a Pick 10 ticket is

$$P(10 \mid 10) = \frac{{}_{10}C_{10} \cdot {}_{70}C_{10}}{{}_{80}C_{20}} \approx 1.1221 \times 10^{-7} \approx \frac{1}{8,911,711},$$

so the house edge is not affected very much by changes in the top payoff, which is itself often close to or equal to the aggregate limit. However, a change in the aggregate limit might lead to a change in the entire payoff structure, not just the top jackpot, and thus a very different house advantage.

When keno was first played legally in Nevada, beginning in the 1930s, a standard keno pay table, Table 3.27, was in common use at casinos throughout the state [91]. This pay table dominated Pick 10 play until 1954, and gave the casino a 26.72% edge.

TABLE 3.27: Pick 10 pay table: 1931–1954

| Catch | Payoff (for 1) |
|---|---|
| 5 | 2 |
| 6 | 18 |
| 7 | 180 |
| 8 | 900 |
| 9 | 1800 |
| 10 | 3600 |

In the 1940s, Harolds Club in Reno offered a $1 Pick 10 game with a $5000 aggregate limit. This ticket was played against Table 3.28, whose payouts match or exceed Table 3.27 on all catches, and so has a lower HA: 21.54%.

TABLE 3.28: Harolds Club Pick 10 pay table

| Catch | Payoff (for 1) |
|---|---|
| 5 | 2 |
| 6 | 18 |
| 7 | 180 |
| 8 | 1200 |
| 9 | 2500 |
| 10 | 5000 |

At the same time, though, Harolds Club enforced the aggregate limit of $5000 as a player's wager increased by lowering the catch threshold for that top prize. Under the heading "How To Win... $5000.00... Limit", pay tables pointed out that a $2 Pick 10 wager could win $5000 by catching 9 or 10 spots, a $4 wager won $5000 for catching 8–10 spots, and if a player was willing to bet $27.80 on a 10-spot ticket, catching 7 or more numbers would result in a $5000 win. Payoffs below the $5000 limit rose in proportion to the amount wagered. This restriction of the top prize raised the HA only slightly for the $2 and $4 tickets, but the edge on the $27.80 ticket jumped to 37.53%.

1954 saw the Pioneer Club in Reno launch a keno game with a $10,000 aggregate limit, which was swiftly matched by other Reno casinos [91]. Other Nevada casinos also quickly adopted this limit; a Pick 10 pay table with a top prize of $10,000 from the Bonanza Club in North Las Vegas is shown in table 3.29. Relative to Table 3.27, this pay table includes some small increases for

TABLE 3.29: Pick 10 pay table with $10,000 top prize

| Catch | Payoff (for 1) |
|---|---|
| 5 | 2 |
| 6 | 18 |
| 7 | 180 |
| 8 | 1300 |
| 9 | 2600 |
| 10 | 10,000 |

catching 8 and 9 spots as well as a near-tripling of the top prize to $10,000, and it has a house advantage of 20.74%.

In the early 1960s, when an aggregate keno limit of $25,000 was the law in Nevada, many Las Vegas casinos including the California Club, Fremont, Golden Nugget, Horseshoe, Las Vegas Club, Mint, Pioneer Club, Sahara, and

Stardust were using Table 3.30 for $1 Pick 10 tickets. This pay table carried a

TABLE 3.30: Pick 10 pay table: Las Vegas, 1960s

| Catch | Payoff (for 1) |
|---|---|
| 5 | 2 |
| 6 | 18 |
| 7 | 180 |
| 8 | 1300 |
| 9 | 2600 |
| 10 | 25,000 |

house advantage of 20.57%, a relatively low rate that indicated an equilibrium of sorts in a competitive market.

When the aggregate limit was raised to $50,000 in 1979, some casinos were quick to raise their top prizes—which had little effect on the HAs. Replacing 25,000 by 50,000 in Table 3.30 only reduces the HA from 20.57% to 20.29%, due to the low probability of catching all 10 numbers. However, this increased top jackpot provided cover for changing payoffs at lower levels to values that raised the HA. Table 3.31 shows a 1979 pay table used at the Sands Casino. Payoffs at the catch-7 and catch-8 levels have been reduced, while catch-9 and

TABLE 3.31: Sands Casino Pick 10 pay table: 1979

| Catch | Payoff (for 1) |
|---|---|
| 5 | 2 |
| 6 | 18 |
| 7 | 120 |
| 8 | 960 |
| 9 | 3800 |
| 10 | 50,000 |

catch-10 payoffs have risen. The net effect of these changes is to raise the HA by over half relative to Table 3.30, to 33.83%. While the top jackpot may be more attractive for having doubled in size, it is no more probable than it was when it paid only $25,000 for $1.

The cap on aggregate limits in keno was abolished in Nevada in 1989 [26], and this act of the Nevada Gaming Commission freed casinos to assume as much risk as they dared in setting keno pay tables. Caesars Palace responded quickly, changing their $2 Pick 10 pay table as shown in Table 3.32.

In 1988, the house advantage on this ticket was 29.88%. The 1989 pay table was slightly better for the player, but most of that improvement comes from increasing the catch-6 payoff, not by changing the top prize. The new HA is a still-healthy 27.42%. Caesars Palace management was likely hoping

TABLE 3.32: Caesars Palace Pick 10 pay tables: 1988 and 1989

| Catch | 1988 | 1989 |
|---|---|---|
| 5 | 2 | 2 |
| 6 | 40 | 44 |
| 7 | 264 | 264 |
| 8 | 1920 | 1920 |
| 9 | 7600 | 7600 |
| 10 | 50,000 | 80,000 |

that increased keno action, from players drawn to the higher top prize, would compensate for their slightly lower advantage.

The Pick 10 pay table at the Orleans (Table 1.4, page 6) has a top prize equal to the casino's aggregate limit: $100,000. Table 3.33 compares this ticket to an equivalent $1 ticket using Caesars Palace's 1989 pay table. Major dif-

TABLE 3.33: Orleans versus Caesars Palace $1 Pick 10 pay tables

| Catch | Orleans | Caesars Palace |
|---|---|---|
| 5 | 1 | 1 |
| 6 | 25 | 22 |
| 7 | 125 | 132 |
| 8 | 1000 | 960 |
| 9 | 10,000 | 3800 |
| 10 | 100,000 | 40,000 |

ferences favoring the Orleans are evident at the catch-9 and catch-10 levels, but the probabilities of these events are so low that they don't affect the HA significantly—paying 25 for 1 instead of 22 for 1 on catching 6 numbers makes the real difference. This change lowers the HA and favors the player by almost 3½%. The HA of the Orleans ticket is slightly better than at Caesars Palace in 1989: 25.23%.

At the Virgin River Casino in Mesquite, Nevada, the aggregate limit of $100,000 affects the Pick 10 pay table only if a player wagers more than $2.50. The pay table, Table 3.34, lists three different pay schedules, for wagers of $1, $5, and $10. Payoffs increase in strict proportion to the wager until the potential win would exceed the aggregate limit. This ticket should pay $200,000 for catching 10 numbers on a $5 ticket and $400,000 for catching 10 on a $10 ticket. The HA of the $1 ticket is 29.92%; this rises slightly to 30.14% on the $5 ticket and 30.26% for a $10 ticket.

TABLE 3.34: Virgin River Pick 10 pay tables

| **Catch** | **$1** | **$5** | **$10** |
|---|---|---|---|
| 5 | 2 | 10 | 20 |
| 6 | 20 | 100 | 200 |
| 7 | 135 | 675 | 1350 |
| 8 | 900 | 4500 | 9000 |
| 9 | 4000 | 20,000 | 40,000 |
| 10 | 40,000 | 100,000 | 100,000 |

## Pick 11: An Uncommon Option

Following the launch of keno in Nevada in 1931, the first ticket introduced besides the familiar 10-spot ticket was an 11-spot ticket [159]. As was the case with the Chinese lottery decades earlier, this ticket got its own special configuration, as an 11-king ticket. Payoffs were prorated to the 55¢ cost, 5¢/number, of the ticket.

**Example 3.17.** For the pick-10 game, catching 5 numbers paid 2 for 1, and catching 6 numbers paid 18 for 1. Catching 6 numbers out of 11 corresponded to five tickets catching 6 of 10 and a further six tickets catching 5 numbers. The total payoff was 102 for 1; at 5¢/number, this was $5.10. ■

The player also had the option of playing an 11-king ticket for 50¢; payoffs for this ticket were handled by multiplying the 55¢ payoff by 10/11 and then rounding down to the nearest nickel.

Despite this historical significance both before and after gambling was re-legalized in Nevada, Pick 11 keno tickets have fewer attractive features than Pick 10 or Pick 12 tickets. Pick 10 is, of course, the original keno bet and carries some popularity down to present times due to that; a number of promotional games use Pick 10 as a base, even if they result in the high HAs we saw on page 88. Some nonstandard games, such as Hi-Low Keno (page 95), involve picking 12 numbers and have drawn some attention to Pick 12 games. Tickets with 12 numbers are also an attractive option because 12 is highly divisible—by 2, 3, 4, and 6—and thus 12-number tickets have great potential as way tickets. Pick 11 occupies an awkward place between the two, and some keno paybooks—for example, the Orleans book excerpted in Table 1.4—simply skip over Pick 11.

However, since offering a particular keno option is a low-cost proposition—once the paybooks are printed, the game can be offered regardless of its popularity, even if no patron ever chooses to play—Pick 11 games are part of the keno menu at other casinos.

**Example 3.18.** The Primadonna Casino in Reno offered two competing Pick 11 games for a time. These had different minimum ticket prices and pay schedules, which are shown in Table 3.35. Which ticket was better?

TABLE 3.35: Primadonna Casino Pick 11 pay tables ($)

| Catch | 65¢ Ticket | 50¢ Ticket |
|---|---|---|
| 0 | 1 | 0 |
| 5 | 0 | 0.50 |
| 6 | 5 | 5 |
| 7 | 40 | 38 |
| 8 | 372 | 240 |
| 9 | 1000 | 800 |
| 10 | 2000 | 2000 |
| 11 | 12,500 | 5000 |

If "better" is defined by the probability of winning, then the 50¢ game holds the edge: 9.83% to 5.70%. However, much of this advantage comes from the break-even payoff for catching 5 of 11 numbers; with this probability removed, the chance of making a profit on the 50¢ ticket drops to 2.43%. Note that every winning payoff on the 65¢ ticket brings a profit. If the house advantage is the criterion for ranking these games, then the 50¢ game still wins out; its 20.24% HA is better than the 28.96% edge that the casino enjoys on the 65¢ ticket. ■

The 11 is, of course, a winning number on the come-out roll in craps, and as such carries some cachet as a lucky number: perhaps a second choice for luck, after 7. Nebraska's Big Red Keno has focused on the lucky reputation of 11 in Pick 11 keno. The "Lucky-N-Wild 7–11" game offered in Lincoln plays into the gamblers' belief that 7 and 11 are particularly lucky numbers by offering special pay tables for Pick 7 and Pick 11 games that highlight the numbers 7 and 11 [10, p. 19]. For the Pick 11 game, the number 7 is prominently featured. Table 3.36 shows the pay tables for both Lucky-N-Wild 7–11, which is based on a $1.25 wager, and the standard Pick 11 pay table as played in Lincoln, which assumes a $1 bet. (Different Nebraska cities served by Big Red Keno may offer slightly different Pick 11 pay tables.)

**Example 3.19.** Lucky-N-Wild 7–11 and the standard Pick 11 game both offer a $7 payoff. For Lucky-N-Wild, this is for catching either 0 or 7 numbers; for the standard game, it's a catch-6 payoff. Which of these three events has the higher probability?

For Lucky-N-Wild, the $7-winning probabilities are

$$P(0\,|\,11) = \frac{{}_{69}C_{20}}{{}_{80}C_{20}} \approx .0327$$

and

$$P(7\,|\,11) = \frac{{}_{11}C_7 \cdot {}_{69}C_{13}}{{}_{80}C_{20}} \approx .0036,$$

while the probability of catching 6 numbers and winning $7 in the standard

TABLE 3.36: Big Red Keno: Pick 11 games in Lincoln, Nebraska [10]

| Catch | **Lucky-N-Wild 7–11** \$1.25 wager | **Standard Pick 11** \$1 wager |
|---|---|---|
| 0 | 7.00 | 0 |
| 1 | 1.25 | 0 |
| 2 | 1.25 | 0 |
| 6 | 1.25 | 7 |
| 7 | 7.00 | 80 |
| 8 | 77 | 425 |
| 9 | 777 | 2250 |
| 10 | 7777 | 20,000 |
| 11 | 77,777 | 50,000 |

game is

$$P(6 \mid 11) = \frac{{}_{11}C_6 \cdot {}_{69}C_{14}}{{}_{80}C_{20}} \approx .0202.$$

While it's more likely that you'll catch 0 of 11 numbers than 6 or 7, you do have to make a larger bet to access that payoff. ■

**Example 3.20.** The house advantage on the Lucky-N-Wild Pick 11 game is 31.39%. Is this tribute to the number 7 hurting or harming the players?

We will answer this question by comparing 31.39% to the HA of the standard Pick 11 game as played in Lincoln. A look at the two pay tables shows a clear pattern: other than the additional payoffs at Lucky-N-Wild 7–11 for catching 0, 1 or 2 numbers, and the payoff for catching all 11, the standard game pays better. The question, as always, is how the payoffs and probabilities combine to produce an expected value.

The house edge for a \$1 bet on the standard Pick 11 bet is found to be 30.93%. With a difference of less than half a percentage point, the games are nearly identical in the casino's advantage.

If you were to wager \$5 at each game, which would cover 4 bets at Lucky-N-Wild or 5 bets at the standard Pick 11 game, your net expected losses would be \$1.57 and \$1.55, respectively. The 2¢ difference scarcely merits mention. ■

## Pick 12: Hi-Low Keno

*Hi-Low Keno* is a betting variation that sounds very simple: mark twelve numbers in three identified groups of four. This sounds like the beginnings of a simple way ticket, but the payoff structure is far more interesting (or complicated) than a standard 4-, 8-, or 12-spot ticket. The difference between Hi-Low and a way ticket lies in how the three blocks are treated.

Payoffs at Hi-Low begin when any 5 of the 12 chosen numbers are caught, and are based on how many numbers *in each group* are drawn by the casino. Catching 8 numbers out of 12, for example, pays off differently based on how those numbers are distributed, with the highest payoff of \$1248 (on a \$3.60 ticket) reserved for catching all four numbers in each of two groups. Other distributions of eight numbers among the three groups pay off smaller amounts, with the lowest payoff of \$508 awarded when the caught numbers split 3–3–2.

We denote the three groups by A, B, and C, and define the numbers $a, b$, and $c$ as the number of caught numbers from groups A, B, and C, respectively. For example, one might mark the twelve numbers from 1–12, dividing them into the three groups $\mathrm{A} = \{1,2,3,4\}, \mathrm{B} = \{5,6,7,8\}$, and $\mathrm{C} = \{9,10,11,12\}$. It follows that $a, b$, and $c$ are integers in the range 0–4. One pay table for Hi-Low Keno, which depends on the precise values of $a, b$, and $c$ as well as their sum, is presented in Table 3.37 [26]. As a rule, full blocks of 4 lead to higher payoffs than distributions without any full groups, and multiple full blocks pay off better, as when 4–4–1 pays more than 4–3–2 in a "catch 9" event. The very name "Hi-Low" is derived from this difference in payoffs: for example, catching 5 numbers distributed 2–2–1 or 3–1–1, which are the most likely distributions, gives a "low" payoff. Catching 5 numbers distributed in the less likely configurations 3–2–0 or 4–1–0 gives a "high" payoff [12].

**Example 3.21.** Compare the probabilities of catching 8 of 12 numbers divided 4–4–0 and 3–3–2.

For a ticket catching 8 numbers, the probability of catching exactly $a$ numbers from group A, $b$ numbers from group B, and $c$ numbers from group C is

$$p(a,b,c) = \frac{{}_4C_a \cdot {}_4C_b \cdot {}_4C_c \cdot {}_{68}C_{12}}{{}_{80}C_{20}}.$$

$p(a,b,c)$ must then be multiplied by a factor $n$ which counts the number of ways to assign the labels A, B, and C to the three groups, and we denote the probability we seek here by $P(a,b,c) = n \cdot p(a,b,c)$.

For a 4–4–0 split, $n = 3$, because there are 3 choices for the group with no catches. For the 3–3–2 split, $n = 3$ again. If the numbers were to split 4–3–1 among the three groups, then $n = {}_3P_3 = 6$. We then have

$$P(4,4,0) = 3 \cdot p(4,4,0) = 3 \cdot \frac{{}_4C_4 \cdot {}_4C_4 \cdot {}_4C_0 \cdot {}_{68}C_{12}}{{}_{80}C_{20}} \approx 6.179 \times 10^{-6},$$

and

$$P(3,3,2) = 3 \cdot p(3,3,2) = 3 \cdot \frac{{}_4C_3 \cdot {}_4C_3 \cdot {}_4C_2 \cdot {}_{68}C_{12}}{{}_{80}C_{20}} \approx 5.932 \times 10^{-4},$$

about 96 times greater—a disparity which is not reflected in the relative payoffs. ■

TABLE 3.37: Hi-Low Keno pay table [26]

| Catch | $a$ | $b$ | $c$ | Payoff |
|---|---|---|---|---|
| 5 | 2 | 2 | 1 | \$2.00 |
| | 3 | 1 | 1 | \$2.40 |
| | 3 | 2 | 0 | \$2.80 |
| | 4 | 1 | 0 | \$3.60 |
| 6 | 2 | 2 | 2 | \$16.80 |
| | 3 | 2 | 1 | \$20.00 |
| | 3 | 3 | 0 | \$26.40 |
| | 4 | 1 | 1 | \$26.40 |
| | 4 | 2 | 0 | \$29.60 |
| 7 | 3 | 2 | 2 | \$102.00 |
| | 3 | 3 | 1 | \$122.40 |
| | 4 | 2 | 1 | \$142.80 |
| | 4 | 3 | 0 | \$183.60 |
| 8 | 3 | 3 | 2 | \$508 |
| | 4 | 2 | 2 | \$656 |
| | 4 | 3 | 1 | \$804 |
| | 4 | 4 | 0 | \$1248 |
| 9 | 3 | 3 | 3 | \$2052 |
| | 4 | 3 | 2 | \$2280 |
| | 4 | 4 | 1 | \$2736 |
| 10 | 4 | 3 | 3 | \$6000 |
| | 4 | 4 | 2 | \$8400 |
| 11 | 4 | 4 | 3 | \$19,200 |
| 12 | 4 | 4 | 4 | \$61,200 |

How do Hi-Low payoffs compare to a standard 12-spot keno game, the pay table for which is given in Table 3.38?

We immediately note that Hi-Low is a better bet if you only catch 5 numbers—the ticket is a net loss unless you break even when the numbers are split 4–1–0, but you lose less than the full price of the ticket, which is your loss in a standard game. For other catches, we need to compute the average return, for 1, given a particular number $k$ of catches. This expected value may be found with the formula

$$E(k) = \frac{\sum_{a+b+c=k} P(a,b,c) \cdot R(a,b,c)}{\sum_{a+b+c=k} P(a,b,c)},$$

where $R(a,b,c)$ is the return on your bet if your $k$ matches split $a-b-c$, from Table 3.37, and the two sums are taken over all partitions of $k$ into nonnegative integers $a$, $b$, and $c$, none greater than 4. Dividing $E(k)$ by the

TABLE 3.38: Pick 12 keno pay table

| Catch | Payoff (for 1) |
|---|---|
| 5 | 0 |
| 6 | 5 |
| 7 | 20 |
| 8 | 200 |
| 9 | 2000 |
| 10 | 7500 |
| 11 | 50,000 |
| 12 | 100,000 |

\$3.60 cost of the ticket gives the equivalent payoff, for 1, of a Hi-Low ticket considered as a 12-spot ticket. See Table 3.39.

TABLE 3.39: Equivalent Hi-Low payoffs as a 12-spot ticket

| Catch | Payoff (for 1) |
|---|---|
| 5 | \$0.64 |
| 6 | \$5.64 |
| 7 | \$32.45 |
| 8 | \$167.27 |
| 9 | \$621.82 |
| 10 | \$1848.48 |
| 11 | \$5333.33 |
| 12 | \$17,000.00 |

Comparing the two games shows that the player is better off playing Hi-Low Keno rather than a standard 12-spot ticket if he or she matches 5–7 numbers—but of course, that is not what 12-spot keno players hope for. A better comparison might be made by looking at the house advantage of each ticket. For a 12-spot ticket with Table 3.38 as its pay table, the HA is 25.44%. For Hi-Low, the house edge is slightly higher, 28.63%.

So the payoffs for the biggest wins are lower on average, and the HA is greater—what is there to recommend Hi-Low Keno over a traditional 12-spot ticket?

Part of the appeal might be the apparently higher payoffs—of course, this is an illusion, because a Hi-Low ticket costs over three and a half times as much as a Pick 12 ticket. More significantly, keno players are, by definition, almost never in the game for the long run, where the average payoffs or HA matter. However, it turns out that there's more to the story of Hi-Low than that. In much the same way that the Union Plaza's 70¢ Pick 9 game was properly regarded as a 71-way Pick 10 game for the purposes of gameplay,

Hi-Low is rightly considered as an 18-way Pick 10 ticket—once again, we see keno returning to its roots even as the game evolved [26].

The 18 ways are determined as follows:

- There are ${}_3C_2 = 3$ ways to choose two of the three complete blocks of four numbers, which are combined into a block of 8.
- Each of these 8-blocks can then be combined with ${}_4C_2 = 6$ choices of two numbers from the four in the third block of three to make a 10-spot way. Multiplying gives $3 \times 6 = 18$ 10-spot ways in total.

Accordingly, making a fair comparison of Hi-Low to a standard ticket is better done by comparing it to a standard Pick 10 pay table rather than Pick 12.

**Example 3.22.** Consider the top prize: \$61,200 for catching all 12 numbers. In this interpretation, you have caught 10 out of 10 numbers on all 18 Pick 10 tickets represented by Hi-Low. It follows that your payoff is \$3400 per winning way. If the ticket price is the usual \$3.60, then you've paid 20¢ per way, and the effective payoff here is 17,000 for 1.

The Pick 10 pay tables at the Excalibur and at Treasure Island (TI) pay 25,000 for 1 when all 10 numbers are caught, making Hi-Low again an inferior ticket at the top prize level.

Catching 11 out of 12 numbers, which must be distributed 4-4-3, is the equivalent of the following 18 Pick 10 catches:

- 3 Catch 10s.
- 3 Catch 9s of the form 4-4-1.
- 12 more Catch 9s of the form 4-3-2 or 4-2-3.

The payoff for catching 11 out of 12 is \$19,200. \$10,200 of this is allocated to the three tickets that catch all 10 spots, leaving \$9000 to be divided among the 15 9-spot catches. Each Catch 9 ticket is thus seen to pay off \$600 for 20¢, for an effective payoff of 3000 for 1.

Again consulting the Excalibur and Treasure Island pay tables for catching 9 numbers on a 10-spot ticket, we see that the payoff is 5000 for 1 at TI and 4000 for 1 at Excalibur. The gap is narrowing, but Hi-Low is still inferior. ■

At this point, Hi-Low seems a misnomer, for we have yet to find a payoff that is higher than for a straight Pick 10 ticket. In an attempt to find evidence supporting the "Hi" part of the name, we look at the Catch 5 payoffs. Catching 5 numbers distributed 2-2-1 pays \$2. How many of the 18 tickets are winners, and what is the effective payoff for 1?

**Example 3.23.** For convenience, suppose that you have marked the numbers 1–12 and grouped them into the three blocks 1–4, 5–8, and 9–12. Suppose further that the 5 numbers you catch are 1, 2, 5, 6, and 9. Your 18 10-spot tickets are listed in Table 3.40, with money-winning catches in bold.

TABLE 3.40: Breakdown of Hi-Low ticket into 18 Pick 10 tickets

| **Ticket** | **Catches** |
|---|---|
| 1,2,3,4,5,6,7,8,9,10 | **5** |
| 1,2,3,4,5,6,7,8,9,11 | **5** |
| 1,2,3,4,5,6,7,8,9,12 | **5** |
| 1,2,3,4,5,6,7,8,10,11 | 4 |
| 1,2,3,4,5,6,7,8,10,12 | 4 |
| 1,2,3,4,5,6,7,8,11,12 | 4 |
| 1,2,3,4,5,6,9,10,11,12 | **5** |
| 1,2,3,4,5,7,9,10,11,12 | 4 |
| 1,2,3,4,5,8,9,10,11,12 | 4 |
| 1,2,3,4,6,7,9,10,11,12 | 4 |
| 1,2,3,4,6,8,9,10,11,12 | 4 |
| 1,2,3,4,7,8,9,10,11,12 | 3 |
| 1,2,5,6,7,8,9,10,11,12 | **5** |
| 1,3,5,6,7,8,9,10,11,12 | 4 |
| 1,4,5,6,7,8,9,10,11,12 | 4 |
| 2,3,5,6,7,8,9,10,11,12 | 4 |
| 2,4,5,6,7,8,9,10,11,12 | 4 |
| 3,4,5,6,7,8,9,10,11,12 | 3 |

The five Catch 5 tickets are the only ones that pay off, so the total win of $2 needs to be divided by 5. This 40¢ win per ticket represents a payoff of 2 for 1. The Excalibur pays 2 for 1 when catching 5 of 10, while TI pays 1 for 1. For smaller catches, Hi-Low is a better ticket—if you're comparing it at the right casino. ■

Combining Hi-Low with a way ticket can result in some massively complicated keno tickets. The ticket in Figure 3.8, with 23 spots marked, is conditioned for *1200* different Hi-Low wagers [91]:

- There are ${}_6C_3 = 20$ ways to choose three groups of three from the six marked groups.
- Each of these can be combined with three of the five kings (10, 20, 30, 40, and 50) to make three groups of four, as a Hi-Low ticket requires. The order in which kings are matched with groups matters here, leading to ${}_5P_3 = 60$ tickets for each of the 20 combinations, or 1200 total tickets.

For example, one Hi-Low combination from the ticket in Figure 3.8 has 4/5/7/**10**, 76/77/78/**50**, and 11/12/13/**20** as its blocks. The kings are indicated in bold; permuting them among the three selected blocks will result in five additional Hi-Low combinations.

At the standard price of $3.60, this ticket costs .3¢ per way, though this

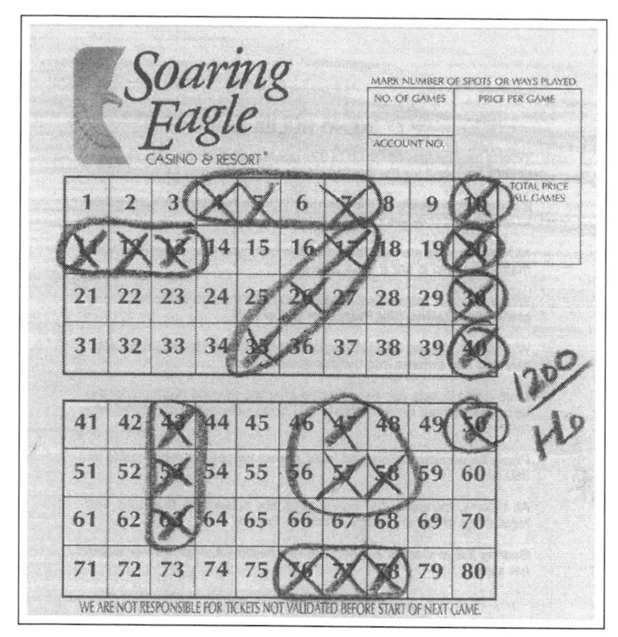


FIGURE 3.8: 1200-way Hi-Low Keno ticket.

price would surely be adjusted to something like 10¢ per way, or $120, if it were accepted for play. Many keno paybooks include a sentence like "All way tickets are subject to approval of the Keno Supervisor."; such a complicated ticket as this would be a prime candidate for rejection on the grounds that it was too complicated for keno writers to process efficiently. With a HA of 28.63% working on every way, the expected value of this complex discounted way ticket is –$34.36.

The Excalibur offers 8- and 9-spot tickets that follow the same principles as Hi-Low: numbers are grouped into several small sets, and payoffs are based on how many caught numbers fall into these various sets. The 8-spot game splits the numbers into 4 groups of 2 and regards the ticket as a 4/2, 6/4, 4/6, 1/8 way ticket, for a total of 15 ways. The pay table starts when any two numbers are caught; the probability of catching at least 2 numbers from 8 is, by the Complement Rule,

$$1 - P(0 \mid 8) - P(1 \mid 8) = 1 - \frac{{}_{72}C_{20}}{{}_{80}C_{20}} - \frac{{}_{8}C_{1} \cdot {}_{72}C_{19}}{{}_{80}C_{20}} \approx .6453.$$

A bet with a greater than 50% chance of winning? Something else must be going on here, and it can be found in the pay table, Table 3.41. Once again, we see that a number of "winning" catches come out as net losses due to the high ticket price of $15, which is $1 per way.

**Example 3.24.** Catching 4 numbers, 1 from each group, pays $10, for a net

TABLE 3.41: Excalibur 8-spot, 15-way pay table ($15 wager)

| Catch | Payoff | Catch | Payoff |
|---|---|---|---|
| 1-1-0-0 | $1 | 2-2-0-0 | $156 |
| 1-1-1-0 | $4 | 2-2-1-0 | $255 |
| 1-1-1-1 | $10 | 2-2-1-1 | $431 |
| 2-0-0-0 | $15 | 2-2-2-0 | $2001 |
| 2-1-0-0 | $19 | 2-2-2-1 | $3669 |
| 2-1-1-0 | $26 | 2-2-2-2 | $26,768 |
| 2-1-1-1 | $46 | | |

loss of $5. The probability of this catch is

$$\frac{({}_2C_1)^4 \cdot {}_{72}C_{16}}{{}_{80}C_{20}} \approx .0186.$$

At the same time, catching 4 numbers distributed 2-2-0-0, an event with probability

$$\frac{{}_4C_2 \cdot {}_{72}C_{16}}{{}_{80}C_{20}} \approx .0070,$$

pays $156 (for a profit of $141)—rather a significant difference.

The probability of 2-2-0-0 is $\frac{3}{8}$ of the probability of 1-1-1-1, but the payoff is considerably more than $\frac{8}{3}$ greater. ■

## Pick 13–14: Back to 10 Spots

As keno game options expanded, Pick 13 through Pick 16 games were introduced. As was the case with Pick 9 at the Union Plaza, the first tickets at these levels were generalized from Pick 10 games, with the prices and pay tables based on the number of 10-spot tickets that could be made from the chosen numbers. For convenience, prices and payoffs were frequently rounded to the nearest nickel.

For a Pick 13 ticket, there are ${}_{13}C_{10} = 286$ ways to select 10 of the 13 numbers. This ticket was priced at $2.85, conveniently close to 1 cent per way. For Pick 14, the number of tickets is ${}_{14}C_{10} = 1001$, and the ticket sold for $10. A 15-spot ticket covered 3003 10-spot ways for $30, and a 16-spot ticket cost $80.05 and encompassed 8008 ways [90].

A 13-spot ticket was played against Table 3.42; this was at a time in the 1940s and 50s when the aggregate limit was $5000.

Catching 5 out of 13 spots corresponded to catching 5 out of 10 spots on ${}_8C_5 = 56$ of the 286 tickets. The number of ways to pick the other 5 numbers on a 10-spot ticket from the 8 numbers not chosen in the drawing is 56. The $1.10 payoff is approximately 2 for 1 on each of these winning ways, consistent with the payoffs in Table 3.27. At the other end, catching all 13 numbers meant

TABLE 3.42: Pick 13 payoffs: $2.85 ticket read as 286 10-spot ways [90]

| Catch | Payoff |
|---|---|
| 5 | $1.10 |
| 6 | $8.55 |
| 7 | $53.10 |
| 8 | $241.40 |
| 9 | $774.00 |
| 10 | $1866.00 |
| 11 | $3591.00 |
| 12 | $5000.00 |
| 13 | $5000.00 |

catching 10 out of 10 numbers on all 286 tickets. This represents about a 1748 for 1 payoff on each winning way, a payoff which is tempered somewhat by the aggregate limit of $5000. The payoff for catching all 10 numbers on a Pick 10 ticket at that time, according to Table 3.27, was 3600 for 1.

## Pick 15: Cover All Challenge

The years since the statewide aggregate limit was eliminated in Nevada and legal gambling spread beyond Nevada and Atlantic City coincided with the rise of inexpensive powerful computers, which make sorting out keno options and identifying winning tickets much easier. As a result, casinos in locations new to keno often introduced new game options that are facilitated by computerized keno operations. The Mirage Casino in Las Vegas, which opened in 1989, offered a game that combined a Pick 15 pay table with a Cover All ticket. This 14-way ticket split the top line of a keno bet slip in half: 1–5 and 6–10. Each half was then combined with each of the 7 other rows of 10 numbers on the bet slip to make a 14-way 15-spot ticket priced at $1 per way. The Pick 15 pay table in use had a 1 for 1 payoff for 0 catches, and its other payoffs started when 6 numbers were caught. Catching 6 numbers paid off at 1 for 1, with prizes increasing until they reached the Mirage's aggregate limit of $250,000 for catching all 15 numbers.

Among the ${}_{80}C_{20}$ different ways to draw the 20 winning numbers, there are many effectively indistinguishable combinations for the purpose of this bet. For example, the selection

$$\{1, 6, 8, 18, 25, 27, 36, 40, 41, 43, 52, 62, 63, 65, 68, 69, 72, 75, 77, 80\}$$

pays off the same as the selection

$$\{2, 6, 8, 18, 25, 27, 36, 40, 41, 43, 52, 62, 63, 65, 68, 69, 72, 75, 77, 80\},$$

which simply exchanges the number 1 for the number 2. Either combination leads to three wins and a \$7 payoff (a net loss of \$7): \$1 for each of 2 catches of 6 numbers: $\{1 \text{ or } 2, 62, 63, 65, 68, 69\}$ and $\{6, 8, 72, 75, 77, 80\}$, and \$5 for one catch of 7 numbers: $\{6, 8, 62, 63, 65, 68, 69\}$.

In general, we can exchange numbers within any of the nine blocks defined by the ticket without changing the payoff. How many distinguishable combinations of 20 numbers are there?

Define the following variables:

$$\begin{aligned}
x_1 &= \text{Number of catches in the range } 1-5 \\
x_2 &= \text{Number of catches in the range } 6-10 \\
x_3 &= \text{Number of catches in the range } 11-20 \\
x_4 &= \text{Number of catches in the range } 21-30 \\
x_5 &= \text{Number of catches in the range } 31-40 \\
x_6 &= \text{Number of catches in the range } 41-50 \\
x_7 &= \text{Number of catches in the range } 51-60 \\
x_8 &= \text{Number of catches in the range } 61-70 \\
x_9 &= \text{Number of catches in the range } 71-80.
\end{aligned}$$

Answering this question amounts to counting the number of integer solutions, $N$, to the equation

$$x_1 + x_2 + x_3 + x_4 + x_5 + x_6 + x_7 + x_8 + x_9 = 20,$$

subject to the nine constraints

$$0 \leqslant x_1, x_2 \leqslant 5;$$

$$0 \leqslant x_3, x_4, x_5, x_6, x_7, x_8, x_9 \leqslant 10.$$

We begin by computing an upper bound for $N$. The number of nonnegative integer solutions to the equation, without considering the upper bounds, is found by modeling the problem as follows: Let the 20 catches be denoted by $\times$s, and consider a row of 20 symbols:

$$\times\times\times\times\times\times\times\times\times\times\times\times\times\times\times\times\times\times\times\times\,.$$

We are trying to count the number of ways to allocate the $\times$s to 9 hypothetical "bins", corresponding to the 9 ranges defined above. Each such way can be illustrated by inserting 8 dividers, denoted here by $|$, into the string of $\times$s. For example, the two draws indicated above have

$$\langle x_1, x_2, x_3, x_4, x_5, x_6, x_7, x_8, x_9 \rangle = \langle 1, 2, 1, 2, 1, 3, 1, 5, 4 \rangle$$

and are both represented by the string of characters

$$\times|\times\times|\times|\times\times|\times|\times\times\times|\times|\times\times\times\times\times|\times\times\times\times.$$

We have $20 + 9 - 1 = 28$ spaces to fill in, and need to count the number of ways to fill 8 of them with the divider $|$. Since order does not matter, this may be done in ${}_{28}C_8 = 3,108,105$ ways—but some of these, for example,

$$\times\times\times\times\times\times|\times|\times\times|\times\times\times\times||\times|\times\times|\times\times|\times\times,$$

do not represent valid keno draws. In this case, $x_1 = 6$, which is outside the specified range 0–5. In general, to count the number of nonnegative integer solutions to the equation

$$x_1 + x_2 + \ldots + x_n = k,$$

with $n$ and $k$ positive integers, we must place $n-1$ dividers among $k$ $\times$s; this may be done in ${}_{n+k-1}C_{n-1}$ ways.

We now know that $N < 3,108,105$. The remaining challenge is to count the various combinations that do not correspond to possible keno draws and remove them from the total, a process known as *inclusion/exclusion*. We begin by excluding all combinations for which $x_1 \geqslant 6$. These can be counted by defining the new variable $x_1' = x_1 - 6$ and counting the number of solutions in nonnegative integers to the equation

$$x_1' + x_2 + x_3 + x_4 + x_5 + x_6 + x_7 + x_8 + x_9 = 14.$$

There are ${}_{14+9-1}C_8 = {}_{22}C_8$ of these. Similarly, there are ${}_{14}C_8$ combinations for which $x_2 \geqslant 6$.

For the remaining seven variables, we adopt the same counting strategy. To count the number of combinations with $x_3 \geqslant 11$, we set $x_3' = x_3 - 11$ and count the number of solutions to

$$x_1 + x_2 + x_3' + x_4 + x_5 + x_6 + x_7 + x_8 + x_9 = 9,$$

which is ${}_{17}C_8$. This quantity must be subtracted 7 times, once for each of the variables $x_3$ through $x_9$.

Our new value for $N$ is now ${}_{28}C_{20} - 2 \cdot {}_{23}C_8 - 7 \cdot {}_{17}C_8$—but we have overcorrected. Certain combinations with two variables exceeding their permissible maximum have been removed twice. For example, one such combination has $x_1 = 6, x_2 = 2, x_7 = 12$, and all other variables 0, and can be illustrated as

$$\times\times\times\times\times\times|\times\times|||||\times\times\times\times\times\times\times\times\times\times\times\times||.$$

To maintain an accurate count, the number of combinations such as this one needs to be added back to $N$—not because these are valid combinations, but because we have oversubtracted and need to correct that error.

Combinations with two variables over their limits are of two types. If $x_1 \geqslant 6$ and $x_2 \geqslant 6$, then we seek to count the number of valid solutions to the equation

$$x_1' + x_2' + x_3 + x_4 + x_5 + x_6 + x_7 + x_8 + x_9 = 8,$$

with $x_1' = x_1 - 6$ and $x_2' = x_2 - 6$ defined as above. There are ${}_{16}C_8$ such solutions.

The remaining solutions that need to be restored have one of $x_1$ and $x_2$ over 5 and one of $x_3$ through $x_9$ over 10. By the FCP, there are 14 different ways to select the two variables that are over their limits; using $x_1' = x_1 - 6$ and $x_3' = x_3 - 11$ to illustrate this idea, we are counting the solutions to

$$x_1' + x_2 + x_3' + x_4 + x_5 + x_6 + x_7 + x_8 + x_9 = 3,$$

and there are ${}_{11}C_8$ of those.

In principle, inclusion/exclusion continues: we would subtract every combination with three variables over their limit, then add back all the combinations with four such variables, and so on until we reach the combinations with all nine variables violating their respective constraints. However, as there are no solutions to the original equation with three or more variables over their limit, the process stops here. Our final value for $N$ is

$$N = {}_{28}C_{20} - 2 \cdot {}_{23}C_8 - 7 \cdot {}_{17}C_8 + {}_{16}C_8 + 14 \cdot {}_{11}C_8 = 1,972,487.$$

Each of these valid combinations corresponds to many actual 20-number keno draws. Our earlier example, with

$$\langle x_1, x_2, x_3, x_4, x_5, x_6, x_7, x_8, x_9 \rangle = \langle 1, 2, 1, 2, 1, 3, 1, 5, 4 \rangle,$$

has

$${}_5C_1 \cdot {}_5C_2 \cdot {}_{10}C_1 \cdot {}_{10}C_2 \cdot {}_{10}C_1 \cdot {}_{10}C_3 \cdot {}_{10}C_1 \cdot {}_{10}C_5 \cdot {}_{10}C_4 \approx 1.429 \times 10^{13}$$

ways to choose its numbers.

The Bourbon Street Casino in Las Vegas offered a different Pick 15 game that seems at least partially intended to generate repeat business. Four of the twelve payoffs at Diamond Lil's Special included a free ticket. In two cases, Catch 0 and Catch 8, the free ticket was included with a cash payment. Table 3.43 shows the payoffs; tickets started at $2.25.

TABLE 3.43: Bourbon Street: Diamond Lil's Special pay table ($2.25 ticket)

| Catch | Payoff | Catch | Payoff |
|---|---|---|---|
| 0 | $25 + free ticket | 10 | $500 |
| 1 | $4.50 | 11 | $3333 |
| 2 | Free ticket | 12 | $8225 |
| 3–6 | $0 | 13 | $30,000 |
| 7 | Free ticket | 14 | $40,000 |
| 8 | $25 + free ticket | 15 | $50,000 |
| 9 | $200 | | |

Though the equation for the expected value $E$ of this ticket includes five terms with an $E$ in them, it nonetheless remains a simple linear equation. The equation with all of the payoffs and probabilities reduces to

$$E = .193716E - .819830,$$

whose solution is E = –$1.02; the resulting HA is 45.16%.

## Pick 16: Expected Catches

At the Red Wind Casino in Olympia, Washington, the familiar phrase "Sweet Sixteen" is pressed into service for a $5 "pay any catch" Pick 16 game based on Table 3.44.

TABLE 3.44: Red Wind Casino Sweet Sixteen pay table ($5 ticket)

| Catch | Payoff | Catch | Payoff |
|---|---|---|---|
| 0 | $50 | 9 | $200 |
| 1 | $5 | 10 | $600 |
| 2 | $3 | 11 | $2500 |
| 3 | $2 | 12 | $12,500 |
| 4 | $1 | 13 | $25,000 |
| 5 | $2 | 14 | $50,000 |
| 6 | $3 | 15 | $50,000 |
| 7 | $5 | 16 | $50,000 |
| 8 | $20 | | |

When picking 16 numbers, how many catches should one expect? The mean number of catches is found by evaluating the sum

$$\sum_{k=0}^{16} k \cdot P(k \,|\, 16),$$

which is 4—or one-fourth of the numbers marked. This is consistent with the fact that the casino draws one-fourth of the 80 numbers in play in any one drawing.

The Red Wind pay table recognizes this most-likely result by assigning it the lowest payoff: $1, which results in a net loss of $4. Catches near 4 also receive break-even or less than break-even payoffs: all catches from 1–7 do no better than break even. The probability of breaking even (8.20%) or losing money (89.77%) on this Sweet Sixteen ticket is 97.97%, making this a very unattractive bet even as it pays out on all catches and offers a fairly reasonable HA of only 29.91%.

At the Crystal Bay Casino, in Crystal Bay, Nevada on the north shore of Lake Tahoe, keno players could choose a 5-way Sweet Sixteen ticket covering all 80 numbers. The ticket was divided into 5 vertical columns, as shown in

Figure 3.9, and could be had for a minimum bet of \$1.50 per way; \$7.50 in total.

| 1 | 2 | 3 | 4 | 5 | 6 | 7 | 8 | 9 | 10 |
|---|---|---|---|---|---|---|---|---|---|
| 11 | 12 | 13 | 14 | 15 | 16 | 17 | 18 | 19 | 20 |
| 21 | 22 | 23 | 24 | 25 | 26 | 27 | 28 | 29 | 30 |
| 31 | 32 | 33 | 34 | 35 | 36 | 37 | 38 | 39 | 40 |
| 41 | 42 | 43 | 44 | 45 | 46 | 47 | 48 | 49 | 50 |
| 51 | 52 | 53 | 54 | 55 | 56 | 57 | 58 | 59 | 60 |
| 61 | 62 | 63 | 64 | 65 | 66 | 67 | 68 | 69 | 70 |
| 71 | 72 | 73 | 74 | 75 | 76 | 77 | 78 | 79 | 80 |

FIGURE 3.9: Crystal Bay Sweet Sixteen 5-Way ticket.

## Pick 20: Can You Catch Them All?

In the history of keno, no player has ever caught all 20 numbers on a Pick 20 ticket. This is no surprise, as the probability of this event is

$$P(20\,|\,20) = \frac{1}{{}_{80}C_{20}} = \frac{1}{3,535,316,142,212,174,320}.$$

Considered in this light, the \$50,000 payoff on a \$5 20 Spot Special ticket for catching 20 out of 20 at the Angel of the Winds Casino hardly seems sufficient reward for such a rare event. Nor is this imbalance between probability and payoff mitigated by the fact that one can win the same \$50,000 for matching 16–19 of 20 numbers.

The full pay table for the 20 Spot Special is given in Table 3.45. As with

TABLE 3.45: Angel of the Winds 20 Spot Special pay table (\$5 wager)

| Catch | Payoff | Catch | Payoff |
|---|---|---|---|
| 0 | \$1000 | 10 | \$50 |
| 1 | \$10 | 11 | \$200 |
| 2 | \$5 | 12 | \$1000 |
| 3 | \$2.50 | 13 | \$4000 |
| 4–6 | \$0 | 14 | \$15,000 |
| 7 | \$2.50 | 15 | \$25,000 |
| 8 | \$10 | 16–20 | \$50,000 |
| 9 | \$25 | | |

Sweet Sixteen, the expected number of catches on a 20-spot ticket is 5: one-fourth of the number drawn, just as picking 20 numbers covers one-fourth of the choices. Accordingly, the 20 Spot Special loses everything if 4, 5, or 6 numbers are caught, covering the casino against the three most probable outcomes. The probability of a \$0 payoff is approximately .6285.

The expected value of a \$5 ticket 20 Spot Special ticket is –\$1.48. In Omaha, Nebraska, Big Red Keno offers a \$5 Pick 20 game called the No Loser 20 Spot (Table 3.46) which attempts to appeal to gamblers by paying out on all catches and raises the top prize from \$50,000 at Angel of the Winds to \$250,000.

TABLE 3.46: Big Red Keno (Omaha) No Loser 20 Spot pay table (\$5 wager)

| Catch | Payoff | Catch | Payoff |
|---|---|---|---|
| 0 | \$650 | 12 | \$1000 |
| 1–3 | \$5 | 13 | \$4000 |
| 4–6 | \$1 | 14 | \$15,000 |
| 7 | \$2 | 15 | \$25,000 |
| 8 | \$5 | 16 | \$50,000 |
| 9 | \$20 | 17 | \$75,000 |
| 10 | \$40 | 18–20 | \$250,000 |
| 11 | \$200 | | |

While the expected value of a No Loser 20 Spot ticket is –\$1.43, slightly better than for the 20 Spot Special, it should be noted that *none* of this difference is attributable to the top payoff. A hypothetical game that paid \$0 for catching 18, 19, or 20 numbers would have the same expected value. Indeed, setting the top payoff, for catching all 20 numbers, to 1 billion dollars—far above the \$300,000 aggregate limit in Omaha—would still result in a game with that same –\$1.43 expectation. The probability of catching 20 out of 20 numbers is simply too small to move the needle on the expectation with any reasonable payoff.

**Example 3.25.** In Hastings, Nebraska, Hastings Keno offers *Korner Keno*, a Pick 20 game where players simply choose one $4 \times 5$ corner of the bet slip and are paid according to how many of the 20 drawn numbers fall into that corner. Multiple corners may be bet on a single ticket; each corner costs \$1. A Korner Keno ticket pays \$0 if 4, 5, or 6 numbers fall into the chosen corner. If you bet the \$4 ticket covering all four corners, what is the chance that all four tickets lose everything?

The four variables counting the number of catches falling into each quadrant are not independent because their sum must be 20, so this question requires more care than simply referring to the result for the 20 Spot Special at Angel of the Winds and computing $(.6285)^4$. The ticket is a \$4 loser if the 20 chosen numbers are divided 4–4–6–6, 4–5–5–6, or 5–5–5–5 among the four corners.

- For 4–4–6–6, there are ${}_4C_2 = 6$ ways to pick the two corners with 4 catches. For each choice, there are then $({}_{20}C_4)^2 \cdot ({}_{20}C_6)^2$ ways to choose the numbers.

- For 4–5–5–6, there are 4 ways to pick the corner with 4 catches, and then 3 ways to pick the corner with 6 catches, for a total of 12 possibilities. Each possibility admits ${}_{20}C_4 \cdot ({}_{20}C_5)^2 \cdot {}_{20}C_6$ choices of numbers.

- There is only one way to catch 5 numbers in each of the 4 corners, and $({}_{20}C_5)^4$ ways to fill in the numbers.

The probability of losing $4 to an even or nearly-even distribution of the 20 catches is then

$$\frac{6 \cdot \left[({}_{20}C_4)^2 \cdot ({}_{20}C_6)^2\right] + 12 \cdot \left[{}_{20}C_4 \cdot ({}_{20}C_5)^2 \cdot {}_{20}C_6\right] + ({}_{20}C_5)^4}{{}_{80}C_{20}} \approx .2294.$$

■

Another variation on Pick 20 keno was *Doublette*, "Developed by a Player for Players", a keno variation that was played for a time at the Peppermill Casino beginning in 1988 [91]. Doublette was a Pick 20 game that was played for two straight draws at a minimum price of $10. The pay table was based on the total number of catches, from 0–40 in the two draws combined, with a payment for every number of catches except 10. Some of these payoffs were less than the cost of the ticket. Table 3.47 shows the payoffs.

TABLE 3.47: Doublette pay table ($10 minimum ticket)

| Catch | Payoff | Catch | Payoff | Catch | Payoff | Catch | Payoff |
|---|---|---|---|---|---|---|---|
| 0 | 50,000 | 11 | 5 | 21 | 1000 | 31 | 27,500 |
| 1 | 1000 | 12 | 5 | 22 | 2000 | 32 | 30,000 |
| 2 | 200 | 13 | 10 | 23 | 5000 | 33 | 32,500 |
| 3 | 50 | 14 | 15 | 24 | 10,000 | 34 | 35,000 |
| 4 | 30 | 15 | 25 | 25 | 12,500 | 35 | 37,500 |
| 5 | 25 | 16 | 30 | 26 | 15,000 | 36 | 40,000 |
| 6 | 15 | 17 | 50 | 27 | 17,500 | 37 | 42,500 |
| 7 | 10 | 18 | 100 | 28 | 20,000 | 38 | 45,000 |
| 8 | 5 | 19 | 200 | 29 | 22,500 | 39 | 47,500 |
| 9 | 5 | 20 | 500 | 30 | 25,000 | 40 | 50,000 |
| 10 | 0 | | | | | | |

**Example 3.26.** An average Doublette ticket would score 5 catches in each drawing, for a total of 10. This most-likely event is the one with zero payoff. What is the probability of catching exactly 10 out of 20 numbers across two draws?

If the first draw catches $x$ of your 20 numbers, $0 \leqslant x \leqslant 10$, we are then interested in the event that the second draw catches $10 - x$ numbers. Since successive keno drawings are independent, the combined probability of these two events occurring in succession is

$$P(10 \mid x) = \underbrace{\frac{{}_{20}C_x \cdot {}_{60}C_{20-x}}{{}_{80}C_{20}}}_{\text{First drawing: } P(x \mid 20)} \cdot \underbrace{\frac{{}_{20}C_{10-x} \cdot {}_{60}C_{10+x}}{{}_{80}C_{20}}}_{\text{Second drawing: } P(10-x \mid 20)}.$$

The probability of catching exactly 10 numbers is

$$\sum_{x=0}^{10} P(10 \mid x) \approx .1660,$$

or slightly less than one chance in 6. ■

## Pick 40: Top and Bottom and Its Variants

Part of the appeal of casino games like roulette and keno is surely the opportunity to bet on one's favorite numbers. Expecting a gambler to identify 40 preferred numbers in the range from 1–80 is a tall order, however, and that's not what Pick 40 games like Top and Bottom are about. Top and Bottom, and its variants such as Left and Right, rely on the layout of the keno bet slip and call for a player simply to choose one side. At the Continental Casino in Las Vegas (now the Silver Sevens), this ticket was introduced as the "Continental Divide" [25].

Payoffs are made according to how many of the casino's drawn numbers fall on the player's half of the ticket. For the conventional keno ticket layout, these bets cover the following numbers:

- **Top**: Numbers from 1–40.
- **Bottom**: Numbers from 41–80.
- **Left**: Numbers ending in a 1 through 5: 1–5, 11–15, and so on up to 71–75.
- **Right**: Numbers ending in a 6 through 0: 6–10 up through 76–80.

The subset of 40 numbers chosen does not affect the probabilities. In any wager, the probability of $k$ numbers falling into the designated half of the ticket is

$$P(k \mid 40) = \frac{{}_{40}C_k \cdot {}_{40}C_{20-k}}{{}_{80}C_{20}}.$$

A typical \$5 Top and Bottom pay table, from the Meskwaki Casino, is given in Table 3.48.

Note that this wager pays off whether you score a lot of catches in your designated half of the ticket, or very few. Only the "intermediate" catches of 8 through 12 are losing tickets, although catching 7 or 13 numbers simply refunds your wager.

TABLE 3.48: Meskwaki Casino's Top and Bottom pay table ($5 wager)

| **Catch** | **Payoff** |
|---|---|
| 0 or 20 | $50,000 |
| 1 or 19 | $5000 |
| 2 or 18 | $2000 |
| 3 or 17 | $500 |
| 4 or 16 | $150 |
| 5 or 15 | $75 |
| 6 or 14 | $15 |
| 7 or 13 | $5 |
| 8–12 | $0 |

**Example 3.27.** Find the probability that your Top and Bottom ticket will lose money.

Let the probability of catching $k$ numbers be denoted here as $P(k)$. The necessary probability is

$$P(8) + P(9) + P(10) + P(11) + P(12) \approx .8039,$$

so roughly 80% of all tickets—regardless of whether they're on Top, Bottom, Left, or Right—will lose $5. The house edge on this wager is 32.65%. ■

Players Keno, which is played at Jerzes Sports Bar and Keno and other locations in Papillion, Nebraska, offers players these four Pick 40 options as well as **Odd** and **Even**, which pay off if 11 or more numbers are either odd or even, as the player may choose [63]. Regardless of which half of the card is chosen, this simple wager at Jerzes pays off at 2 for 1, with a $5 minimum bet, if 11 or more numbers are caught. For these six bets, the probability of winning is

$$P(11) + P(12) + \ldots + P(20) \approx .3984,$$

and the house advantage is 20.32%.

In Norfolk, Nebraska, Big Red Keno offers a different Top and Bottom or Left and Right game which doubles the payoff, from 2 for 1 to 4 for 1, if 16 or more numbers fall into the chosen half of the ticket. The probability of a doubled payoff is then

$$\sum_{k=16}^{20} P(k \mid 40) = \sum_{k=16}^{20} \frac{{}_{40}C_k \cdot {}_{40}C_{20-k}}{{}_{80}C_{20}} \approx \frac{1}{526.52},$$

so roughly 1 drawing in 527 will witness a doubled payoff when one half of the ticket catches 16 or more numbers. At 12 games per hour, an average of about 44 hours will elapse between successive double payoffs. Owing to this extreme unlikeliness, this game's HA is barely less than in Papillion: 19.94% versus 20.32%.

However, each Big Red Keno outlet offers a monthly special game, which improves the payoffs on one of its keno games and so decreases the house advantage. In April 2017, Norfolk's special was a revised pay table for Top and Bottom, shown in Table 3.49, which reduced the house advantage to 8.64%.

TABLE 3.49: Big Red Keno: Norfolk April 2017 special Top and Bottom pay table ($1 wager)

| **Catch** | **Standard** | **Special** |
|---|---|---|
| 10 | 0 | .50 |
| 11–15 | 2 | 2 |
| 16–20 | 4 | 10 |

Of the two payoff changes, the bigger effect comes from paying 50¢ for catching 10 out of 20 numbers. While this is a net loss after deducting the $1 ticket price, this change alone drops the HA more than 10%, to 9.78%; the increased payoff for the very unlikely Catch 16 through Catch 20 events only cuts the HA by about 1.14%.

*Pattern Play* is a Pick 40 keno option offered by the British Columbia provincial lottery. In addition to top, bottom, left, and right options, players have the opportunity to customize a 40-number wager by picking any 4 rows, any 5 columns, or any 3 rows and 2 columns [17]. Players may also opt for a Quick Pick ticket, which picks 40 numbers at random.

Additionally, the lottery offers three other preprogrammed 40-number patterns: Crazy Corners, Diamond Daze, and Bulls-eye (Figures 3.10–3.12; page 114).

Pattern Play has the same probabilities as Top and Bottom, but the payoffs for a given fixed wager are different. A $2 Pattern Play wager pays off in accordance with Table 3.50.

TABLE 3.50: Pattern Play pay table ($2 wager) [17]

| **Catch** | **Payoff** |
|---|---|
| 0 or 20 | $50,000 |
| 1 or 19 | $25,000 |
| 2 or 18 | $2000 |
| 3 or 17 | $400 |
| 4 or 16 | $50 |
| 5 or 15 | $15 |
| 6 or 14 | $5 |
| 7 or 13 | $3 |
| 8–12 | $0 |

| 1 | 2 | 3 | 4 | 5 | 6 | 7 | 8 | 9 | 10 |
|---|---|---|---|---|---|---|---|---|---|
| 11 | 12 | 13 | 14 | 15 | 16 | 17 | 18 | 19 | 20 |
| 21 | 22 | 23 | 24 | 25 | 26 | 27 | 28 | 29 | 30 |
| 31 | 32 | 33 | 34 | 35 | 36 | 37 | 38 | 39 | 40 |
| 41 | 42 | 43 | 44 | 45 | 46 | 47 | 48 | 49 | 50 |
| 51 | 52 | 53 | 54 | 55 | 56 | 57 | 58 | 59 | 60 |
| 61 | 62 | 63 | 64 | 65 | 66 | 67 | 68 | 69 | 70 |
| 71 | 72 | 73 | 74 | 75 | 76 | 77 | 78 | 79 | 80 |

FIGURE 3.10: Pattern Play Crazy Corners bet.

| 1 | 2 | 3 | 4 | 5 | 6 | 7 | 8 | 9 | 10 |
|---|---|---|---|---|---|---|---|---|---|
| 11 | 12 | 13 | 14 | 15 | 16 | 17 | 18 | 19 | 20 |
| 21 | 22 | 23 | 24 | 25 | 26 | 27 | 28 | 29 | 30 |
| 31 | 32 | 33 | 34 | 35 | 36 | 37 | 38 | 39 | 40 |
| 41 | 42 | 43 | 44 | 45 | 46 | 47 | 48 | 49 | 50 |
| 51 | 52 | 53 | 54 | 55 | 56 | 57 | 58 | 59 | 60 |
| 61 | 62 | 63 | 64 | 65 | 66 | 67 | 68 | 69 | 70 |
| 71 | 72 | 73 | 74 | 75 | 76 | 77 | 78 | 79 | 80 |

FIGURE 3.11: Pattern Play Diamond Daze bet.

| 1 | 2 | 3 | 4 | 5 | 6 | 7 | 8 | 9 | 10 |
|---|---|---|---|---|---|---|---|---|---|
| 11 | 12 | 13 | 14 | 15 | 16 | 17 | 18 | 19 | 20 |
| 21 | 22 | 23 | 24 | 25 | 26 | 27 | 28 | 29 | 30 |
| 31 | 32 | 33 | 34 | 35 | 36 | 37 | 38 | 39 | 40 |
| 41 | 42 | 43 | 44 | 45 | 46 | 47 | 48 | 49 | 50 |
| 51 | 52 | 53 | 54 | 55 | 56 | 57 | 58 | 59 | 60 |
| 61 | 62 | 63 | 64 | 65 | 66 | 67 | 68 | 69 | 70 |
| 71 | 72 | 73 | 74 | 75 | 76 | 77 | 78 | 79 | 80 |

FIGURE 3.12: Pattern Play Bulls-eye bet.

The house advantage for a $2 wager is 30.18%—right in line with the HA for the $5 Top and Bottom ticket at the Meskwaki Casino.

In the early 1980's, the Royal Americana Casino in Las Vegas put a twist on the Top and Bottom wager with the "Over-10" bet [123]. Over-10 is a simple bet that more than half of the 20 numbers drawn will appear on the selected half of the keno ticket. In Over-10, the player's wager is carried over to the next drawing if the numbers split 10/10 between the top and bottom halves; in a standard Top and Bottom bet, a 10/10 split would lose for both sides.

The probability of a 10/10 split is

$$p = \frac{({}_{40}C_{10})^2}{{}_{80}C_{20}} = \frac{847,660,528^2}{3,535,316,142,212,174,320} \approx .2032$$

The remaining approximately 80% of drawings will result in a win for one side or the other; the symmetry of the wager means that the probability of winning an Over-10 bet on Top is approximately .3984. The wager paid off at 10–11, or 21 for 11, a payoff structure commonly seen in Nevada sports books. Computing the expected return on an $11 bet (the minimum) involves solving the equation

$$E = (21) \cdot .3984 + (E + 11) \cdot .2032 - 11,$$

where the second term reflects the fact that a 10/10 split simply awards the player a free ticket with the same payoff structure, but without the $11 price.

The solution to this equation is $E = -\$.50$. Dividing by the $11 cost of the ticket shows that the house edge is low: only 4.55%.

At about the same time, the Marina offered a Top and Bottom game that eliminated the need for the player to choose a side. The Marina's game, called "Coverall Keno", paid off if at least 13 of the 20 drawn numbers fell on one half, the top or bottom, of the keno ticket [167]. The pay table for a $1 ticket on this game was Table 3.51.

TABLE 3.51: Marina Coverall Keno pay table

| Catch | Payoff |
|---|---|
| 13 | $1 |
| 14 | $3 |
| 15 | $10 |
| 16 | $37.50 |
| 17 | $175 |
| 18 | $750 |
| 19 | $3750 |
| 20 | $10,000 |

The only way that this bet lost was if the 20 drawn numbers split 10/10, 11/9, or 12/8 between the two halves of the ticket. The probability of a loss was then

$$\frac{[{}_{40}C_{10} \cdot {}_{40}C_{10}] + [2 \cdot {}_{40}C_{11} \cdot {}_{40}C_{9}] + [2 \cdot {}_{40}C_{12} \cdot {}_{40}C_{8}]}{{}_{80}C_{20}} \approx .8039.$$

This bet has only about a 20% chance of winning, and most of that (12.7%) comes from the break-even payoff for a 13/7 split. With the probability of actually winning any money a low 6.9%, it comes as little surprise that this bet held a 31.41% house advantage.

---

## 3.2 Game Variations

*Keno is a game with a dual personality. It can be simple, uncomplicated and relaxing, or the most complex game in the casinos. It's up to the player to choose the variety.*—[Walter I. Nolan, in [112].]

### Progressive Jackpots

Like progressive slot machines and lotteries such as Powerball, progressive keno jackpots accumulate by adding a portion of the price of every ticket sold into a top prize fund, which starts at a specified fixed value and resets to that value when someone wins the prize, typically when their ticket catches all of the marked numbers. Progressive jackpots were originally introduced to keno in an effort to offer larger payoffs to players than gaming regulations that imposed aggregate limits on drawings permitted. Progressive amounts added to jackpots were not subject to the aggregate limit restrictions [130]. While the elimination of aggregate limits in Nevada beginning in 1989 rendered this justification unnecessary, progressive keno jackpots may still be found in Nevada and in other states where aggregate limits remain in force.

At La Vista Keno, for instance, 4% of each wager on a progressive keno game is added to the prize pool [74]. The progressive games in La Vista pay *only* on tickets that catch all marked numbers: the Pick 5 game jackpot starts at $2800 when all 5 chosen numbers are caught and the Pick 6 jackpot starts at $14,000 for catching 6 out of 6. These games can be played within the constraints of a way ticket; both require a minimum $2.50 wager per way, which contributes 10¢ per way to the progressive prize, to be eligible for the jackpot. Since, for a given drawing, the jackpot is fixed and does not rise in proportion to the wager, there is no value in betting more than $2.50 per way.

For the Pick 5 game, the expected value of a $2.50 ticket starts at

$$E = (2800) \cdot \frac{{}_{5}C_{5} \cdot {}_{75}C_{15}}{{}_{80}C_{20}} - 2.50 \approx -\$.6942,$$

but it rises as the jackpot builds. As with most casino games with a progressive feature, there comes a point when the jackpot rises so high that the expectation turns positive and the player has the advantage. Denoting the jackpot by $J$, this game has a positive expectation whenever

$$E(J) = J \cdot \frac{{}_5C_5 \cdot {}_{75}C_{15}}{{}_{80}C_{20}} - 2.50 > 0,$$

or when $J > \$3876.42$, which can be accumulated with the 4% contributions from 10,765 losing \$2.50 tickets.

The chance of the jackpot reaching this amount is simply the probability that 10,765 straight 5-spot tickets fail to catch all 5 numbers. This probability is

$$\left(1 - \frac{{}_{75}C_{15}}{{}_{80}C_{20}}\right)^{10,765} \approx (.9994)^{10,765} \approx 9.366 \times 10^{-4},$$

which means that a positive-expectation drawing is highly unlikely.

**Example 3.28.** Nevada's biggest progressive keno game is Mega10 Keno, a \$1.50 Pick 10 game which is played at a collection of networked casinos including the El Cortez, Jerry's Nugget, Plaza, and the Red Rock Resorts properties in metro Las Vegas.

The Mega10 progressive jackpot starts at \$1,000,000. While progressive games remain subject to the casino's aggregate limits on payouts, only an initial amount of \$25,000 is counted against the limit. If there are multiple winners on one drawing, the rules stipulate that each winner receives \$25,000 and an equal share of the accumulated additional money. The pay table is shown in Table 3.52.

TABLE 3.52: Mega10 progressive pay table

| Catch | Payoff |
|---|---|
| 5 | \$1.50 |
| 6 | \$10 |
| 7 | \$100 |
| 8 | \$1000 |
| 9 | \$7500 |
| 10 | Progressive |

With the initial \$1,000,000 top prize, the HA runs very high, at 56.89%. The Mega10 expectation turns positive when the top prize rises to \$8,605,380. This number is close to the denominator of the probability of catching 10 out of 10 numbers, which is approximately 1/8,911,711. ■

XpertX, the company managing Mega10 Keno, assumes responsibility for paying the amount of the jackpot that exceeds \$25,000; that initial amount is paid by the casino where the winning ticket is sold. The progressive jackpot is awarded as a sequence of 11 annual payments over 10 years [166].

## Changing the Casino's Draw

In the previous section, we examined a host of betting variations and game enhancements that arise from expanding the player's options beyond the original Pick 10 ticket. Another option for keno innovators might be to change the number of balls drawn by the casino from 20. This path has been explored far less thoroughly; perhaps the most extensive experiment took place at the Cal-Neva Casino, which literally stood on the Nevada-California border near Lake Tahoe. The Cal-Neva experimented for a time with keno games where the casino chose 10, 20, 30, and 40 numbers; players had to pick the game they wanted when purchasing their tickets.

Denote the number of numbers that the casino draws by $N$. As $N$ changes from 20, we note the following changes in keno's probabilities:

- The general formula for catching $k$ numbers on an $n$-spot ticket includes the additional variable $N$, and so becomes

$$p(k \mid n, N) = \frac{{}_nC_k \cdot {}_{80-n}C_{N-k}}{{}_{80}C_N}.$$

- At $N = 10$, the expected number of catches on an $n$-spot tickets drops from $n/4$ (page 107) to $n/8$, as the casino is drawing only $1/8$ of the total numbers. Additionally, drawing fewer numbers increases the probability of no catches and decreases the probability of catching a large number of marked spots—precisely the ones that win money.

  **Example 3.29.** On a 5-spot ticket, the probability of catching all 5 numbers drops from $6.45 \times 10^{-4}$ with 20 numbers drawn to $1.05 \times 10^{-5}$ when only 10 are drawn, a result about 61.5 times less likely. ■

- If $N$ increases, the situation is reversed: it's less likely to catch 0 spots, and more likely that a ticket will catch some or all of its numbers.

  **Example 3.30.** For the 5-spot ticket, the chance of catching all 5 numbers is .0059 with a 30-ball draw, and rises to .0274, or about 1 chance in 36.5, when the casino draws 40 numbers. ■

The Cal-Neva's composite table for this collection of games, Table 3.53, reflected both of these mathematical realities.

The games listed in Table 3.53 are very carefully designed, with HAs ranging within a fairly narrow band from 20.12% to 25.95%. Better games could be found at higher levels, where the Pick 5 through Pick 10 games included a progressive jackpot payout for catching all marked numbers when played against a 20-number casino draw.

The Cal-Neva also offered a 22-ball game. As with the games shown in Table 3.53, the casino imposed a $2 minimum wager on tickets played against a 22-number draw. A variety of games at the Pick 1 and Pick 2 levels were offered, as shown in Table 3.54.

TABLE 3.53: Cal-Neva variable draw pay tables ($2 wager)

| | | Casino Draws ($N$) | | | |
|---|---|---|---|---|---|
| **Pick** | **Catch** | **10** | **20** | **30** | **40** |
| 1 | 0 | 0 | 0 | 0 | 0 |
| | 1 | 12 | 6 | 4 | 3 |
| 2 | 0 | 0 | 0 | 1 | 2 |
| | 1 | 2 | 0 | 0 | 0 |
| | 2 | 74 | 26 | 8 | 4 |
| 3 | 0 | 0 | 0 | 2 | 5 |
| | 1 | 2 | 0 | 0 | 0 |
| | 2 | 10 | 2 | 1 | 0 |
| | 3 | 400 | 94 | 17 | 8 |
| 4 | 0 | 0 | 0 | 2 | 3 |
| | 1 | 0 | 0 | 0 | 0 |
| | 2 | 6 | 2 | 0 | 0 |
| | 3 | 140 | 9 | 4 | 2 |
| | 4 | 3000 | 250 | 45 | 16 |

TABLE 3.54: Cal-Neva 22-ball keno Catch All pay table

| Pick | $2 Special | $3 Bonus | $5 Bonus Plus |
|---|---|---|---|
| 1 | 6 | 9 | 15 |
| 2 | 23 | 36 | 61 |

The Catch 1 payoff for the standard 20-ball draw at the Cal-Neva was the usual 3 for 1, with its 25% house advantage. The Catch 1 games listed in Table 3.54 keep that same 3 for 1 ratio while increasing the number of ways to catch that one number, which decreases the HA in each game, to 17.5%. Standard Catch 2 paid 13 for 1, resulting in a 21.84% HA. While the three Pick 2 games listed paid off at 11.5, 12, and 12.2 for 1 respectively, the 21st and 22nd balls were enough to lower these HAs as well, as low as 10.82% in the case of the $5 Super Bonus.

The Cal-Neva took several steps to highlight the 21st and 22nd numbers drawn: they flashed on the keno board instead of steadily glowing, and were designated on the confirmation slips by punched square holes. These distinctions were necessary for two Flashing $pecial games that paid off handsomely if the player's numbers were among the last two drawn.

- The Flashing 1 Spot $pecial game paid off at 30 for 1 if the chosen number was among the last two drawn, and nothing otherwise. The probability of winning this game was

  $$P(\text{Number not drawn in the first 20}) \cdot P(\text{Number drawn as \#21 or 22}).$$

We know that the first factor in this product is ¾. The remaining probability is

$$\frac{{}_{59}C_1}{{}_{60}C_2} = \frac{1}{30},$$

making the probability of winning this bet 1/40. The expected value of a \$1 ticket turns out to be the same –25¢ that we have seen for a regular Pick 1 ticket; the HA remains 25%.

- The Flashing 2 Spot \$pecial was advertised with the sentence "2 Spots Gets You \$5000 to \$50,000 Only At Cal-Neva Keno". The minimum ticket price was \$2; the game paid 2500 for 1 if the player caught the last two numbers, so a \$20 wager could return \$50,000.

  The probability of catching 0 out of 2 numbers in the first 20 draws is 177/316, as we computed on page 62. The probability of then catching the last two numbers is

$$\frac{1}{{}_{60}C_2} = \frac{1}{1770},$$

  and so the expectation on a \$2 ticket is

$$E = (5000) \cdot \frac{177}{316} \cdot \frac{1}{1770} - 2 \approx -\$.4177,$$

  making the house edge 20.89%. This is equivalent to a standard Pick 2 ticket with 20 numbers drawn paying 13.16 for 1.

## Catch 22: Impossible, Yet Not Unplayable

As the Cal-Neva offered a game where the casino chose 22 numbers, so Caesars Tahoe developed a game where players could choose 22 numbers. Much like the Sweet Sixteen wager, *Catch 22* keno took a common numerical phrase and repurposed it as the name of a new way to play the game. Of course, with the casino drawing its usual 20 numbers, the pay table only needed to consider the standard catch-0 through catch-20 outcomes, and it is shown in Table 3.55.

The probability of catching $k$ numbers when 22 are chosen by the player and 20 are selected by the casino is

$$p(k \mid 22) = \frac{{}_{22}C_k \cdot {}_{58}C_{20-k}}{{}_{80}C_{20}}.$$

Comparing this to the standard keno probability function $p(k \mid 20)$ shows that $p(k \mid 20) > p(k \mid 22)$ only if $0 \leqslant k \leqslant 5$.

With a house advantage of 25.74%, the standard pay table gave a result right in line with keno payoffs, but a player incentive was available to tip the scales a bit. A player buying into Catch 22 for 20 or more consecutive

TABLE 3.55: Caesars Tahoe Catch 22 pay table ($5 wager)

| Catch | Payoff | Catch | Payoff |
|---|---|---|---|
| 0 | $375 | 12 | $25 |
| 1 | $25 | 13 | $62.50 |
| 2 | $12.50 | 14 | $375 |
| 3 | $7.50 | 15 | $2250 |
| 4–8 | $2.50 | 16 | $6250 |
| 9 | $5 | 17 | $50,000 |
| 10 | $7.50 | 18–20 | $100,000 |
| 11 | $12.50 | | |

drawings saw the payoffs for catching 0 or greater than 18 numbers increased: a catch-0 ticket paid $1500 instead of $375—this was helpfully labeled in the brochure as "No ID check", meaning that the net win of $1495 was not subject to automatic IRS notification—and a ticket catching 18 or more numbers paid $250,000. Changing the payoff at the top end induces an insignificant change in the HA; the probability of the top three catches is, once again, too low to change the expectation by even a penny. The quadrupling of the catch-0 payoff, however, is sufficient to drop the HA down to 13.98%—of course, this requires an initial investment of at least $100 and the willingness to root against catching one's numbers.

## Betting the Edge: Merlin's Magical Keno Kube

The Excalibur Casino is built in the shape of a castle and has an Arthurian England theme. This theme extends as far as the keno lounge, where the Merlin's Magical Keno Kube wager offers the player a $5 ticket covering all 32 numbers around the edge of a keno ticket. These numbers are shaded in Figure 3.13.

| 1 | 2 | 3 | 4 | 5 | 6 | 7 | 8 | 9 | 10 |
|---|---|---|---|---|---|---|---|---|---|
| 11 | 12 | 13 | 14 | 15 | 16 | 17 | 18 | 19 | 20 |
| 21 | 22 | 23 | 24 | 25 | 26 | 27 | 28 | 29 | 30 |
| 31 | 32 | 33 | 34 | 35 | 36 | 37 | 38 | 39 | 40 |
| 41 | 42 | 43 | 44 | 45 | 46 | 47 | 48 | 49 | 50 |
| 51 | 52 | 53 | 54 | 55 | 56 | 57 | 58 | 59 | 60 |
| 61 | 62 | 63 | 64 | 65 | 66 | 67 | 68 | 69 | 70 |
| 71 | 72 | 73 | 74 | 75 | 76 | 77 | 78 | 79 | 80 |

FIGURE 3.13: Keno bet slip with Merlin's Magical Keno Kube bet shaded.

This wager pays off whenever fewer than 7 or more than 9 of the edge numbers are caught by the player; the maximum payoff of $100,000 for catching 19 or 20 numbers is also the Excalibur's aggregate limit on any one game. The full pay table is given in Table 3.56.

TABLE 3.56: Merlin's Magical Keno Kube pay table ($5 wager)

| Catch | Payoff | Catch | Payoff |
|---|---|---|---|
| 0 | $27,500 | 11 | $5 |
| 1 | $1500 | 12 | $7 |
| 2 | $275 | 13 | $25 |
| 3 | $25 | 14 | $175 |
| 4 | $7 | 15 | $1000 |
| 5 | $5 | 16 | $7500 |
| 6 | $4 | 17 | $35,000 |
| 7–9 | $0 | 18 | $50,000 |
| 10 | $3 | 19–20 | $100,000 |

The expected value of a $5 Keno Kube wager is –$1.39, giving a HA of 27.77%. This long-term expectation may not be the best indicator of how the game plays out, so we look at some single-game probabilities. The probability of losing $5 by catching 7, 8, or 9 of 32 numbers is .5702—greater than 50%. The probability of losing money encompasses catches of 6 through 10 numbers, and is .8132. Adding in the two break-even payoffs shows that the probability of coming out even or behind on a single draw is .9367, leaving the chance of winning money a minuscule 6.33%, or approximately 1 in 15.8 tickets.

**Example 3.31.** Merlin's Magical Keno Kube is easily adapted to other casino themes. At the FireKeepers Casino in Battle Creek, Michigan, this ticket is called the "Ring Of Fire" and pays off according to Table 3.57 for a $5 wager.

TABLE 3.57: Ring Of Fire pay table ($5 wager)

| Catch | Payoff | Catch | Payoff |
|---|---|---|---|
| 0 | $25,000 | 11 | $5 |
| 1 | $2500 | 12 | $10 |
| 2 | $250 | 13 | $25 |
| 3 | $25 | 14 | $250 |
| 4 | $10 | 15 | $1000 |
| 5 | $5 | 16 | $7500 |
| 6 | $3 | 17 | $40,000 |
| 7–9 | $0 | 18 | $50,000 |
| 10 | $3 | 19–20 | $100,000 |

Does Ring Of Fire give the player a more favorable HA than Merlin's Magical Keno Kube, whose HA is 27.77%?

The probabilities have not changed, only the payoff amounts have. Calculating the HA shows that Ring Of Fire is slightly more player-favorable, as the HA is only 23.35%. The probability of losing money or breaking even is the same as for Merlin's: approximately .9367. ■

In 1991, Caesars Palace offered two Pick 32 tickets: the Outside bet akin to Merlin's Magical Keno Kube and an analogous Inside bet, which stays away from the edge. This wager used the pattern in Figure 3.14.

| | | | | | | | | | |
|---|---|---|---|---|---|---|---|---|---|
| 1 | 2 | 3 | 4 | 5 | 6 | 7 | 8 | 9 | 10 |
| 11 | 12 | 13 | 14 | 15 | 16 | 17 | 18 | 19 | 20 |
| 21 | 22 | 23 | 24 | 25 | 26 | 27 | 28 | 29 | 30 |
| 31 | 32 | 33 | 34 | 35 | 36 | 37 | 38 | 39 | 40 |
| 41 | 42 | 43 | 44 | 45 | 46 | 47 | 48 | 49 | 50 |
| 51 | 52 | 53 | 54 | 55 | 56 | 57 | 58 | 59 | 60 |
| 61 | 62 | 63 | 64 | 65 | 66 | 67 | 68 | 69 | 70 |
| 71 | 72 | 73 | 74 | 75 | 76 | 77 | 78 | 79 | 80 |

FIGURE 3.14: Keno bet slip with Caesars Palace Inside bet shaded.

Both of these bets used the same pay table. Caesars Palace's pay table was somewhat more restrictive than those of the other casinos, returning $0 on the $5 minimum bet for catching 6–10 of the 32 numbers rather than 7–9. Raising the top prize to $300,000, for catching 19 or 20 numbers, did not compensate for this change. The probability of losing money or breaking even was the same 93.67%, but the HA at Caesars Palace was higher: 29.19%.

## Two-Game Parlays

In considering casino games for space on the gaming floor, one factor to be considered is a game's "hold": the percentage of money gambled by players that is ultimately retained by the casino. Hold is defined as the amount won divided by the "drop": the total amount of money the casino has in its cash boxes at the end of a specified time interval. This differs from the house advantage in that players may hit unlucky streaks and lose 100% of their buy-in over time, rather than only dropping the percentage that the HA calls for the casino to retain.

**Example 3.32.** In a recent year in Nevada, the hold percentage for blackjack was 12.44%—very different from that game's low HA for perfect basic strategy play. The hold percentage for keno that year was 27.85%, which is much more in line with the game's HA [53]. ■

One technique that has some value in increasing casinos' hold is to provide opportunities for players to play recent winnings immediately into another wager, as with a "Let It Ride" button that pops up on a video poker or blackjack machine after a player win. In keno lounges, one such option is the *two-game parlay.* The general structure of a two-game parlay is that the player bets the same combination of numbers on two consecutive drawings, forgoing some or all of the prize on the first draw for a chance at an increased payoff if he or she catches their selected numbers in both games. The casino, of course, sees this as two chances at winning the player's initial wager.

Keno Casino, purveyors of Bellevue, Nebraska's keno games, offers a game called *2-Game Parlay Keno.* This is a game variation that can be played on two successive drawings, marking either 1 or 2 spots [65]. The 1-spot game pays \$12 if both tickets catch their number and 1¢, a payout required by state law, if only one ticket wins. The one winning ticket may be either the first or second, making the probability of winning a penny

$$\frac{1}{4} \cdot \frac{3}{4} + \frac{3}{4} \cdot \frac{1}{4} = \frac{6}{16}.$$

Since the two drawings are independent, the probabilities of winning each prize may be found using the Multiplication Rule. The expected value of a \$2 wager covering two drawings is

$$E = (12) \cdot \frac{1}{4} \cdot \frac{1}{4} + (.01) \cdot \frac{6}{16} - 2 = -\$1.24625,$$

and the HA is 62.31%—very high even for keno.

A gambler who walks away without waiting to collect the penny on a single winning ticket forfeits only .19% more to the house.

By contrast, if this 1-spot ticket is played twice, as two separate \$1 tickets, the probability distribution in Table 3.58 is in effect.

TABLE 3.58: Probability distribution for two standard \$1 Pick 1 keno tickets

| Total Win | Probability |
|---|---|
| \$6 | 1/16 |
| \$3 | 6/16 |
| \$0 | 9/16 |

The corresponding HA is then 25%—just as it is for a single Pick 1 ticket. This two-game parlay is no bargain.

At the Riviera, the two-game parlay was called "The Daily Double". A 1-spot \$2 Daily Double ticket (\$1 per game) paid off only if the number was caught in both the Red and Blue games running alternately. With a payoff of \$24, or 12 for 1, the expected value of a \$2 ticket was

$$E = (24) \cdot \left(\frac{1}{16}\right) - 2 = -\$.50,$$

and the HA was the same 25% as a standard Pick 1 game.

The Union Plaza offered a $2 two-game parlay called "The Big Q", which was said to offer "Keno's Biggest Payoff". The 1-spot game paid $25 if the number was caught in both games, slightly better than the Riviera, but that biggest payoff came on the 5-spot ticket. Catching 5 out of 5 numbers in two straight games paid $250,000—double the Union Plaza's aggregate limit at the time, since this was a payoff for two games. The game paid $500 for catching 5 out of 5 on only one game, making the expectation

$$E = (250,000) \cdot P(5\,|\,5)^2 + (500) \cdot \{2 \cdot P(5\,|\,5) \cdot [1 - P(5\,|\,5)]\} - 2,$$

or about –$1.25. The HA of this bet is in line with Bellevue's parlay: 62.58%.

The high payoffs may be very tempting, but most parlays like this amounted to giving the casino a better shot at retaining a player's money.

## Lagniappe Keno

The Orleans Casino promotes a novel keno wager with a tip of its hat to the resort's New Orleans Mardi Gras theme:

> Shopkeepers in the French Quarter of New Orleans have a long-standing practice of showing appreciation to their customers by presenting them with something extra, such as an additional praline or beignet when a dozen are purchased or a free accessory to match a garment. The thank-you gift is called a *lagniappe* (pronounced *lan-yap*).
>
> In keeping with this venerable tradition, The Orleans Keno Department offers a special series of keno tickets. Lagniappe keno is played at 50¢ per game (minimum of 20 games). Each ticket features its own lagniappes.
>
> To receive a lagniappe, catch all of your marked numbers.

A lagniappe, then, represents a bonus paid on a keno bet that catches all of the numbers chosen by the player. For a Pick 5 game, the lagniappe was $100 from Monday through Thursday; the weekend lagniappe was a free buffet dinner for two, with a retail value of $39.98. Similar prize structures were in place for Pick 6 and Pick 7 games, with the Monday-Thursday cash prize and the value of the weekend dinner (and the prestige of the restaurant) increasing.

One possible explanation for the different lagniappes lies in the Orleans' dual role as a locals' casino and a tourist casino. Gamblers from Monday through Thursday are more likely to be Nevada residents and perhaps more regular customers than tourists who appear in greater numbers on the weekends, and the Orleans management was probably trying to offer a game that would be attractive to locals who might be turned into frequent customers.

There is reason to be suspicious when any casino is offering something apparently for nothing. The Pick 5 lagniappe keno game—when played Monday through Thursday—had the following pay table:

| Catch | Payoff |
|---|---|
| 4 | \$9 |
| 5 | \$500 (includes \$100 lagniappe) |

We can incorporate the value of the lagniappe as we find the expectation of a lagniappe keno ticket:

$$E = (9) \cdot \left( \frac{{}_5C_4 \cdot {}_{75}C_{16}}{{}_{80}C_{20}} \right) + (500) \cdot \left( \frac{{}_{75}C_{15}}{{}_{80}C_{20}} \right) - .50 \approx -\$.1414.$$

Note that the 20-game minimum called for in the Orleans' publicity materials has no effect on the expectation of a single game, and so it need not be included in this analysis. Since a lagniappe keno ticket costs 50¢, we divide this expectation by .50 to get a house edge of 28.28%, which is high for a casino game but not surprising for keno.

Without the lagniappe option, a Pick 5 keno game at the Orleans offered the following pay table for a \$1 minimum wager:

| Catch | Payoff (for 1) |
|---|---|
| 3 | 1 |
| 4 | 15 |
| 5 | 750 |

If we divide the payoffs in this table by 2 to equalize the reward per dollar wagered, we find that the lagniappe has been accommodated by the following changes to the original pay table:

- Elimination of the break-even payoff for catching 3 of 5 numbers.
- Increases in the payoffs for catching 4 and 5 numbers, including the lagniappe on a 5-number catch.

The expectation for this bet is

$$E = (1) \cdot P(3 \,|\, 5) + (15) \cdot P(4 \,|\, 5) + (750) \cdot P(5 \,|\, 5) - 1 \approx -\$.2510,$$

and the HA would be 25.10%. Much of the player value in this keno game comes from the break-even payment; once that's eliminated, raising the other payoffs by the amounts specified still results in a net increase in the HA.

We can see that lagniappe keno, played from Monday-Thursday, represents a worse game—in the form of a slightly higher house edge—for the player than the Orleans' standard Pick 5 keno game.

If, however, you're playing lagniappe keno on a weekend, the value of the lagniappe drops by 60%. The new expectation is

$$E = (9) \cdot \left( \frac{{}_5C_4 \cdot {}_{75}C_{16}}{{}_{80}C_{20}} \right) + (439.98) \cdot \left( \frac{{}_{75}C_{15}}{{}_{80}C_{20}} \right) - .50 \approx -\$.1801,$$

and the new HA is 36.02%. This game is even worse for the player; we, of course, knew this simply from seeing the lagniappe drop in value.

## Pop 80

*Pop 80* (short for "Popular 80¢ rate") refers to a keno pay table that is usually based on tickets priced at 80 cents per way. The payback rate on Pop 80 tickets, regardless of their actual price, is typically 75–80% or more—somewhat higher than most keno games. The D Casino offers a Pop 80 rate in its 40¢ "Deano" tickets, which are cited as the best keno pay table in Las Vegas [73]. We saw one of these games in our consideration of variable payoffs on Pick 2 games (page 65), where the HA on "Simply The Best Special Rate Ever!" Pick 2 tickets was an appealingly-low 18.83%. The full Pop 80 pay table from the D is shown in Table 3.59.

TABLE 3.59: Deano pay table ($) at the D, Las Vegas (40¢ wager)

| | Numbers marked | | | | | | | | | |
|---|---|---|---|---|---|---|---|---|---|---|
| **Catch** | 1 | 2 | 3 | 4 | 5 | 6 | 7 | 8 | 9 | 10 |
| 1 | 1.40 | | | | | | | | | |
| 2 | | 5.40 | .40 | .40 | | | | | | |
| 3 | | | 20 | 1 | .40 | .40 | .20 | | | |
| 4 | | | | 65 | 3 | 1 | .40 | .40 | .60 | .40 |
| 5 | | | | | 400 | 30 | 7 | 3 | 2 | .40 |
| 6 | | | | | | 1200 | 70 | 30 | 15 | 5 |
| 7 | | | | | | | 7000 | 700 | 100 | 50 |
| 8 | | | | | | | | 10,000 | 1000 | 500 |
| 9 | | | | | | | | | 12,000 | 5000 |
| 10 | | | | | | | | | | 20,000 |

On its face, the pay table does not suggest a better game for players, in part because the 40¢ ticket price requires some extra calculation to find the player return rate. Computing the house advantages for each game shows that this pay table offers a better rate of return to players than other games. Table 3.60 compares the HA of these games to the HA for the corresponding $1 games at the D.

The choice is clear: The Pop 80 rates are more player-friendly than the standard rates, anywhere from 7½–12½%. The rules for the Deano/Pop 80 rate offer 10¢ ways, but call for a minimum wager of 40¢ per game and $4

TABLE 3.60: Comparison of standard and Pop 80 house edges at The D

| Pick | Standard HA | Pop 80 HA |
|---|---|---|
| 1 | 25.00% | 12.50% |
| 2 | 27.85% | 18.83% |
| 3 | 27.85% | 16.75% |
| 4 | 25.94% | 18.14% |
| 5 | 27.84% | 18.04% |
| 6 | 28.46% | 17.97% |
| 7 | 28.70% | 15.40% |
| 8 | 29.74% | 21.43% |
| 9 | 29.20% | 20.02% |
| 10 | 28.57% | 20.50% |

per ticket. These rules are perhaps most easily followed by betting the Pop 80 rate as a multi-way king ticket [71].

**Example 3.33.** Suppose that you want to chase the $1200 payday for catching 6 of 6 numbers. An easy way to do this at a low price while giving yourself a chance at winning is to choose 7 numbers, play them all as kings, and configure your ticket as a 7/6, 1/7 way ticket: 8 tickets in all. 40¢ per way, for a total wager of $3.20, will not hit the $4 ticket minimum; that minimum may be easily reached by playing the same ticket over 2 or more consecutive drawings.

The probability of winning the $1200 jackpot by catching exactly 6 of 7 numbers is

$$\frac{{}_{7}C_{6} \cdot {}_{73}C_{14}}{{}_{80}C_{20}} \approx \frac{1}{1366}.$$

By contrast, the probability of catching 6 numbers out of 6 is

$$\frac{{}_{74}C_{14}}{{}_{80}C_{20}} \approx \frac{1}{7753},$$

about one-sixth the probability above. ■

## Double-Up Keno

**Double-Up Keno!!!**
**Pick** 2–10 numbers.
**Pay** just $1 per number.
**Win** on *all* catches—your prize *doubles* with each additional catch!

This might be an advertisement for a hypothetical game called *Double-Up Keno*. The pay table for this game is simple: catching $k$ numbers gives a payoff of $\$2^k$. For example, the pay table for a Pick 5 game (ticket price \$5) is given in Table 3.61.

TABLE 3.61: Double-Up Keno Pick 5 pay table

| Catch | Payoff |
|---|---|
| 0 | \$1 |
| 1 | \$2 |
| 2 | \$4 |
| 3 | \$8 |
| 4 | \$16 |
| 5 | \$32 |

In practice, payoffs for small catches are increased, while payoffs for the far-less-likely large catches are smaller. The net effect of these changes is a significant reduction in the house advantage for larger tickets.

**Example 3.34.** For a Pick 1 game, though, the advantage would lie with the player. The payoffs on a \$1 ticket would be \$1 for matching 0 numbers and \$2 for matching 1; the resulting expectation is

$$E = (1) \cdot \frac{3}{4} + (2) \cdot \frac{1}{4} - 1 = \$.25 > 0.$$

This is why Double-Up Keno starts its pay table at Pick 2. ■

Figure 3.15 shows the house advantage for Double-Up Keno as a function of the number of numbers chosen.

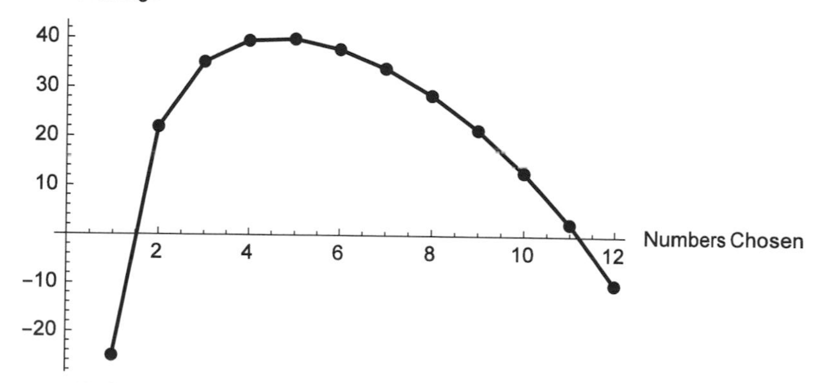


FIGURE 3.15: House advantage for Double-Up Keno.

The HA reaches its maximum of 39.88% on a Pick 5 game and then slowly decreases. A Pick 10 game has a house edge of 12.93%, and a Pick 11 ticket's HA is only 2.51%—lower than for European roulette. A Pick 12 game under this set of rules would return the edge, at 9.91%, to the player.

## Kingo

*Kingo* is a cross between keno and bingo that is on offer in Bellevue, Nebraska. A Kingo player receives a ticket with 24 randomly-selected numbers arranged in a 5 by 5 grid with a free space in the center square, as in Figure 3.16. Kingo tickets are Quick Pick only, so players need not spend time trying to pick and arrange 24 favored numbers.

| | | | | |
|---|---|---|---|---|
| 60 | 21 | 66 | 31 | 57 |
| 26 | 74 | 32 | 56 | 67 |
| 19 | 69 | **Free** | 15 | 8 |
| 62 | 43 | 13 | 11 | 3 |
| 59 | 4 | 47 | 30 | 61 |

FIGURE 3.16: Sample Kingo ticket.

As with bingo cards, order matters in laying out a Kingo ticket. Since any of the 80 numbers may be placed into any square on the ticket, there are ${}_{80}P_{24} \approx 1.007 \times 10^{44}$ possible tickets. Some of these are, for all practical game purposes, identical—taking Figure 3.16 and rotating it 90, 180, or 270 degrees, or reflecting it over any of its four axes of symmetry, produces a ticket with a different arrangement of its 24 numbers but which scores the same number of winning combinations for a given set of 20 drawn numbers.

Kingo tickets are priced at $1. Three possible winning ways are available; all are based on catching numbers that complete a particular pattern on the bingo card.

- A **Regular Kingo** fills in a single horizontal, vertical, or diagonal line including the center free square, and pays $25. This prize is also awarded to a ticket catching all four of its corner numbers (60, 57, 59, and 61 in Figure 3.16).

- A **Hardway Kingo** completes a horizontal or vertical line *without* using the center free square, and pays \$60.
- **X Kingo** is the top prize: \$10,000 awarded for catching all 8 numbers that comprise both diagonals.

A Kingo ticket is best interpreted as a 14-way ticket, with five 4-spot ways (Regular Kingo), eight 5-spot ways (Hardway Kingo), and one 8-spot way (X Kingo). The probability of winning \$10,000 is then

$$\frac{{}_{8}C_{8} \cdot {}_{72}C_{12}}{{}_{80}C_{20}} \approx 4.346 \times 10^{-6},$$

although this low value also includes the possibility of catching additional winning ways.

**Example 3.35.** What is the maximum possible win on a Kingo ticket?

The best possible Kingo ticket would certainly include the X Kingo configuration, which accounts for 8 catches. This wins only the \$10,000 top prize, and does not also qualify for three Regular Kingo payoffs that it covers. The challenge is then to find the best possible way to distribute the other 12 catches so as to complete the maximum number of lines.

Since Hardway Kingos pay more than Regular Kingos, we should seek to fill in 5-spot lines over 4-spot lines which use the free square. Completing any one of those lines, either rows or columns, uses up three catches, but these may also contribute to filling lines in the other direction.

Completing rows (or columns) 1, 2, 4, and 5 brings the total win to \$10,240, but includes no winning combinations going the other way. A larger win can be achieved by catching all numbers in rows 1 and 2 (6 additional catches beyond the X Kingo squares), together with the remaining numbers in columns 1, 2, and 3. These columns use up the last 6 catches, and complete three columns. One maximal configuration, which would win a total of \$10,265, is shown in Figure 3.17.

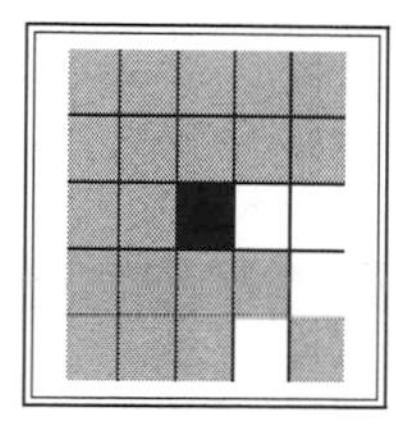

FIGURE 3.17: Kingo ticket with maximum win highlighted.

■

There are several equivalent patterns of 20 catches that lead to the same payoff. For example, we could simply shift the two catches in column 1, rows 3 and 4, to the corresponding spots in column 5. In rearranging the catches to find other maximal configurations, it is necessary not to disturb the 8 catches making up the central X.

The most probable event in a single game of Kingo is catching 6 numbers, one-fourth of the 24 on the ticket. There are ${}_{24}C_6 = 134,596$ ways to distribute 6 catches around a Kingo ticket. How many of these are winners?

- By catching all four corners and a Regular Kingo along a diagonal, there are 2 ways to score two wins while catching 6 numbers. Here, every catch contributes to a win, and so there is no need to consider catches that are not part of a winning pattern.

- Five-spot wins come from Hardway Kingos: there are 4 horizontal and 4 vertical wins possible. Each may be combined with 19 places to locate the sixth catch, which gives a total of $8 \cdot 19 = 152$ winning patterns.

- A single 4-spot win requires choosing a winning arrangement and then placing the other two catches, which can be done in ${}_{20}C_2 = 190$ ways. There are 5 ways to score a Regular Kingo win: two diagonal, one vertical, one horizontal, and the four corners. In the case of diagonal and corner wins, we must exclude the double-win patterns already counted. The total number of Regular Kingo wins with 6 catches is then

$$\underbrace{2 \cdot ({}_{20}C_2 - 1)}_{\text{Diagonal}} + \underbrace{{}_{20}C_2}_{\text{Vertical}} + \underbrace{{}_{20}C_2}_{\text{Horizontal}} + \underbrace{{}_{20}C_2 - 2}_{\text{4 Corners}} = 5 \cdot {}_{20}C_2 - 4 = 946.$$

We conclude that there are 1100 possible winning arrangements with 6 catches, and we have the following probabilities:

$$P(\text{Win something} \mid \text{Catch 6 numbers}) = \frac{1100}{134,596} \approx .0082.$$

$$P(\text{Catch 6 numbers}) = \frac{{}_{24}C_6 \cdot {}_{52}C_{14}}{{}_{80}C_{20}} \approx .2210.$$

$$P(\text{Win with 6 catches}) \approx .0082 \cdot .2210 \approx \frac{1}{554}.$$

## 100 Dimes

If 40 cents, the minimum keno bet at the El Cortez, is too much to risk on a keno bet, the four Las Vegas casinos that offer the Just 7777s game (page 78) also offered a 10¢ game that may be more economically appealing. The catch is that this ticket must be played for a minimum of 100 games, and so represents an initial outlay of at least \$10. This Pick 7 game generates a phenomenal casino edge by eliminating the standard payoff if 4 or 5 numbers are matched; while the game includes a break-even payoff if none of your 7 numbers are drawn, the probability of this event is only .1216, and so you can, in the long run, only count on about 12 break-even tickets in a typical run of 100 games.

The 100 Dimes game has the following pay table:

| Catch | Payoff |
|---|---|
| 0 | \$.10 |
| 6 | \$20 |
| 7 | \$1200 |

The expectation on a 10¢ ticket is

$$E = (.10) \cdot P(0 \,|\, 7) + (20) \cdot P(6 \,|\, 7) + (1200) \cdot P(7 \,|\, 7) - .10 \approx -\$.0439,$$

which makes the house advantage 43.9%.

At this point, you might as well buy a state lottery ticket (HA approximately 50%) and perhaps derive some satisfaction from supporting education—of course, Nevada has no state lottery.

## Nevada Numbers

Part of the reason why Nevada has no state lottery is that casino industry lobbyists can be counted on to oppose any legislative efforts to introduce this form of gambling competition into their state. California state lottery tickets sell quite briskly in stores near the California-Nevada border, indicating a high level of interest in lotteries among some Nevada residents. For a brief while in the early 2000s, several Nevada casinos cooperated on a keno game called *Nevada Numbers*, which had many of the same features as a state lottery.

Nevada Numbers was a Pick 5 keno game with a modified pay table. While standard Pick 5 keno pays off only when 3 or more numbers are caught, the pay table for Nevada Numbers added payoffs for catching 1 or 2 numbers, while increasing the payoffs for 3–5 catches. The 5 out of 5 payoff was a progressive jackpot that started at \$5,000,000.

What made Nevada Numbers resemble a lottery? First, this single drawing determined the fate of tickets available throughout Nevada, from Laughlin in the south to Reno in the north. Second, only 5 numbers from 1–80, rather than 20, were drawn by the host casino, Bally's in Las Vegas, and broadcast to the keno lounge at each participating casino [134]. Finally, as with state lottery games such as Powerball, this was not a continuously-running game; the drawing was conducted only once daily, at 6:00 P.M. Pacific time.

A Nevada Numbers ticket cost \$2 and paid off in accordance with Table 3.62.

**Example 3.36.** Find the probability of hitting the progressive jackpot.

The denominator in this calculation is ${}_{80}C_5$ rather than the usual ${}_{80}C_{20}$; there is only one way to win. The probability of catching 5 out of 5 numbers is then

$$\frac{1}{{}_{80}C_5} = \frac{1}{24{,}040{,}016},$$

and we see that the denominator dwarfs the $5,000,000 payoff. ■

TABLE 3.62: Nevada Numbers pay table [134]

| Catch | Payoff |
|---|---|
| 1 | $1 |
| 2 | $2 |
| 3 | $20 |
| 4 | $2000 |
| 5 | Progressive (min. $5,000,000) |

If the progressive jackpot is at its minimum value, Nevada Numbers can be shown to have a house advantage of 71.44% (Exercise 3.8), placing it well beyond keno wagers and even beyond most state lotteries. Unlike the latter, Nevada Numbers was never intended to fund education or other public works projects. If the jackpot were to rise to $39,347,298, however, Nevada Numbers would have a positive expectation.

Nevada Numbers was discontinued in 2009 and replaced in 2011 by Nevada Numbers Lite [106]. Nevada Numbers Lite is a Pick 9 keno game with the standard 20 numbers drawn by the casino and payoffs as in Table 3.63.

TABLE 3.63: Nevada Numbers Lite pay table [107]

| Catch | Payoff |
|---|---|
| 4 | $1 |
| 5 | $5 |
| 6 | $10 |
| 7 | $100 |
| 8 | $2000 |
| 9 | Progressive (min. $200,000) |

Nevada Numbers Lite was initially launched at three Nevada casinos: Treasure Island in Las Vegas, Jerry's Nugget in North Las Vegas, and the Aquarius in Laughlin. The progressive jackpot for catching all 9 numbers on a $1 ticket starts at $200,000. Unlike Nevada Numbers, the new game is played every 10 minutes [107].

**Example 3.37.** Which game, Nevada Numbers or Nevada Numbers Lite, offers a player the higher probability of making a profit on a single ticket?

For Nevada Numbers, the desired probability is the probability of matching 3 or more numbers out of 5, since the match 1 payoff is a net loss and matching 2 numbers merely breaks even. As in previous sections, we denote the probability of matching $n$ numbers by $P(n)$; then

$$P(n) = \frac{{}_{5}C_{n} \cdot {}_{75}C_{5-n}}{{}_{80}C_{5}}$$

and we then have

$$P(\text{Profit}) = P(3) + P(4) + P(5) \approx .0012.$$

A Nevada Numbers Lite wager yields a profit if you match at least 5 numbers; in this game

$$P(n) = \frac{{}_{9}C_{n} \cdot {}_{71}C_{20-n}}{{}_{80}C_{20}},$$

and we then have

$$P(\text{Profit}) = P(5) + P(6) + P(7) + P(8) + P(9) \approx .0389$$

—so it's about 33 times more likely that you'll come out ahead on Nevada Numbers Lite. ■

As with Nevada Numbers, the house advantage in Nevada Numbers Lite depends on the size of the progressive jackpot. If the jackpot is at its \$200,000 minimum, the casino's edge is 39.65%—intermediate between what we expect from keno and from a lottery. If the jackpot rises to \$747,420, then the gambler has the advantage.

## Penny Keno

Way tickets give the keno player a way to take the price per wager down even lower than 10¢. At the FireKeepers Casino, *Penny Keno* lives up to its name by offering keno bets at 1¢ each. These special way tickets are generated by a pattern of spots on the ticket forming a letter of the alphabet: K, E, N, O, C, A, S, H, and F are all possible letters for this game. Figure 3.18 shows a "K" configuration with the active numbers shaded.

| 1 | 2 | 3 | 4 | 5 | 6 | 7 | 8 | 9 | 10 |
|---|---|---|---|---|---|---|---|---|---|
| 11 | 12 | 13 | 14 | 15 | 16 | 17 | 18 | 19 | 20 |
| 21 | 22 | 23 | 24 | 25 | 26 | 27 | 28 | 29 | 30 |
| 31 | 32 | 33 | 34 | 35 | 36 | 37 | 38 | 39 | 40 |
| 41 | 42 | 43 | 44 | 45 | 46 | 47 | 48 | 49 | 50 |
| 51 | 52 | 53 | 54 | 55 | 56 | 57 | 58 | 59 | 60 |
| 61 | 62 | 63 | 64 | 65 | 66 | 67 | 68 | 69 | 70 |
| 71 | 72 | 73 | 74 | 75 | 76 | 77 | 78 | 79 | 80 |

FIGURE 3.18: Penny Keno bet slip with shaded "K" bet.

This ticket covers 50 of the 80 spaces on a keno ticket; these 50 spaces are divided into 21 blocks of 2 or 3 numbers each, which are labeled alphabetically as shown in Figure 3.19.

| A | A | I | 4 | 5 | 6 | T | T | T | U |
|---|---|---|---|---|---|---|---|---|---|
| B | B | I | 14 | 15 | S | S | S | 19 | 20 |
| C | C | J | 24 | R | R | R | 28 | 29 | 30 |
| D | D | J | M | M | M | 37 | 38 | 39 | 40 |
| E | E | K | N | N | N | 47 | 48 | 49 | 50 |
| F | F | K | 54 | O | O | O | 58 | 59 | 60 |
| G | G | L | 64 | 65 | P | P | P | 69 | 70 |
| H | H | L | 74 | 75 | 76 | Q | Q | Q | U |

FIGURE 3.19: Penny Keno bet slip with "K" bet groups labeled alphabetically.

These blocks are then combined to generate a way ticket consisting of 190 Catch All groups of 3, 4, or 5 numbers, divided as follows:

- Eight 3-spot tickets, consisting of the 8 3-number blocks (M-T in Figure 3.19). For example, block M covers the numbers 34, 35, and 36. A 3-spot way pays off 55¢ if all three numbers are caught.
- Seventy-eight 4-spot tickets, which are formed by all choices of 2 of the 13 2-number blocks (A-L and U): ${}_{13}C_2 = 78$. By combining blocks C and K, we get the 4-spot ticket 21, 22, 43, and 53. Winning 4-spot ways pay $2.50.
- One hundred four 5-spot tickets, obtained by combining each 2-number block with each 3-number block: $8 \times 13 = 104$. N and U, taken together, give the sample 5-number block consisting of 44, 45, 46, 10, and 80. 5-spot ways are paid $11.50 if they are drawn.

The price of this ticket is 1¢ per way, or $1.90—effectively bringing the cost per ticket down to a penny. In buying such a ticket, you lose the ability to personalize the ticket by betting on your favorite numbers, but it's highly unlikely that anyone has 50 favorite numbers in the range 1–80. If you have a favorite letter, though, this is the time to bet it.

**Example 3.38.** One notes that the K bet covers 5/8 of the numbers on a keno slip, and so an "average" draw would consist of 5/8 of the K bet numbers, or 12.5 out of 20. What is the probability that *none* of the numbers covered by the K card are drawn by the casino? There are only 30 numbers not involved in this bet, so the probability is

$$\frac{{}_{30}C_{20}}{{}_{80}C_{20}} \approx 8.4985 \times 10^{-12},$$

making that a very improbable event. ■

**Example 3.39.** Suppose that the casino's 20 numbers cover 4 of the blocks of 3 and 4 blocks of 2—an unusual occurrence where all 20 drawn numbers are contained within the K. The probability of this event is

$$p = \frac{{}_8C_4 \cdot ({}_3C_3)^4 \cdot {}_{13}C_4 \cdot ({}_2C_2)^4}{{}_{80}C_{20}} = \frac{50,050}{{}_{80}C_{20}} \approx 1.4157 \times 10^{-14}.$$

What's the payoff?

This combination of caught numbers covers 4 3-spot ways, which pay off 55¢, ${}_4C_2 = 6$ 4-spot ways, at \$2.50 each, and $4 \times 4 = 16$ 5-spot ways, paying \$11.50 apiece. The total win is \$201.20; on subtracting the \$1.90 ticket cost, the profit is \$199.30. ■

Can we do better? What is the (admittedly very unlikely) maximum profit?

Based on the pay table, we intuitively suspect that the highest payoff corresponds to filling the maximum possible number of 5-spot ways; each of those come from a full 3-spot block and a full 2-spot block. Define the following variables:

$$m = \text{Number of 2-spot blocks completed.}$$
$$n = \text{Number of 3-spot blocks completed.}$$

It follows from these definitions that a ticket with specified values of $m$ and $n$ would win on $n$ 3-spot ways, ${}_mC_2$ 4-spot ways, and $mn$ 5-spot ways. We seek to maximize the profit function $F(m, n)$ given by

$$F(m, n) = 11.50mn + 2.50\ {}_mC_2 + .55n - 1.90.$$

Since

$${}_mC_2 = \frac{m(m-1)}{2},$$

we have

$$F(m, n) = 1.25m^2 + 11.50mn + .55n - 1.25m - 1.90.$$

We denote this function by $F$ rather than $P$ to avoid confusion, since $F$ is not a probability function. Our intuitive guess above amounts to making $mn$ as large as possible under the following constraints on $m$ and $n$:

- **Nonnegativity**: $m \geqslant 0$ and $n \geqslant 0$.
- **20-spot limit**: $2m + 3n \leqslant 20$.
- **Upper bounds**: The layout of the K ticket gives $m \leqslant 13$ and $n \leqslant 8$. We can improve these bounds to $m \leqslant 10$ and $n \leqslant 6$ by using the 20-spot limit inequality.

- **Integers**: $m$ and $n$ must be whole numbers.

Maximizing the quadratic function $F(m, n)$ subject to the first three constraints listed above is a task for the branch of applied mathematics known as *nonlinear programming*; the fourth requirement, that the variables $m$ and $n$ be integers, turns this into a problem in the related field of *integer programming*. Integer programming is typically more challenging than nonlinear programming; one option in small problems is to sidestep this difficulty by simply evaluating the target function for *all* of the pairs of integers $(m, n)$ satisfying the other constraints. These points are called *feasible points*, and there are 44 such points satisfying this problem's constraints. They lie in a right triangle in the first quadrant of the $mn$-plane, and are plotted in Figure 3.20.

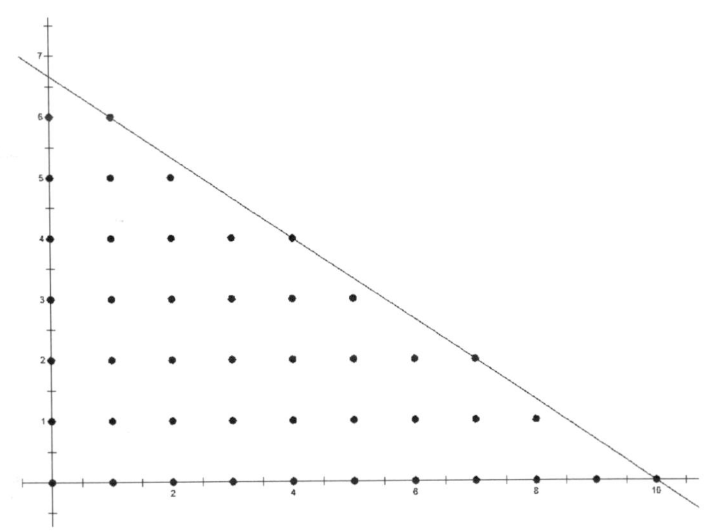


FIGURE 3.20: Feasible points for Penny Keno's K bet

We can make this brute-force approach to finding the maximum profit easier by applying one additional constraint that greatly simplifies the problem by eliminating many of these 44 feasible points. If $2m + 3n \leqslant 18$, we can catch another 2-spot or 3-spot block (increasing $m$ or $n$, respectively) and raise the total number of winning combinations while still satisfying the constraint $2m+3n \leqslant 20$. There will always be blocks available, since the K ticket contains 13 2-spot blocks and 8 3-spot blocks, and a point $(m, n)$ for which $2m+3n \leqslant 18$ cannot use all of either type of block.

Accordingly, we recognize that the best possible solution will also satisfy the following constraint:

- **Maximality**: $2m + 3n \geqslant 19$.

The two constraints $2m + 3n \leqslant 20$ and $2m + 3n \geqslant 19$, together with the nonnegativity criteria, generate a narrow band in the first quadrant where our maximum profit must lie. Using the maximality criterion reduces the number of feasible integer points from 44 to 7, which are graphed in Figure 3.21.

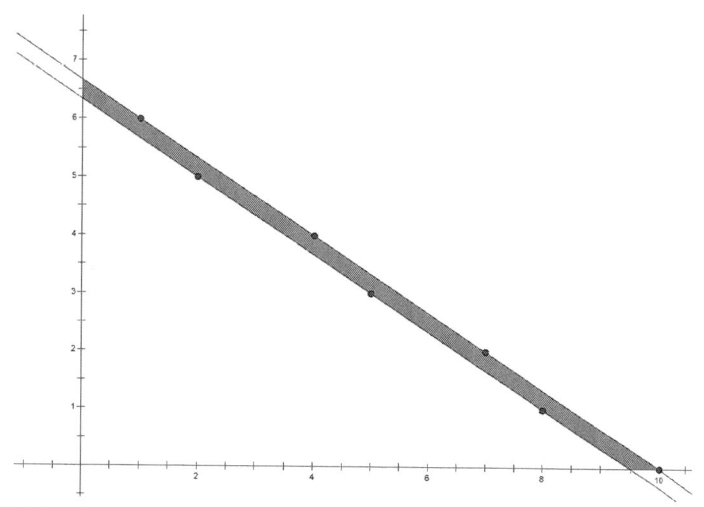


FIGURE 3.21: Feasible region with maximal integer points plotted for Penny Keno's K bet.

A simple spreadsheet calculation of all 7 returns gives the following profits:

| $m$ | $n$ | $F(m,n)$ |
|---|---|---|
| 1 | 6 | \$70.40 |
| 2 | 5 | \$118.35 |
| 4 | 4 | \$199.30 |
| 5 | 3 | \$197.25 |
| 7 | 2 | \$212.70 |
| 8 | 1 | \$160.65 |
| 10 | 0 | \$110.60 |

The maximum profit of \$212.70 is found at the point (7,2), corresponding to catching 7 full 2-blocks and 2 full 3-blocks. (4,4), the configuration in Example 3.39, gives the second-highest profit. It should be noted that (7,2) gives rise to only 14 winning 5-spot ways, while (4,4) yields 16; this is in conflict with our initial intuition. The two extra 5-spot ways in a (4,4) configuration are easily overcome by the excess in winning 4-spot ways: (7,2) has 21 of those, while (4,4) has only 6.

The *third*-highest profit, \$197.25, is at $(m, n) = (5, 3)$, which includes only 19 of the player's 50 numbers among the casino's 20 spots. This combination includes 3 3-way wins, 10 4-way wins, and 15 5-way wins.

We consider next the Penny Keno "S" ticket, which covers 63 of the 80 numbers and is shown, with its blocks labeled, in Figure 3.22.

| A | A | A | B | B | B | C | C | C | D |
|---|---|---|---|---|---|---|---|---|---|
| G | G | F | F | F | E | E | E | D | D |
| G | H | 23 | 24 | 25 | 26 | 27 | 28 | 29 | 30 |
| H | H | J | J | J | L | L | L | N | N |
| I | I | I | K | K | K | M | M | M | N |
| 51 | 52 | 53 | 54 | 55 | 56 | 57 | 58 | 59 | O |
| U | U | T | S | S | R | Q | Q | P | O |
| U | T | T | S | R | R | Q | P | P | O |

FIGURE 3.22: Penny Keno bet slip with "S" bet groups labeled alphabetically.

The active numbers on an S ticket are divided into 21 groups of 3, which can be combined to form ${}_{21}C_2 = 210$ 6-spot tickets at a total price of \$2.10. The payoff is \$1 if 5 of 6 numbers on a way ticket are caught and \$35 if all 6 are caught.

**Example 3.40.** Find the probability that the 20 numbers drawn by the casino consist of one number from each of 20 of the 3-spot blocks.

There are ${}_{21}C_{20} = 21$ ways to choose the blocks that are hit, and then 3 choices for the number chosen from each block, making the requested probability

$$p = \frac{21 \cdot 3^{20}}{{}_{80}C_{20}} \approx 2.0712 \times 10^{-8}.$$

This is an unlikely occurrence. ■

To find the maximum profit as we did for the K ticket, we begin by noting that a difference between the S and K tickets is that a 6-spot way on the S pays off if either 5 or 6 numbers are drawn; accordingly, we must account for incomplete as well as complete block fills. If $m$ = the number of 3-spot blocks containing 2 drawn numbers and $n$ = the number of 3-spot blocks where all 3 numbers are drawn, we find ourselves looking for the maximum value of the profit function

$$F(m,n) = 1 \cdot mn + 35 \cdot {}_nC_2 - 2.10,$$

or

$$F(m,n) = mn + 17.50n^2 - 17.50n - 2.10.$$

Before proceeding to identify feasible integer points, we step back and look at the big picture. Catching 6 numbers of a 6-spot way pays off 35 times more than matching 5 numbers, so we suspect that the maximum payoff will be achieved by filling as many 6-spot ways as possible. This is done by making $n$ as large as we can.

The feasible region for the S ticket is the same as it was for the K ticket (Figure 3.21), which gives the same set of 7 feasible points that potentially lead to the maximum payoff. Evaluating the profit function $F(m, n)$ at these 7 points gives the following set of outcomes:

| $m$ | $n$ | $F(m,n)$ |
|---|---|---|
| 1 | 6 | \$528.90 |
| 2 | 5 | \$357.90 |
| 4 | 4 | \$223.90 |
| 5 | 3 | \$117.90 |
| 7 | 2 | \$46.90 |
| 8 | 1 | \$5.90 |
| 10 | 0 | −\$2.10 |

Our intuition is confirmed; the maximum profit, of \$528.90, is achieved when hitting 6 3-spot blocks completely and 2 of the 3 numbers in a 7th block. This ticket collects on 15 6-way tickets catching all 6 numbers and 6 6-way tickets with 5 catches. We note further that as $n$ increases, so too does the payoff, again as we suspected before solving the problem.

The probability of this windfall is quite small:

$$p = \frac{{}_{21}C_6 \cdot 15 \cdot {}_3C_2}{{}_{80}C_{20}} = \frac{2,441,880}{3,535,316,142,212,174,320} \approx 6.9071 \times 10^{-13},$$

or slightly less often than once every 1.44 trillion drawings.

At Jerry's Nugget, Penny Keno omits the letter F but includes the letters I and D, as well as a Cover All option incorporating all 80 numbers in 16 groups of 5, producing a 120-way 10-spot ticket, and an Eagle wager (Figure 3.23).

| C | 2 | 3 | 4 | 5 | 6 | 7 | 8 | 9 | L |
|---|---|---|---|---|---|---|---|---|---|
| A | A | 13 | 14 | I | I | 17 | 18 | L | K |
| B | B | C | 24 | H | H | 27 | J | J | K |
| 31 | D | D | E | E | F | F | G | G | 40 |
| 41 | 42 | M | M | N | N | O | O | 49 | 50 |
| 51 | 52 | 53 | 54 | P | P | 57 | 58 | 59 | 60 |
| 61 | 62 | 63 | Q | Q | R | R | 68 | 69 | 70 |
| 71 | 72 | S | S | 75 | 76 | T | T | 79 | 80 |

FIGURE 3.23: Penny Keno bet slip with Eagle bet shaded and groups labeled alphabetically.

The 40 active numbers in the Eagle design are divided into 20 blocks of two numbers each, and the ticket is configured as a way ticket consisting of 190 blocks of 4. The price is then \$1.90, and the payoff for filling in a block of 4 is \$2.50. If we denote the number of 2-spot blocks that are completely caught by $m$, the profit function for the Eagle ticket is quite simple:

$$F(m) = (2.50) \cdot {}_mC_2 - 1.90 = 1.25m^2 - 1.25m - 1.90.$$

Since the profit function for the Eagle ticket involves only one variable, no advanced mathematical techniques are necessary to find its maximum payoff. The best possible draw for the player is when $m = 10$; when the 20 numbers drawn by the casino fill in exactly 10 of the blocks of 2. This generates ${}_{10}C_2 = 45$ winning blocks of 4, for a total win of \$112.50 and a profit of \$110.60.

## High Roller Tickets

*The less you bet, the more you lose when you win.*—[Fremont Casino Keno Paybook]

Penny Keno is perhaps the ultimate expression of the idea that keno can be a low-rollers' game, with gambling available at very low rates per ticket or per way. At the other end of the price spectrum, casinos have been known to offer very high-priced tickets, perhaps in an effort to draw high rollers from other games into the keno lounge.

As we noted in our examination of Pick 2 payoffs (page 65), it is often the case in casinos that table games with a higher minimum wager offer better odds or playing conditions. One might reasonably expect, then, that gamblers willing to risk more money on a single keno ticket would be rewarded with better odds or a lower house advantage. The odds of catching a given number of spots don't change, of course, which leaves us looking at the HA—or equivalently, at the payouts offered.

**Example 3.41.** The Dunes Casino in Las Vegas offered a \$30 1-spot ticket paying \$100 for a catch. This $3\frac{1}{3}$ for 1 payoff gave the player a game with a 16.67% HA, down from the standard 25%. ■

The Gold Coast Casino in Las Vegas currently operates a \$10 game, *Hot \$10 Ticket*, which ranks as one of the highest-priced single-way tickets available there. Whether or not \$10 players are getting a better game for their money than dollar bettors can be checked by comparing pay tables, which are shown in Table 3.64. This table contains the payoff odds, for 1, on each winning catch; this allows for easier comparison of the two games. The dollar amounts of Hot \$10 Ticket payoffs are ten times the number shown in the H columns.

It's not necessary to compute the respective HAs: a Hot \$10 Ticket pays off at least as well as a standard ticket on every catch, and so represents a better gamble. The player's reward for risking more money is a slightly lower HA. The standard game carries a house edge ranging from 21.07–25.10%; the house advantage on the Hot \$10 Ticket runs from 15.85–18.85%.

TABLE 3.64: Gold Coast standard and Hot $10 Ticket pay table (for 1)

| | **Pick 3** | | **Pick 4** | | **Pick 5** | | **Pick 6** | |
|---|---|---|---|---|---|---|---|---|
| **Catch** | **S** | **H** | **S** | **H** | **S** | **H** | **S** | **H** |
| 2 | 1 | 1 | 1 | 1 | | | | |
| 3 | 45 | 50 | 2 | 2.50 | 1 | 1 | 1 | 1 |
| 4 | | | 160 | 170 | 15 | 17.50 | 4 | 6.50 |
| 5 | | | | | 750 | 800 | 80 | 85 |
| 6 | | | | | | | 2000 | 2000 |

S: Standard ticket. H: Hot $10 Ticket.

Up a bit from $10 tickets was a sequence of $50 High Roller Special tickets on offer at Bally's in Las Vegas. These Catch All tickets were available for 2, 3, and 4 spots, paying off as in Table 3.65.

TABLE 3.65: Bally's Las Vegas High Roller Special pay table

| **Pick** | **Payoff** |
|---|---|
| 2 | $650 |
| 3 | $2700 |
| 4 | $11,900 |

The Pick 2 ticket effectively paid 13 for 1, which we have seen leads to a 21.84% HA, slightly better than a standard 12 for 1 payoff. At the Pick 3 level, the 54 for 1 payoff led to a 25.07% HA—roughly in the middle of Table 3.5's range of house edges. The Pick 4 game had the highest HA among these high-priced options: 238 for 1 leads to an edge of 27.09%. High rollers weren't getting much of a break for their money on this game.

The Barbary Coast Casino in Las Vegas (now the Cromwell) was home to a set of Jackpot Special Catch All tickets that provided keno players with a chance to hit the $50,000 aggregate limit on a single ticket. Tickets were available in Pick 3 through Pick 6 versions; the game parameters are shown in Table 3.66.

TABLE 3.66: Barbary Coast Jackpot Special Catch All pay tables

| | **Pick 3** | **Pick 4** | **Pick 5** | **Pick 6** |
|---|---|---|---|---|
| Ticket price | $1000 | $200 | $50 | $10 |
| Payoff | $50,000 | $50,000 | $50,000 | $50,000 |
| HA | 30.62% | 23.42% | 35.51% | 35.51% |

The best game for a player seeking to win the maximum quickly and run turns out to be the Pick 4 ticket, but there are no real bargains to be had among the HAs on these tickets.

At the Frontier in Las Vegas, a $250 Pick 4 ticket offered an appealing HA by dropping the Catch All format seen at Bally's and the Barbary Coast. Big Bucks Keno paid on 2–4 catches, as is typical at lower ticket prices. The pay table, Table 3.67, gave players a game with a house edge of only 8.82%. This

TABLE 3.67: Frontier Big Bucks Keno Pick 4 pay table ($250 ticket)

| Catch | Payoff |
|---|---|
| 2 | $250 |
| 3 | $500 |
| 4 | $50,000 |

house advantage is competitive with some of the less-favorable craps bets, and is better than the Tie bet at baccarat, a game favored by many high-stakes gamblers.

Big Bucks Keno was also offered in Pick 5 ($50 ticket) and Pick 6 ($20 ticket) versions; the HAs there were also under 10%. High rollers were finally getting a better keno deal, at the Frontier.

## What Makes a Rate "Special"?

> *Special Keno—In this exciting version of Keno, you need to hit more numbers to win...but your payoffs are much larger.*—[Desert Inn Keno Paybook]

Many of these high-priced tickets have names including the word "special", suggesting that something exciting is on offer. "Special" keno rates were introduced in Nevada casinos beginning in 1974, when the basic keno rate was raised from 60¢ to 70¢ and most payoffs were adjusted—in the casinos' favor [112]. Many casinos offer an array of "special" keno pay tables in parallel with their standard pay tables, leaving the choice of the game to the player. We have seen that "special" need not mean "better": the special Catch All Pick 10 tickets at Caesars Palace and Harrah's in Las Vegas (p. 88) had house edges exceeding 90%. Do all "special" tickets share this high HA? Are any such tickets a better choice for the keno player?

**Example 3.42.** At the Four Queens Casino, "Super Special" pay tables are available for Pick 3 through Pick 6 games. These games differ from the standard pay tables at the Four Queens in the following ways:

- *Higher price.* The standard pay tables are configured for a $1 wager, while Super Special tickets require a minimum bet of $1.50.

- *Narrower range of winning catches.* The minimum number of catches required to win any money on a Super Special ticket is one more than that for the standard game.
- *Different payoffs—some higher, some lower—when corrected for the higher ticket price.* A standard Pick 4 ticket carries a 115 for 1 payoff if all 4 numbers are caught; the same catch on a ticket played at the Super Special rate pays off at 200 for 1. On the other hand, the payoff for catching 5 out of 5 numbers is 720 for 1 on a standard ticket, but falls to 666.67 for 1 on the Super Special Pick 5 game.

Making a fair comparison of standard and Super Special tickets requires a look at the expected value of a ticket. Table 3.68 compares the regular and Super Special Pick 4 and 5 tickets, with the ticket prices indicated.

TABLE 3.68: Four Queens: Standard and Super Special pay tables

| | **Pick 4** | | **Pick 5** | |
|---|---|---|---|---|
| | Standard | Super Special | Standard | Super Special |
| **Catch** | $1 rate | $1.50 rate | $1 rate | $1.50 rate |
| 2 | 1 | 0 | 0 | 0 |
| 3 | 3 | 2 | 1 | 0 |
| 4 | 115 | 300 | 13 | 30 |
| 5 | — | — | 720 | 1000 |

The house advantages of the four tickets can then be easily calculated:

| **Ticket** | **HA** |
|---|---|
| Pick 4 Standard | 21.07% |
| Pick 4 Super Special | 32.97% |
| Pick 5 Standard | 25.10% |
| Pick 5 Super Special | 32.82% |

In each case, the standard ticket carries a lower HA than the Super Special. This is also true for the Four Queens' Pick 3 and Pick 6 games. ■

"Super Special" here appears to be a way to highlight the higher payoffs, without calling too much attention to the increase in ticket cost necessary to gain access to those payoffs. Once again, we see that increasing payoffs for greater numbers of catches does not fully compensate the player for taking away a small payoff for a small catch.

**Example 3.43.** The Golden Nugget Casino, located across the street from the Four Queens, offers $1.25 Specials which it advertises as "Best Value!". Table 3.69 compares the Pick 4 through Pick 6 pay tables.

TABLE 3.69: Golden Nugget: Standard and Special pay tables

| | Pick 4 | | Pick 5 | | Pick 6 | |
|---|---|---|---|---|---|---|
| | $1 | $1.25 | $1 | $1.25 | $1 | $1.25 |
| **Catch** | Standard | Special | Standard | Special | Standard | Special |
| 2 | 1 | 0 | 0 | 0 | 0 | 0 |
| 3 | 3 | 6 | 1 | 1 | 1 | 1 |
| 4 | 112 | 200 | 13 | 8 | 4 | 3 |
| 5 | — | — | 720 | 1100 | 85 | 90 |
| 6 | — | — | — | — | 1480 | 3000 |
| **HA** | 31.45% | 30.23% | 29.45% | 28.79% | 30.20% | 29.52% |

Except for the Pick 4 level, the lowest for which the Specials are available, these tickets differ in one significant way from the Four Queens' Super Specials: the number of catches necessary to qualify for a payoff is the same for both games; there is no higher threshold that must be cleared to win money. For the Pick 4, 5, and 6 games, the HAs on $1.25 Special tickets are more favorable to the player, but the difference is never more than 2%. ■

### Shortcake Keno

In the 1960s, Binion's Horseshoe Club in downtown Las Vegas (now Binion's) promoted its special keno pay tables with a paybook that borrowed from gamblers' lingo:

> *"To go for the Strawberry Shortcake" is an old gambling expression meaning a chance for the big money.*

This is apparently a very old expression; a 21st-century Internet search for this phrase turns up only restaurant Web sites.

Such a chance did not come without cost, though. At the Horseshoe Club, *Shortcake Keno* referred to a separate set of pay tables for Pick 4 through Pick 11 games that diminished or eliminated payoffs for small catches while increasing the payoff for catching more numbers.

**Example 3.44.** Table 3.70 compares the pay tables for the Horseshoe Club's regular and Shortcake Pick 6 games.

While the difference in ticket prices makes a direct comparison tricky, the top payoff has indeed increased on a Shortcake ticket, while the Catch 3 payoff has been eliminated. Computing house advantages shows that the standard game is the better of the two wagers, as it carries a HA of 20.99% while the Shortcake ticket's HA is 24.84%. ■

TABLE 3.70: Binion's Horseshoe Club Pick 6 and Shortcake Pick 6 pay tables

| Catch | Standard 50¢ ticket | Shortcake 55¢ ticket |
|---|---|---|
| 3 | \$0.50 | \$0 |
| 4 | \$2.80 | \$4 |
| 5 | \$55 | \$55 |
| 6 | \$620 | \$1000 |

## Opportunity Knocks at the Continental and Sahara Tahoe

In 1990, the Continental Casino offered a "Super Special" pick-6 game with a *player advantage* that exceeded 20% [26]. The \$3 ticket was played against Table 3.71.

TABLE 3.71: Continental Pick 6 pay table (\$3 ticket) [163]

| Catch | Payoff |
|---|---|
| 4 | \$15 |
| 5 | \$800 |
| 6 | \$5500 |

Note that catching 3 of 6 numbers paid nothing, in contrast with standard Pick 6 pay tables that typically return 1 for 1 on a catch of 3 spots. The expected value of a single \$3 ticket was positive:

$$E = (15) \cdot \frac{{}_6C_4 \cdot {}_{74}C_{16}}{{}_{80}C_{20}} + (800) \cdot \frac{{}_6C_5 \cdot {}_{74}C_{15}}{{}_{80}C_{20}} + (5500) \cdot \frac{{}_{74}C_{14}}{{}_{80}C_{20}} - 3 \approx \$.614,$$

and the players enjoyed a 20.47% advantage over the casino.

Here, then, was a case when a special ticket was truly better for players. There are several reasons why the Continental might have offered this game.

- **An error.** This seems like the most likely explanation. The casino may have made a computational error in setting the pay tables, or an error in printing the keno booklets might not have been caught. Setting the payoff for 5 catches at \$300 rather than \$800 would have given a house advantage of 31.13%, which favors the casino. It's easy to see how a 3 could become an 8 in the printing process.

- **A loss leader.** The probability of winning any money on this ticket is only 3.18%. While the Law of Large Numbers would take its inevitable toll on all wagers—this time to the casino's detriment—someone in management might have thought that the risk was insignificant. Players betting this game might also be induced to bet less favorable games. It is certainly not obvious from the pay table that players have a 20% edge.

- **A teaser rate.** The game may have been designed to attract customers who would stay even after the game was discontinued. A small change to the pay table might have gone unnoticed by casual players even as it returned the edge to the casino, especially if a 1 for 1 payoff for 3 catches was included.

Another game with a player advantage was briefly offered in 1978, at the Sahara Tahoe Casino in Stateline, Nevada. The casino was offering South African krugerrands, gold coins worth about $200 at the time, as a bonus beyond their prize to players catching 7 or more numbers on a 70¢ Pick-10 ticket [142]. This didn't generate a lot of winners or excitement, as the probability of winning a krugerrand was only about .00175, so casino officials extended the offer to players catching 6 or more numbers.

This small change, extending the krugerrand bonus to Catch 6 tickets, tipped an already-favorable wager even more toward the players. The pay table in use at the time, for a 70¢ wager, was one like Table 3.72. This does not include the value of the krugerrand.

TABLE 3.72: 70¢ Pick 10 pay table, circa 1978

| Catch | Payoff |
|---|---|
| 5 | $1.40 |
| 6 | $14.00 |
| 7 | $99.40 |
| 8 | $700.00 |
| 9 | $3150.00 |
| 10 | $13,300.00 |

Without the krugerrand bonus, this ticket has a typical HA of 27.37%. If the top four prizes are increased by $200, the edge shifts to the player, whose advantage is then 22.71%. By extending the krugerrand to the catch-6 level, the player's expected return jumped to $2.45—3½ times the price of the ticket. The player's edge was then 350.69%.

What makes a rate "special", then? It could be nothing more than a marketing gimmick, or it could indicate that the game is slightly—or considerably—better. If you want to find out which, there's no substitute for doing the math.

## Keno Coupons

Some casinos, as an incentive to get customers to try the game, have issued coupons which entitle the bearer to a discount on a keno wager. These can cut down the HA considerably, making keno bets with a coupon competitive with other casino games. One such coupon was issued by Harolds Club, and can be seen in Figure 3.24.

FIGURE 3.24: Harolds Club keno coupon.

This coupon offered a 50¢ discount on any ticket costing $1.50 or more, and so could effectively multiply the payoffs on a $1 ticket by 1.5 with no additional risk. The increased payoffs were subject to the aggregate limit of $50,000 printed on the ticket.

How much does this coupon reduce the house edge? Consider the Pick 10 pay table in Table 3.73. The payoffs for a $1 ticket in the second column are

TABLE 3.73: Pick 10 pay table: 1979–1989 (Aggregate limit $50,000)

| **Catch** | **$1 ticket** | **$1.50 ticket** |
|---|---|---|
| 5 | 2 | 3 |
| 6 | 18 | 27 |
| 7 | 120 | 180 |
| 8 | 960 | 1440 |
| 9 | 3800 | 5700 |
| 10 | 50,000 | 50,000 |

converted to the equivalent payoffs for a $1.50 ticket in the third column. Since the aggregate limit is $50,000, the payoff for 10 catches does not increase with the higher wager.

The house advantage on a $1 ticket played against the second column is 33.83%. By redeeming a coupon, the same $1 buys a ticket with the third column as its pay table, and the HA drops to a mere 1.03% because most payoffs increase while the amount subtracted to find the expected value does not change. Armed with this coupon, a gambler could turn 10-spot keno into a game with a house edge slightly below the Banker bet at baccarat, which is, at 1.06%, among the best bets in any casino.

The Harolds Club coupon book from which Figure 3.24 is taken included two 50¢ keno coupons. Given the sharp reduction in the HA seen above, it is no surprise that the coupon carries the restriction "Maximum of one coupon per Keno Ticket". Allowing two coupons on a $1.50 ticket creates a ticket with only a 6.47% chance of winning, but with a positive expected value of 48.97¢—a 97.95% player edge.

**Example 3.45.** Barney's Casino in Stateline, Nevada offered a 30¢-off keno coupon for a special Pick 3 game. Playing the coupon with 50¢ garnered a $35 payoff if all three numbers were caught.

At the full 80¢ price, this ticket carried a 39.29% house edge. With the coupon, the expected return was

$$E = (35) \cdot \frac{{}_{3}C_{3} \cdot {}_{77}C_{17}}{{}_{80}C_{20}} - .50 = -\$.0143,$$

which produced a low HA of 2.87%. ■

## Free Keno Tickets

The ultimate keno coupon would be a free ticket, which would guarantee a player the edge as no money would be risked. The Barney's coupon described in Example 3.45 was printed at the top of a page of coupons. The bottom coupon on the page was good for 50¢ in free nickels, and instructions on the coupon sheet stated that this coupon had to be redeemed before any of the other three on the page. With 50¢ in hand, a player would have the cost of the discounted keno ticket handed to him or her, and could then play the keno coupon with that money—and no risk. Combining coupons in this fashion would result in an expected return of 48.6¢. Keeping the nickels, of course, gave an expected return of 50¢.

Throughout the 1980s and early 1990s, Vegas World (now the Stratosphere) was running one of Las Vegas' most ambitious promotions. One offer charged $396 per couple for a package including two nights' accommodation, $200 in cash, $800 in free casino action in the form of slot machine tokens and nonnegotiable table game chips, unlimited drinks at casino bars, and a variety of souvenirs. A challenge to management was keeping patrons attracted by this offer on-site and gambling. While Vegas World was located a quarter-mile north of the other Strip casinos, there was a very real risk that that distance would not be a barrier to customers availing themselves of the incentives and then leaving the resort to spend their cash elsewhere. Since the "free" vacation program relied on some patrons spending a lot of money in the casino to be viable, this posed a problem.

One idea management devised to keep people at Vegas World was a sequence of free keno drawings exclusively for hotel guests, taking the price per way from the 1¢ in Penny Keno down to nothing. These games were held five times daily, every three hours: at 11:15 A.M. and 2:15, 5:15, 8:15, and 11:15 P.M. The catch was that guests had to be present to win. Attendance at the daily drawings was encouraged by posting the names of absent winners together with the amount of money they'd lost by missing the drawing [143].

The free game was an 9-spot game with a modified pay table. For convenience, numbers were printed on in-room telephones and permanently assigned to each room; they were not freely chosen by hotel guests. Table 3.74 compares the payoffs from a standard $1 9-spot game with the same top prize and the Vegas World promotional drawing.

TABLE 3.74: Standard and Vegas World promotional 9-spot pay tables

| **Standard ($1)** | | **Vegas World** | |
|---|---|---|---|
| Catch | Payoff | Catch | Payoff |
| 5 | $6 | 5 | $0 |
| 6 | $30 | 6 | $10 |
| 7 | $300 | 7 | $100 |
| 8 | $5000 | 8 | $5000 |
| 9 | $50,000 | 5 | $50,000 |

By eliminating the payoff for catching 5 numbers, the probability of winning something on the free ticket drops from 3.89% to .63%, but the expectation is positive since it is not necessary to deduct the cost of a ticket from the return. The return on the regular 9-spot ticket is –25.6¢, for a house advantage of 25.6%. For the promotional bet, the expected return is

$$E = (10) \cdot P(6 \mid 9) + (100) \cdot P(7 \mid 9) + (5000) \cdot P(8 \mid 9) + (50,000) \cdot P(9 \mid 9) \approx \$.3155,$$

or about 31.6¢. The edge—small though it may be—rests with the player, as we would expect with any no-cost wager.

This may not seem like a particularly attractive incentive, but in practice, giving guests about $1.58 per day in expected value—coupled, perhaps, with the fear of missing out on a big win and seeing their name posted as a winner who lost a prize by their absence—turned out to be enough to get Vegas World patrons to stick around the casino for the free drawings and to gamble their own money between games [143].

A different Vegas World keno promotion involved the nonnegotiable casino chips included as part of the vacation amenities package. Ordinarily, these chips were designated for use only on even-money bets on table games, such as blackjack, Pass and Don't Pass wagers at craps, or the various 1–1 bets at roulette. Gamblers who might have been reluctant to play live table games with their chips had the option to use them to play *7–11 Keno*, a special Pay Any Catch game that was only open to players using their "action chips". As the name suggests, this game came in two versions: Pick 7 and Pick 11. The pay tables are shown in Table 3.75.

The minimum wager at 7–11 Keno was $25 in action chips. How should this be valued in assessing the game? On one hand, the chips were free—excluding the cost of the vacation package, which a player could have assigned to his or her room cost, making the chips genuinely free. On the other hand,

TABLE 3.75: Vegas World 7–11 Keno pay tables

| Catch | Pick 7 | Pick 11 |
|---|---|---|
| 0 | 10 | 50 |
| 1 | 5 | 10 |
| 2 | 5 | 5 |
| 3 | 5 | 5 |
| 4 | 25 | 5 |
| 5 | 100 | 10 |
| 6 | 1000 | 25 |
| 7 | 50,000 | 100 |
| 8 | — | 1000 |
| 9 | — | 10,000 |
| 10 | — | 75,000 |
| 11 | — | 100,000 |

the chips could have been used at the tables. These were one-time chips: a winning player received "live" standard casino chips, but the action chips were retained by the dealer; losing bets were, of course, also taken. Such chips have an effective value of approximately half their face value, since even-money bets are won just under 50% of the time [14]. If we take the cost of a 7–11 Keno ticket to be $12.50, we find that the Pick 7 game had a house edge of 24.6% and the Pick 11 game's HA was 27.3%.

The Landmark Casino in Las Vegas offered a different free promotional keno ticket [164]. The term "loss leader" was perhaps never more appropriate to a casino promotion than to this one, as landing a player who was initially drawn to keno by a free ticket but stayed around to pay to play was a worthy goal for the casino.

The free ticket was a 5-spot paying $50 if all five numbers were drawn, and nothing otherwise. The probability of winning on this ticket was then

$$P(5 \mid 5) = \frac{{}_5C_5 \cdot {}_{75}C_{15}}{{}_{80}C_{20}} \approx 6.449 \times 10^{-4} \approx \frac{1}{1551},$$

and the expected value of this ticket was 3.22¢ per drawing.

This promotional ticket was good for all keno draws in a 5-hour period [75]. At approximately 5 minutes per game, the probability of winning at least once was

$$1 - P(\text{Lose every game}) = 1 - \left(\frac{1550}{1551}\right)^{60} \approx .0380,$$

or about 3.8%.

Perhaps the best strategy in playing the Landmark's promotional ticket was to take it from the clerk and offer to sell it to the person waiting behind you in line for $1.94, the approximate expected value of a 60-game ticket. You might even get $2 if that patron wasn't carrying exact change.

The Riviera Casino offered its own free 5-spot ticket, during a time in the mid-1990s when the casino was part of a pushback against Las Vegas' emerging family-friendly image and marketing itself as "The Alternative for Grown-Ups" [125]. The Pick 5 ticket was offered only to hotel guests, and the prize for catching all 5 numbers was two free nights at the Riviera—on a return visit within one year. While the free ticket was valid for as long as a guest was registered at the hotel, winning tickets had to be claimed, as is standard in keno, before the start of the next drawing. As with Vegas World's promotion, this could be seen as a way to keep hotel guests gambling in the Riviera's own casino. There was a limit of one win per patron.

Let the cost of a night at the Riviera be $\$R$. (The promotion was not valid during holidays, when hotel rates traditionally rise with demand.) The expected value of this ticket would then be

$$(2R) \cdot \frac{{}_5C_5 \cdot {}_{75}C_{15}}{{}_{80}C_{20}} \approx \frac{R}{775}.$$

For most standard nights at the Riviera in those years, where $R < \$200$ was the norm and rooms could frequently be had for under \$100 per night, this would be less than a quarter.

From 1977–1981, *Gambling Times* magazine offered its readers a free monthly keno game played by mail. The game was a Pick-10 challenge for readers; the winning numbers were determined by the last live keno drawing of the month at the Las Vegas partner casino: the Sahara from 1977–1980 and the Hacienda in 1981. The winning player was the first one catching the most numbers from that drawing; their prize was a 3-day, 2-night, all expense-paid vacation for two people at the casino resort. This was not exactly free, since players had to mail in their entries, but *Gambling Times* encouraged readers to send their picks in on a postcard, which would allow them entry for only 10¢.

Most months, catching 7 out of 10 was enough to win—the challenge, of course, was to be both early and lucky. The Sahara package had a retail value of \$126, making the expected value of an entry, assuming that you could be the first person to catch 7 or more numbers,

$$E = (126) \cdot [P(7 \mid 10) + P(8 \mid 10) + P(9 \mid 10) + P(10 \mid 10)] - .10 \approx \$.1208,$$

making this a positive expectation game for early responders with luck on their side.

---

## 3.3 The Big Picture

A review of the house edge in pretty much every live keno game indicates very clearly that this is not a game worth playing, not when compared to other far less disadvantageous options in any casino with table games or reasonable video poker and slot machines. A Pick 1 ticket carries a 2 to 1 payoff, with a HA of 25%; one can get these same 2–1 odds by betting one of the columns at an American roulette table while only fighting a house edge of 5.26%. This disadvantage, of course, is not a mandatory feature of keno—it would be easy to design a keno game with a single-digit HA by simply increasing the payoffs until a desirable HA is reached.

Consider a Pick 2 game. At the FireKeepers Casino, this game pays $12 on a $1 wager if both chosen numbers are caught. The expected value of this bet is

$$E = (12) \cdot \frac{_{78}C_{18}}{_{80}C_{20}} - 1 = (12) \cdot \frac{19}{316} - 1 = -\$\frac{22}{79} \approx -\$.2785.$$

If we change the payoff from $12 for 1 to $X for 1, the new expected value of a $1 ticket as a function of $X$ is

$$E(X) = \frac{19X - 316}{316},$$

as we saw on page 63. Values of $E(X)$ are tabulated in Table 3.76.

TABLE 3.76: Variable Pick 2 Keno expectation ($)

| $\boldsymbol{X}$ | 11 | 12 | 13 | 14 | 15 | 16 | 17 |
|---|---|---|---|---|---|---|---|
| $\boldsymbol{E(X)}$ | −.3386 | −.2785 | −.2184 | −.1582 | −.0981 | −.0380 | +.0222 |

We can see that raising the payoff by $1 corresponds to approximately a 6% decrease in the house edge, to the point where a 15 for 1 payoff for matching both numbers brings the HA down to just under 10%, and 16 for 1 results in a house edge under 4%—higher than many table games, but lower than American roulette. At 17 for 1, this bet favors the gambler.

If lowering the HA to a level more on a par with table games is possible, why then, one might ask, is it not done? One possible explanation lies in the relative popularity of games like blackjack, craps, and roulette: their low house advantages relative to keno are nonetheless multiplied by thousands of dollars in daily action at popular casinos, and so they win money by taking a small percentage of the high total handle. Keno, by contrast, is somewhat less popular, and so the rules are set to draw in players with the hope of a large payoff relative to the cost of a ticket rather than to attract many players. Since fewer gamblers risk money on keno, the games must retain a higher percentage of the lower amount wagered in order to justify their expense. Moreover, their expense can be considerable. Historically, keno has been one of the most labor-intensive casino games, employing ticket writers and dealers as well as runners to collect tickets from patrons elsewhere in the casino.

It's plausible that a casino in a competitive market might try experimenting with a more player-favorable pay table in its keno games, perhaps as an advertised rate to draw keno activity away from nearby casinos. To the extent that keno lounges compete on payoffs, which is an uncertain proposition, this might succeed. However, since keno pay tables are easily changed, competitors could quickly match any new payoff structure, and it wouldn't be long before everyone in a market was offering the same better payoffs. While this is good for players, the effect for casinos would be that everyone would be writing about the same amount of keno action as with the more lucrative pay tables, but making less money while doing so. While small variations in payoffs might be found by looking carefully, there is close agreement among keno house edges in competitive markets. Competition, where it is present, is typically found in the variety of games on offer rather than in how the standard games pay off.

A related question might be "Why do people continue to play keno?" despite these unfavorable conditions. It is certainly true that many, perhaps most, keno players are not aware of the tremendous house advantage on virtually every ticket. There are certainly better gaming options in any casino; each of the four most popular table games has reasons for finding favor with gamblers:

- **Baccarat**, in its appeal to high rollers and association with sophisticated casinos and wealthy players, carries undeniable glamor. A game frequently played by James Bond requires little else to recommend it—but in addition, the HAs on two of the three primary wagers are among the lowest to be found in any casino.
- **Blackjack** has a low house advantage when played using basic strategy under favorable rules, and the knowledge that the game can be turned in the player's favor by counting cards draws many aspiring gamblers, at varying degrees of skill.
- **Craps** offers, in addition to low HAs on the Pass and Don't Pass lines, the excitement of rolling the dice. Players aren't allowed to deal the cards in blackjack, nor to spin the wheel in roulette, but they may participate actively in craps.
- **Roulette** shares with baccarat a glamorous reputation and with keno the chance to choose one's own lucky numbers. These factors are enough to draw players seeking to overcome the standard 5.26% HA on most bets at American roulette.

Control over part of the game is common to keno and roulette, and is surely part of the reason for keno's continued niche appeal in casinos. The ability to pick their own numbers is a way for gamblers to exert some small measure of influence over the game; an opportunity that's not present in most other game options. The value of control, or perceived control, in attracting players to a random process such as casino gaming was explored at some length as a psychological phenomenon in [122].

Another part of keno's appeal, in the face of unfavorable odds, is surely the chance to win a large sum of money for a small initial outlay. This opportunity is also present in progressive slot machines, which promise a shot at life-changing sums of money for a small initial investment—at very long odds that parallel those for keno. Blackjack and baccarat may offer lower house advantages than keno, on the order of 1%, but neither game offers the chance to win $50,000 or more for a single $1 wager. Starting with $1 and turning it into $50,000 at a craps table requires multiple consecutive wins and the willingness to let one's accumulated winnings—which can be substantial in the runup to $50,000—ride through multiple bets.

How do these money-making opportunities compare? A Pick 8 pay table from Texas Station offered a $50,000 payoff for a $1 wager catching all 8 numbers. The probability of this event is

$$P(8\,|\,8) = \frac{{}_8C_8 \cdot {}_{72}C_{12}}{{}_{80}C_{20}} \approx \frac{1}{230,115}.$$

By contrast, the probability of winning on the Pass line at a craps table is approximately .4929. To win $50,000 by starting with $1, making an even-money bet on Pass, and letting the winnings ride (assuming that the table maximum permits) requires winning 16 consecutive decisions, an event with probability

$$(.4929)^{16} \approx \frac{1}{82,309}$$

—approximately 2.79 times the chance of going from $1 to $50,000 on a single Pick 8 keno ticket [47]. The road to fortune may lead through a casino, but it doesn't pass near the keno lounge.

Nonetheless, keno endures—not with the popularity it once had, but well enough to attract players as it joins state lottery offerings and maintains some casino floor space.

---

## 3.4 Video Keno

Is a faster and cheaper version of keno really better?

With *video keno*, keno's high house advantage can be lowered while the casino maintains its profit by driving up game volume through increased speed. Video keno is much like video poker, in that it can be played by a single player at his or her own pace on a dedicated gaming machine. By embedding keno in a video terminal, each game is played much more quickly than standard keno. While a typical keno game takes about 5 minutes to play, including time for processing tickets, video keno games, like video poker or slot machines, can be played in a tiny fraction of that time: as little as 5–10 seconds for a committed player. On one hand, this is more opportunity for the house advantage to work against the gambler; on the other hand, the base house advantage in video keno is often considerably less than the HA in a traditional game using paper tickets. Additionally, video keno games are frequently available at very low rates—5¢ games are common, and 1¢ games exist—so that lower HA may be taking a small percentage of a smaller individual bet.

The challenge to game designers and programmers is to balance these factors; the challenge for players is then to find the best game, which, as we have seen, is not a transparent process.

**Example 3.46.** Table 3.77 contains the payoff schedule for a common video keno game: Game King, one of several games in a single video terminal manufactured by International Game Technology (IGT).

TABLE 3.77: Game King video keno pay table (for 1)

| | Numbers marked | | | | | | | | |
|---|---|---|---|---|---|---|---|---|---|
| **Catch** | **2** | **3** | **4** | **5** | **6** | **7** | **8** | **9** | **10** |
| 2 | 15 | 2 | 2 | | | | | | |
| 3 | | 48 | 5 | 3 | 3 | 1 | | | |
| 4 | | | 100 | 13 | 4 | 2 | 2 | 1 | |
| 5 | | | | 838 | 75 | 22 | 13 | 6 | 5 |
| 6 | | | | | 1660 | 422 | 100 | 44 | 24 |
| 7 | | | | | | 7000 | 1670 | 362 | 146 |
| 8 | | | | | | | 10,000 | 4700 | 1000 |
| 9 | | | | | | | | 10,000 | 4500 |
| 10 | | | | | | | | | 10,000 |

Our analysis from page 61 on Pick 2 payoffs and page 68 on Pick 3 pay tables shows that the house edge on Game King's Pick 2 game is only 9.81%, a rate achieved in live keno only by betting $10 or more, and the edge at Pick 3 is a very-low-for-keno 5.65%.

Figure 3.25 illustrates the HA of each Game King video keno option. This graph shows that the Pick 2 game is the worst for the gambler; indeed, its 9.81% HA is something of an outlier among the options, though lower than the HA of any $1 live keno ticket. It should be noted that raising the payoff for catching 2 of 2 numbers to 16 for 1 would cut the HA to 3.80%, making it the lowest HA available. While this would be much better for players, it is unlikely that a casino would do well with that payoff. Knowledgeable keno players would be drawn almost exclusively to that game option, with the effect that the house might have difficulty covering its expenses from the low return. A 15.5 for 1 payoff might be a good compromise, as it would cut the HA to 6.80%, much more in line with the range of HAs for Game King's other games.

The lowest HA, 5.01%, is found on the Pick 6 game, though Pick 5, 7, and 8 are all within .1% of this edge. ■

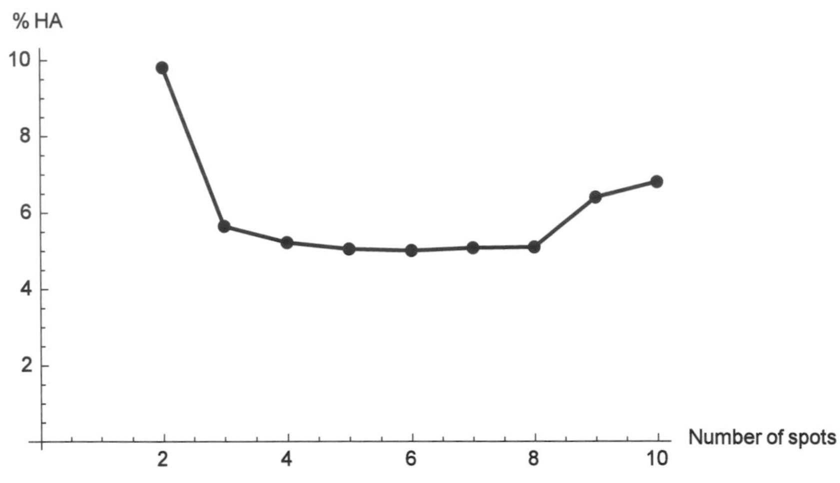


FIGURE 3.25: House advantage at Game King video keno.

### The Best Game?

Choosing the "best" video keno game, from a player's perspective, ought to be a simple matter of comparing pay tables and looking for the highest numbers—after all, while payoffs may change from machine to machine, the underlying probabilities are the same across games. This simplifies the question, perhaps too far. Is it, for example, better to play fewer spots and trade a lower payoff structure for an increased probability of winning something, or to pick more spots and take a low-probability run at a bigger jackpot?

While these questions are certainly also valid for live keno, they take on a greater significance in video keno, where it's possible to find very different pay tables on adjacent machines, even machines from the same manufacturer. Finding a better game doesn't have to involve finding a different casino. From Example 3.46, we see that the average HA of the nine games on offer in Game King is 6.01%, which means that the game could reasonably be marketed as having a 94% return rate. However, the individual HAs of these games are not 6.01%: some are higher, and some are lower. Figure 3.25 suggests that picking a number of spots toward the middle of the 2–10 range of games gives the lowest HA.

A guide to making the best choice from among video keno options was promoted by Brad Fredella, general manager of Stetson's Casino (now part of the Dotty's chain of neighborhood casinos) in Henderson, Nevada. Fredella's advice focused on the Pick 5 game and the payoff for catching all 5 numbers [40]. In casinos throughout metro Las Vegas, this catch-5 payoff, on a standard game paying out on catches of 3 or more numbers, ranges from 740 for 1 to 838 for 1. (Some video keno pay tables offer a small payoff for catching 0 of 5 marked numbers and pay out as low as 400 for 1 on a 5-for-5 catch; such games require separate analysis.)

As one might expect, finding the best game simply amounts to looking for the highest payoff for catching 5 numbers out of 5. In a competitive market such as metro Las Vegas, these top values vary within the 750–838 range. In a 2017 survey, payoffs from 750 for 1 to 838 for 1 were found in Las Vegas-area casinos. The most commonly observed payoff was 810 for 1, at 34 casinos. The complete data may be found in Table 3.78.

TABLE 3.78: Las Vegas area video keno Pick 5/Catch 5 payoffs: 2017

| Payoff for 1 | Number of casinos |
|---|---|
| 750 | 20 |
| 804 | 14 |
| 810 | 34 |
| 838 | 9 |

**Example 3.47.** In March 2016, the Alamo Casino was offering 740 for 1 on a catch-5 event; this low payoff had risen to 750 for 1 at the time of Table 3.78. The Alamo is not a casino resort, but a truck stop with fast food restaurants and a gas station. The casino is small, with two blackjack tables and a sports book among the slot and video poker machines. As the casino is not the primary focus, offering competitive odds is less of a concern for management. Similarly, the rental car center at McCarran Airport in Las Vegas, where gaming machines are placed as an attraction for travelers either just arriving in Nevada or about to depart, was only offering 750 for 1—this facility neither depends strongly on gambling revenue nor competes with other gaming outlets, so has little or no need to pay competitive rates. ■

The California was home to machines paying both 804 and 810 for 1; these same payoffs can both be found at the Soaring Eagle Casino in Mount Pleasant, Michigan. This suggests that shopping around a bit within a casino might be a useful pursuit.

**Example 3.48.** Shopping around within a casino need not involve much walking. At the FireKeepers Casino, the Pick 5 pay tables in Table 3.79 were found on *adjacent* video keno machines in July 2016.

TABLE 3.79: FireKeepers Casino Pick 5 video keno pay tables

| Catch | Game A | Game B |
|---|---|---|
| 3 | 3 | 3 |
| 4 | 11 | 12 |
| 5 | 804 | 750 |

To determine which game is preferable, we can derive a formula similar to those we developed earlier for Pick 2 and Pick 3 games. If the pay table offers $X$ for 1 for catching 3 numbers, $Y$ for 1 for catching 4, and $Z$ for 1 for catching all 5, the expected value of a single \$1 wager is

$$\begin{aligned} E(X,Y,Z) &= P(3\,|\,5)\cdot X + P(4\,|\,5)\cdot Y + P(5\,|\,5)\cdot Z - 1 \\ &= \frac{13,275}{158,158}X + \frac{3825}{316,316}Y + \frac{51}{79,079}Z - 1. \end{aligned}$$

For the two games described above, we find that Game A has $E = -\$.0967$ for a house advantage of 9.67%, and Game B offers $E = -\$.1194$, so its HA is 11.94%. Game A is the better choice. ■

If the top payoff in Game B were 785.25 or higher for 1, though, its expectation would exceed the expectation for Game A.

## 369 Way Keno

*369 Way Keno* is a video keno game that calls for a player to choose 9 numbers in three groups of 3 each, and arranges them into a seven-way 3/3, 3/6, 1/9 way ticket. In Figure 3.26, these groups are labeled A, B, and C, and for convenience, are chosen in close proximity to other group numbers.

Group A: 49, 60, 68
Group B: 14, 19, 37
Group C: 51, 52, 54

| 1 | 2 | 3 | 4 | 5 | 6 | 7 | 8 | 9 | 10 |
|---|---|---|---|---|---|---|---|---|---|
| 11 | 12 | 13 | B | 15 | 16 | 17 | 18 | B | 20 |
| 21 | 22 | 23 | 24 | 25 | 26 | 27 | 28 | 29 | 30 |
| 31 | 32 | 33 | 34 | 35 | 36 | B | 38 | 39 | 40 |
| 41 | 42 | 43 | 44 | 45 | 46 | 47 | 48 | A | 50 |
| C | C | 53 | C | 55 | 56 | 57 | 58 | 59 | A |
| 61 | 62 | 63 | 64 | 65 | 66 | 67 | A | 69 | 70 |
| 71 | 72 | 73 | 74 | 75 | 76 | 77 | 78 | 79 | 80 |

FIGURE 3.26: Sample 369 Way Keno bet slip with groups A, B, and C.

The 369 Way pay table is shown in Table 3.80. How does this compare to a standard Pick 9 pay table?

A quick comparison of the top payout in 369 Way Keno to the standard pay table on page 6 suggests that 369 Way Keno is by far the inferior game: the payoff for catching all 9 numbers is \$50,000 in the standard game, but pays only \$10,000 here. However, catching all 9 numbers in 369 Way Keno also pays off for catching 3 numbers out of 3 and 6 numbers out of 6, 3 times

TABLE 3.80: 369 Way Keno pay table (for 1)

| Catch | Pick 3 | Pick 6 | Pick 9 |
|---|---|---|---|
| 2 | 2 | | |
| 3 | 47 | 3 | |
| 4 | | 4 | 1 |
| 5 | | 67 | 6 |
| 6 | | 1600 | 42 |
| 7 | | | 320 |
| 8 | | | 4700 |
| 9 | | | 10,000 |

each. This raises the maximum possible payoff to $14,941—which is still not competitive with an ordinary Pick 9 ticket.

A 369 Way Keno ticket is, however, not an ordinary Pick 9 ticket—it is possible, for example, to win back some money with as few as 2 catches. As with Hi-Low Keno, a full pay table for 369 Way Keno would have to take the distribution of the caught numbers among the three groups into account. This pay table is shown in Table 3.81.

Our work with Hi-Low Keno may be easily modified to find the expected return on a 369 Way Keno bet. Once again, we must account for the number of ways to assign the three groups of numbers to the player's identified groups A, B, and C. If a given draw catches $a$ numbers from one group, $b$ from the second, and $c$ from the third, then there are 6 ways to assign these numbers to the three groups if all three numbers are different, 3 ways if two numbers are the same, and 1 way if all three numbers are the same. Denoting this factor by $N$, we have the following probability for catching $n = a + b + c$ numbers distributed $a, b,$ and $c$ among the three groups:

$$p(a, b, c) = \frac{{}_3C_a \cdot {}_3C_b \cdot {}_3C_c \cdot N \cdot {}_{71}C_{20-n}}{{}_{80}C_{20}}.$$

**Example 3.49.** The probability of catching $n = 3$ numbers depends on how they're distributed. We have $p(3, 0, 0) \approx .0088$, $p(2, 1, 0) \approx .1582$, and $p(1, 1, 1) \approx .0791$. ∎

The most probable outcome is a catch of 2 numbers, distributed 1-1-0; this has a probability of approximately .2373. It is surely no surprise to learn that this event pays nothing.

Combining this formula with the payoffs in Table 3.81 reveals an expected return of –$0.80 on a $7 ticket ($1 per way); the house advantage is 11.36%. Since 369 Way Keno can be played multiple times per minute, a player can subject a lot of money to the casino's edge in a very short time, and so this game can be as lucrative for a casino as a live Pick 9 game, where the HA runs about 25.61%.

TABLE 3.81: 369 Way Keno: Distribution of catches with payoffs

| Catch | A | B | C | Payoff |
|---|---|---|---|---|
| **0** | 0 | 0 | 0 | 0 |
| **1** | 1 | 0 | 0 | 0 |
| **2** | 2 | 0 | 0 | 2 |
| | 1 | 1 | 0 | 0 |
| **3** | 3 | 0 | 0 | 50 |
| | 2 | 1 | 0 | 5 |
| | 1 | 1 | 1 | 0 |
| **4** | 3 | 1 | 0 | 52 |
| | 2 | 2 | 0 | 9 |
| | 2 | 1 | 1 | 7 |
| **5** | 3 | 2 | 0 | 122 |
| | 3 | 1 | 1 | 61 |
| | 2 | 2 | 1 | 20 |
| **6** | 3 | 3 | 0 | 1742 |
| | 3 | 2 | 1 | 165 |
| | 2 | 2 | 2 | 60 |
| **7** | 3 | 3 | 1 | 2022 |
| | 3 | 2 | 2 | 509 |
| **8** | 3 | 3 | 2 | 6530 |
| **9** | 3 | 3 | 3 | 14,941 |

## Caveman Keno

Video keno is a forum for different keno options, including bonus game or bonus wager options such as are found on many video slot machines, that can also serve to decrease the house edge since the casino is gaining its money through rapid gameplay. *Caveman Keno* is a version of video keno developed by IGT. This game uses a graphics set with prehistoric scenes and implements a multiplier bonus through the use of dinosaur eggs—this despite the fact that cavemen and dinosaurs never lived on Earth at the same time. The bonus involves the following steps:

- Once the player has selected her numbers, the computer randomly selects three numbers from the set the player did not choose and marks them with dinosaur eggs.

  **Example 3.50.** In a Pick 2 game, suppose the player chooses the numbers 6 and 28. The computer might then pick 35, 51, and 69 as the egg numbers. ■

  These three numbers function as bonus numbers if they are drawn in the play of the game.

- If any of these bonus numbers are drawn among the 20 game numbers, the corresponding egg hatches.
- If two or three of the eggs are hatched, any player win is multiplied: by 4 if two eggs hatch and by 8 or 10 if all three hatch.

In order to accommodate this bonus, the standard pay table for Caveman Keno typically a carries a higher house edge than for a live keno game. The following table compares Caveman Keno to the $1 Pick 4 game played at the El Cortez:

| **El Cortez** | | **Caveman** | |
|---|---|---|---|
| Catch | Payoff | Catch | Payoff |
| 2 | 1 | 2 | 1 |
| 3 | 3 | 3 | 5 |
| 4 | 125 | 4 | 81 |
| HA: 27.47% | | HA: 32.30% | |

The HA for Caveman Keno in this chart does not include the effect of the bonus eggs, though. To incorporate the bonus eggs and determine the true HA, we must determine the probability of 2 or 3 bonus eggs being drawn *along with* a winning combination.

We begin by considering the simplest Caveman Keno game: Pick 2, which pays off at 11–1 if both of a player's numbers are caught. The probability of winning at this game is

$$\frac{{}_{78}C_{18}}{{}_{80}C_{20}} = \frac{19}{316} \approx .0601.$$

This 6% chance of a win must then be combined with the probability that 2 or 3 of the bonus numbers will be caught among the 18 numbers that are not the player's, triggering a payoff multiplier. The probability of catching $k$ eggs, $0 \leqslant k \leqslant 3$, among 18 numbers chosen from 78 is

$$p_k = \frac{{}_3C_k \cdot {}_{75}C_{18-k}}{{}_{78}C_{18}},$$

which gives the following probabilities:

| $k$ | 0 | 1 | 2 | 3 |
|---|---|---|---|---|
| $p_k$ | .4498 | .4188 | .1207 | .0107 |

The expectation of a $1 bet is then

$$E = (11) \cdot \left(\frac{19}{316}\right) \cdot [1 \cdot (p_0 + p_1) + 4p_2 + 8p_3] - 1,$$

where the term in brackets is the "egg multiplier", which indicates the change in the payoff as 0–3 eggs are caught. Substituting the four probabilities above into this expression gives an egg multiplier of 1.4371 and an expectation of −$.04952, for a considerably more reasonable HA of 4.95%.

When playing a game where more than 2 numbers are chosen, the pay table allows for more than one payoff, which makes the calculations slightly more complicated. Consider a general game when $n$ numbers are picked by the player and a payoff is activated when $r$ of those numbers are caught. In addition to the probability of catching $r$ of the $n$ numbers, we also need to calculate the probability that the 20 numbers chosen will include $k$ eggs, where $k$ ranges from 0 to 3. These probabilities depend on $k, n$, and $r$, and so we denote them by the conditional expression $p(k \mid n, r)$. Using this notation, the probability $p_k$ above would be written $p(k \mid 2, 2)$.

In general, the process of playing and evaluating the outcome of a round of Caveman Keno involves the following:

- Count the number of catches $r$ from the $n$ numbers chosen. This leaves $20 - r$ slots to be filled.
- Count the number of eggs $k$ in the final set of 20 numbers.
- Count the number of ways to pick these $k$ eggs from the 3 on the screen; this may be done in ${}_3C_k$ ways.
- Fill in the remaining $20 - r - k$ slots from the numbers that were neither chosen by the player nor selected as eggs. There are $77 - n$ of these to choose from.

Once the $n$ numbers picked by the player are selected, $80 - n$ remain, and we want to choose $20 - r$ of them. Collecting all of these choices gives

$$p(k \mid n, r) = \frac{{}_3C_k \cdot {}_{77-n}C_{20-r-k}}{{}_{80-n}C_{20-r}}.$$

For the Pick 3 game, Caveman Keno pays off at 2–1 if 2 numbers are chosen and 26–1 if all 3 are selected. The probabilities $p(k \mid 3, r)$ are tabulated here:

| $p(k \mid 3, r)$ | | $r$ = 2 | $r$ = 3 |
|---|---|---|---|
| $k$ | 0 | .4444 | .4678 |
| | 1 | .4210 | .4113 |
| | 2 | .1234 | .1116 |
| | 3 | .0112 | .0093 |

We then compute the expectation by summing across all possible winning outcomes. The egg multiplier referred to above depends on how many numbers $r$ are matched; once $r$ is known, we compute $p(0 \mid 3, r) + p(1 \mid 3, r) + 4p(2 \mid 3, r) + 8p(3 \mid 3, r)$. This sum appears twice in the expression for $E$: for $r = 2$ and $r = 3$.

$$E = \underbrace{2 \cdot \frac{{}_3C_2 \cdot {}_{77}C_1}{{}_{80}C_{20}} \cdot [p(0 \mid 3,2) + p(1 \mid 3,2) + 4p(2 \mid 3,2) + 8p(3 \mid 3,2)]}_{\text{Catch 2 eggs}}$$
$$+ \underbrace{26 \cdot \frac{{}_{77}C_{17}}{{}_{80}C_{20}} \cdot [p(0 \mid 3,3) + p(1 \mid 3,3) + 4p(2 \mid 3,3) + 8p(3 \mid 3,3)]}_{\text{Catch 3 eggs}} - 1$$

Numerically, the egg multiplier for $r = 2$ is 1.4483, and for $r = 3$ it's 1.3997. This may be interpreted as the "typical" multiplier given that $r$ of the player's chosen numbers are caught. For $r = 0$ or $r = 1$, the egg multiplier can be calculated, but it has no meaning in the game since the player wins nothing with fewer than 2 catches.

Evaluating this expression reveals another low house edge: only 4.75%.

It's possible to simplify this expression somewhat by writing the egg multiplier $M$ as a function of $k$, the number of eggs caught. Since there are four points to be considered—$(0,1), (1,1), (2,4)$, and $(3,8)$—a cubic polynomial will give an exact fit to these data, and the simple equation

$$M(k) = -\frac{k^3}{3} + \frac{5k^2}{2} - \frac{13k}{6} + 1$$

connects $M$ and $k$. We can then rewrite the expectation above as

$$E = 2 \cdot \frac{{}_3C_2 \cdot {}_{77}C_1}{{}_{80}C_{20}} \cdot \sum_{k=0}^{3} M(k) \cdot p(k \mid 3,2) + 26 \cdot \frac{{}_{77}C_{17}}{{}_{80}C_{20}} \cdot \sum_{k=0}^{3} M(k) \cdot p(k \mid 3,3) - 1,$$

where the egg multipliers are written as finite sums.

If the multiplier for 3 eggs is 10 instead of 8, $M(k)$ can be recalculated, and is found to be the simpler function

$$M(k) = \frac{3k^2}{2} - \frac{3k}{2} + 1.$$

This will very slightly increase the average return to the gambler; the egg multiplier jumps to 1.4585 in a Pick 2 game, and rises all the way to 1.4706 ($r = 2$) and 1.4183 ($r = 3$) in Pick 3.

These examples give the expectation when $n = 2$ or 3 numbers are marked and hatching all three eggs carries a multiplier of 8. If we compute the corresponding values for $n = 4$–10 (Table 3.82), we find that the house advantage hovers around 5% for all nine wagers. This is less than one-fourth of the typical HA for live keno.

In practice, how does Caveman Keno compare to regular keno? While the HA is lower, this benefit can be quickly canceled out by the increased game pace. A standard Pick 3 game has a house edge of 23.69%, but plays through at only about 12 games per hour, whereas the Pick 3 Caveman Keno game

TABLE 3.82: House advantage at Caveman Keno with $n$ numbers marked

| $n$ | 2 | 3 | 4 | 5 | 6 | 7 | 8 | 9 | 10 |
|---|---|---|---|---|---|---|---|---|---|
| **HA** | 4.95% | 4.75% | 4.64% | 5.02% | 4.80% | 5.04% | 5.02% | 4.92% | 4.86% |

described above has a 4.75% HA and can reasonably run through 6 games in a minute, or 360 per hour. This imbalance can be mitigated somewhat by decreasing the amount wagered per game, say from $1 to 25¢. Hour for hour, this amounts to risking $12 at live keno, or $90 at Caveman Keno.

Standard keno then results in a loss of

$$\$1 \cdot 12 \cdot .2369 = \$2.84$$

per hour, while an hour of Caveman Keno costs, on average,

$$\$0.25 \cdot 360 \cdot .0475 = \$4.28.$$

While the loss per hour is higher at Caveman Keno, it's possible to decrease the loss by intentionally playing slower or by betting less money on each game. By slowing the game pace to 3 games per minute, the loss is halved, to $2.14—lower than for traditional keno. Playing at 6 games per minute, but lowering the bet to 10¢ per game (where available), cuts the hourly loss to $1.71. Combining these strategies, 3 games per minute at 10¢ each, gives an expected hourly loss of $0.86 on a total wager of $18.

However, if you invest that same $18 in a single hand of blackjack rather than Caveman Keno, using basic strategy and playing a game with moderately favorable rules, your expected loss is about ½% of your wager, or 9¢. For Caveman Keno, 360 games at 25¢ represents an investment of $90, and the corresponding expected loss on a single $90 blackjack bet is 45¢.

For an objectively low HA and a game speed that doesn't negate it, the keno lounge—or its electronic counterpart—is not the place to look.

## Speed Keno

*Speed Keno* is a gameplay variation marketed by Gaming Arts. In Speed Keno, the gambler has some control over the cost of a ticket, which is connected to the number of games played and can be as low as 3¢. This modification would not be possible without computer assistance: Speed Keno offers 100 drawings in 15 minutes, and allows players to buy in to all or part of the 100 games, as they wish, with a sliding scale of prices per game based on the number of games entered:

| **Number of games** | 10 | 25 | 50 | 100 |
|---|---|---|---|---|
| **Price per game** | 10¢ | 8¢ | 5¢ | 3¢ |
| **Total cost** | $1.00 | $2.00 | $2.50 | $3.00 |

A single game takes about 9 seconds to complete: 2 to draw the numbers and 7 between consecutive drawings to give players a fighting chance at processing the results for their ticket. There's a 5-minute break between runs of 100 games to allow players to cash in winning tickets and bet the next set of games. Speed Keno offers a full array of standard keno games, from Pick 1 through Pick 10. Way tickets are also a Speed Keno option, but require playing all 100 games at 5¢ per game.

**Example 3.51.** Since, as in standard keno, the payoff schedule for Speed Keno is linked to the amount wagered through the payoff odds, it follows that the expectation is identical to that for standard keno. We shall illustrate this by looking at the Pick 6 game, whose payoff structure is shown in Table 3.83.

TABLE 3.83: Speed Keno pay table

| Catch | Payoff odds |
|---|---|
| 3 | 1–1 |
| 4 | 2–1 |
| 5 | 75–1 |
| 6 | 2500–1 |

For example, if you play 100 games of Pick 6 Speed Keno and the first drawing catches 4 out of your 6 numbers, the payoff is 6¢, for a 3¢ profit on that ticket—you still have 99 other tickets in play and $2.97 invested in them, of course.

Suppose that you've bet $x$ cents on a single round of Speed Keno and picked 6 numbers. Your expected return in cents is, simply stated,

$$E = (x) \cdot P(3) + (2x) \cdot P(4) + (75x) \cdot P(5) + (2500x) \cdot P(6) - x,$$

where $P(n)$ represents the probability of catching $n$ of your 6 numbers. Factoring out an $x$ gives

$$E = (x) \cdot [P(3) + 2 \cdot P(4) + 75 \cdot P(5) + 2500 \cdot P(6) - 1].$$

To find the expectation as a percentage of the original wager, we simply divide this last expression by $x$, the amount of the bet, which yields the quantity in brackets—precisely the expectation of a $1 bet on a Pick 6 keno game.

Speed Keno may put you at no further disadvantage relative to standard keno, but since the house edge on a Pick 6 bet is 25.85%, this wager, while cheap, is no bargain. ■

**Example 3.52.** Consider a 100-game Speed Keno Pick 2 wager. The total outlay is $3, or 3¢ per ticket, and the payoff is 12–1 if both of your numbers are drawn. Find the probability that your $3 investment returns a profit.

A winning ticket pays 36¢, so you win money if you win at least 9 of the

100 games, which seems like a simple task. Denote by $p$ the probability of winning on a single ticket; we then have

$$p = \frac{_{78}C_{18}}{_{80}C_{20}} \approx .0601.$$

Let $q = 1 - p$. Since the number of winning tickets $X$ is a binomial random variable, the probability of turning a profit on this bet is

$$P(X \geqslant 9) = \sum_{k=9}^{100} {_{100}C_k} \cdot p^k \cdot q^{100-k}.$$

This sum has 92 terms. It can be simplified to a sum of 9 terms by using the Complement Rule:

$$\sum_{k=9}^{100} {_{100}C_k} \cdot p^k \cdot q^{100-k} = 1 - \sum_{k=0}^{8} {_{100}C_k} \cdot p^k \cdot q^{100-k}.$$

Evaluating this sum shows that the probability of winning money is approximately .1476, so about one 100-game run in seven will result in a profit. ■

---

## 3.5 Keno Side Bets

Keno is notorious for having high house advantages.

Side bets, such as the Super 31 blackjack bet or the Fire Bet in craps, are known for having high house advantages: 18–33% and 24.9% respectively.

So why not put the two together?

It's been done—not often, and with only intermittent interest—but there have been a few attempts to enliven keno with some high-risk opportunities for extra action added to a conventional bet. These side bets involve something other than a simple bet based on catching some numbers; one such wager pays off for catching *no* numbers. Many keno side bets introduce order into the game; while basic keno bets do not require that the caught numbers are drawn in any spot among the casino's 20 numbers, these side bets offer a bonus payout if the right numbers arrive at the right spots in the draw.

### Insurance

One of the earliest extra wagers added to keno was instituted in 1947, at the Nevada Club in Reno [91]. The *insurance* bet was an additional wager on tickets marking four or more numbers that paid off if none of the numbers were caught—thus the player was purchasing insurance against a particularly

TABLE 3.84: Keno insurance pay table [91]

| Numbers marked | Payoff (for 1) |
|---|---|
| 4 | 2 |
| 5 | 3 |
| 6 | 4 |
| 7 | 6 |
| 8 | 9 |
| 9 | 12 |
| 10 | 17 |
| 11 | 24 |
| 12 | 34 |
| 13 | 48 |
| 14 | 70 |
| 15 | 101 |

unlucky draw. The cost of insurance was \$1 no matter how many numbers were marked; the payoff rose based on that number, as shown in Table 3.84.

The probability of winning an insurance bet with $k$ numbers selected is simply the probability that the casino's 20 numbers all come from the $80-k$ numbers not chosen. This probability is

$$P(0\,|\,k) = \frac{{}_{80-k}C_{20}}{{}_{80}C_{20}}.$$

**Example 3.53.** The probability of winning an insurance bet protecting a Pick-7 ticket is then

$$P(0\,|\,7) = \frac{{}_{73}C_{20}}{{}_{80}C_{20}} = \frac{429,803,415,410,341,896}{3,535,316,142,212,174,320} \approx .1216.$$

The corresponding expectation is $E = (6)\cdot .1216 - 1 \approx -\$.2706$, giving the casino a 27.06% edge. ■

Similar calculations for other insurance bets show a HA ranging from 19.04% (Pick 15) to 38.34% (Pick 4). However, there was a loophole. Occasionally, the Nevada Club offered free insurance on all tickets written for a specified drawing, typically once every two or three hours [91]. This could turn a \$1 ticket bearing insurance into a positive-expectation wager. For a 5-spot ticket, the effective pay table on a \$1 ticket is given by Table 3.85.

The expected value of a \$1 ticket with free insurance was \$1.4032, giving the player about a 40% edge. On small wagers, the casino was taking minimal risk, since the chance of catching either 1 or 2 numbers, and thus losing both the main bet and the insurance, was .6761. On the other hand, when one player showed up at the keno desk with 2800 Pick 5 tickets ready for play with free

TABLE 3.85: Pay table for 5-spot ticket with insurance wager

| Catch | Payoff (for 1) |
|---|---|
| 0 | 3 (insurance) |
| 1 | 0 |
| 2 | 0 |
| 3 | 1 |
| 4 | 9 |
| 5 | 820 |

insurance, management disallowed the bet, saying that free insurance was only to promote goodwill among the casino's patrons [91]. A gambler making this wager would expect to win the insurance bet 22.7% of the time, or on approximately 636 tickets. The expected profit on 2800 $1 Pick 5 tickets, with free insurance, was $1128.84—and the casino's goodwill understandably did not extend quite that far.

Soon after that, keno insurance as a separate wager largely disappeared, though the idea of a payoff for catching no numbers has found its way onto some conventional keno tickets. For example, the Pick 15 pay table at the Mirage (page 103) paid 1 for 1 on a zero-catch ticket.

At the Riverside Casino in Laughlin, keno insurance survives in a collection of special $2.50 tickets dubbed "Bet Against Numbers". Table 3.86 shows the pay table.

TABLE 3.86: Riverside Casino's Bet Against Numbers pay table. All tickets pay only on 0 catches.

| Pick ($n$) | Payoff ($x$) |
|---|---|
| 7 | 10 + free ticket |
| 8 | 15 + free ticket |
| 9 | 25 + free ticket |
| 10 | 35 + free ticket |
| 15 | 200 + free ticket |

Since the payoff for each win includes a free ticket, the expected value of a single $2.50 ticket can be found by solving the following equation for $E$:

$$E = (x + [E + 2.50]) \cdot P(0 \mid n) - 2.50,$$

where $x$ is the cash portion of the prize. Rearranging gives

$$E = \frac{P(0 \mid n) \cdot x}{1 - P(0 \mid n)} - 2.50.$$

The lowest HA, 31.91%, is on the Pick 9 ticket, and the highest, 44.64%, comes from the Pick 7 ticket. This modern version of keno insurance is at or beyond the high end of its original counterpart.

**Example 3.54.** The Peppermill Casino in Mesquite, Nevada took a different approach to keno insurance. Any \$5 ticket played on the casino's \$100,000 Bonus Keno progressive game for at least 5 consecutive games (a total investment of \$25) was augmented with free insurance paying off \$25 on 8-spot tickets, \$50 on 9-spot tickets, and \$100 on 10-spot tickets, provided that *none* of the 5 tickets produced a winner. The catch at the Peppermill was that this insurance payoff was made in "Keno money", or free keno tickets, not in cash.

These games offered progressive jackpots starting at \$50,000 and potentially rising to \$100,000. For the 10-spot ticket, the pay table started at 3 catches, paying \$3, so the probability of collecting on the insurance wager was

$$[P(0 \mid 10) + P(1 \mid 10) + P(2 \mid 10)]^5 \approx .0382,$$

approximately once in 26 5-ticket wagers. Without the free insurance, and assuming a \$100,000 progressive jackpot for catching all 10 numbers, the HA on this game was 29.09%. Reconfiguring the 5-ticket purchase as a "pay any catch" wager—for a player not qualifying for the insurance payout had to have won something on at least one of the five tickets—the HA only drops to 18.70% [26]. ■

## Keno Quinela

Recall that modern keno can be traced back to racehorse keno, which separated keno from bingo and made it feasible for the game to be played with as few as one player. In that spirit, the *quinela* keno bet, seen in Macau in the mid-1970s, took a common horse racing term and moved it into the keno lounge [58].

A quinela bet at the racetrack calls for the bettor to pick the top two horses without regard to order. A quinela bet in keno asks the player to correctly identify two numbers: the first and last numbers drawn. This bet pays off at 2500 for 1, reflecting the clear difficulty of picking two correct numbers out of 80 possibilities.

Analyzing a keno quinela bet seems at first glance to be a challenging task: Sure, the probability of picking the first ball correctly is an easy $\frac{2}{80}$ proposition, but how do you properly consider everything that might happen in the 18 balls between the first and last? Perhaps after a successful first catch, the second number you chose was drawn somewhere along the way, taking you out of the bet completely—how to account for that?

It turns out that you don't need to. Consider the 20 keno balls as drawn and arranged in sequence. Winning the quinela bet simply calls for identifying the first and last number, in either order. This is equivalent to correctly selecting two numbers from 80, without regard to order. Since there are ${}_{80}C_2 = 3160$

possible ways to do this, the probability of winning is $\frac{1}{3160}$. Suddenly this doesn't seem like a very good bet.

The expected value of a $1 quinela wager is

$$E = (2500) \cdot \frac{1}{3160} - 1 = -\$.2089,$$

giving the casino a 20.9% edge on this bet.

## Bulls-eye Keno

The *Bulls-eye* keno wager is a variation on the keno quinela bet that is played in several state lotteries, including Kansas and Missouri, as well as in Caribbean Keno, which is played across six Caribbean island groups: Anguilla, Antigua, Bermuda, St. Kitts and Nevis, Sint Maarten, and the U.S. Virgin Islands. When all 20 keno numbers have been drawn, one—not necessarily the first or final number—is designated the bulls-eye number, and winning players who have chosen the Bulls-eye bet option receive increased payoffs, albeit for an increased ticket price.

**Example 3.55.** In Caribbean Keno, the Pick 1 game pays 2 for 1 if the chosen number is drawn, but 50 for 1 if that number is the bulls-eye number. Making the bulls-eye bet requires doubling the wager, as from $1 to $2. The probability of a given number being chosen as the bulls-eye is a simple 1 in 80, which gives an expectation for the 1-spot bulls-eye wager of

$$E = (100) \cdot \frac{1}{80} - 2 = -\frac{60}{80} = -\$0.75.$$

The resulting HA is 37.5%. By comparison, the HA on the non-bulls-eye wager is 50%. ■

In the Missouri lottery, two bulls-eye numbers, denoted red and green, are drawn, and players may choose the red game with 1 bulls-eye, or the green game with 2. In each case, opting for the bulls-eye option raises the total wager: doubling it for the red game and tripling it for the green.

Table 3.87 shows the Pick 5 pay table for the Missouri Lottery's Club Keno game with all three bulls-eye options listed.

We note immediately that the payoff for catching 2 numbers with one bulls-eye is less than that for catching one bulls-eye alone, which seems counterintuitive. We shall develop a model for the probability of catching bulls-eyes which will explain this payoff structure. Define the following variables:

- $k$ = The number of catches. In the Missouri Lottery game, we have $1 \leqslant k \leqslant 10$.
- $m$ = The number of bulls-eyes selected on the ticket. Call these *active* bulls-eyes.

TABLE 3.87: Missouri Lottery Club Keno Pick 5 Bulls-eye pay table (for 1)

| | **Bulls-eyes caught** | | |
|---|---|---|---|
| **Catch** | **0** | **1** | **2** |
| 1 | 0 | 5 | — |
| 2 | 0 | 4 | 10 |
| 3 | 2 | 12 | 30 |
| 4 | 20 | 80 | 200 |
| 5 | 330 | 930 | 2325 |

For the Missouri game, $m = 1$ or 2, while other lotteries with only a single bulls-eye option are restricted to the case $m = 1$.

- $b =$ The number of bulls-eyes actually caught in the drawing: $0 \leqslant b \leqslant m$.

The language of conditional probability will be useful here. We are looking for the probability of catching $b$ bulls-eyes, given that $k$ numbers have been caught and $m$ bulls-eyes are active, and we denote that probability by $p(b \mid k, m)$. In a lottery where only one bulls-eye is possible and $m$ must be 1, we might choose to drop the explicit reference to $m$ and simply write $p(b \mid k)$. This must be multiplied by the probability of catching $k$ numbers to determine the probability of catching $k$ numbers with $b$ bulls-eyes on a given ticket with $m$ active bulls-eyes.

If $m = 1$, this situation is quite straightforward. If $k$ numbers have been caught and only one bulls-eye is active, the probability of one of those $k$ numbers being selected as the bulls-eye number is simply

$$p(1 \mid k, 1) = \frac{k}{20},$$

and the probability of not catching the bulls-eye is

$$p(0 \mid k, 1) = \frac{20 - k}{20}.$$

With two active bulls-eyes, the calculations are a bit more involved. With $k$ numbers caught, we have

$$p(0 \mid k, 2) = \frac{{}_{20-k}C_2}{{}_{20}C_2} = \frac{(20-k)(19-k)}{380},$$

$$p(1 \mid k, 2) = \frac{{}_{k}C_1 \cdot {}_{20-k}C_1}{{}_{20}C_2} = \frac{k(20-k)}{190}, \text{ and}$$

$$p(2 \mid k, 2) = \frac{{}_{k}C_2}{{}_{20}C_2} = \frac{k(k-1)}{380}.$$

Returning to our original question, we see that the probability of marking

5 numbers and catching only the bulls-eye number is

$$p(\text{Catch } 1) \cdot p(1 \mid 1, 1) = \frac{{}_5C_1 \cdot {}_{75}C_{19}}{{}_{80}C_{20}} \cdot \frac{1}{20} \approx .0203.$$

The probability of catching two numbers including the bulls-eye is

$$p(\text{Catch } 2) \cdot p(1 \mid 2, 1) = \frac{{}_5C_2 \cdot {}_{75}C_{18}}{{}_{80}C_{20}} \cdot \frac{2}{20} \approx .0270,$$

so we see that this second probability is indeed greater than the first, and so deserving of a lower payoff.

## Golden Keno Ball

Moving back into the casino, the *Golden Keno Ball* game option marketed by Gaming Arts, the company also responsible for Speed Keno (page 166), brings the bulls-eye concept under player control. Players choosing this option designate one of their numbers as the Golden Keno Ball, and if that number is chosen as the casino's final number, the winning payoff is sharply increased. The probability of the Golden Keno Ball coming last among the casino's numbers is simply $\frac{1}{80}$.

**Example 3.56.** In a Pick 1 game, the payoff on a \$1.25 ticket is \$3 if the chosen number is not the last number drawn, and \$20 if it's drawn last. The expected value of a ticket is

$$E = (3) \cdot \frac{19}{80} + (20) \cdot \frac{1}{80} - 1.25 = -\$\frac{23}{80} \approx -\$.2875,$$

giving a house edge of 23%—nearly equal to the 25% HA for a standard 1-spot ticket paying 3 for 1 on a catch. Note that the payoff is only 2.4 for 1 if the chosen number is not drawn last. ■

More generally, the probability of catching $k$ numbers out of $n$, including one as the last number drawn, is

$$\frac{{}_nC_{k-1} \cdot {}_{80-n}C_{20-k}}{{}_{79}C_{19}} \cdot \frac{1}{61}.$$

The first factor in this expression is the probability of catching $k-1$ of the first 19 numbers; the factor of $\frac{1}{61}$ represents the probability of the designated Golden Ball number being drawn last, given that it has not been drawn among the first 19 numbers. The probability of catching $k$ numbers out of $n$ without catching the last number is then

$$\frac{{}_nC_k \cdot {}_{80-n}C_{20-k}}{{}_{80}C_{20}} - \left( \frac{{}_nC_{k-1} \cdot {}_{80-n}C_{20-k}}{{}_{79}C_{19}} \cdot \frac{1}{61} \right).$$

This expression can be simplified, resulting in

$$\frac{{}_{80-n}C_{20-k} \cdot (61 \cdot {}_nC_k - 4 \cdot {}_nC_{k-1})}{61 \cdot {}_{80}C_{20}},$$

where the factor of 4 in the numerator arises from dividing ${}_{80}C_{20}$ by ${}_{79}C_{19}$.

How does the Golden Ball option affect the casino's advantage? Are the increased payoffs sufficient to balance the lower probability of catching the 20th number drawn and qualifying for them?

**Example 3.57.** Consider a Pick 6 game with the Golden Ball bonus. The payoffs and probabilities for this game are shown in Table 3.88.

TABLE 3.88: Golden Keno Ball Pick 6 pay table ($1.25 ticket)

| | **Standard payoffs** | | **Golden Ball payoffs** | |
|---|---|---|---|---|
| **Catch** | **Payoff** | **Probability** | **Payoff** | **Probability** |
| 3 | 1 | .1234 | 2 | .0064 |
| 4 | 2 | .0260 | 5 | .0025 |
| 5 | 75 | .0026 | 150 | .0005 |
| 6 | 200 | $7.8237 \times 10^{-5}$ | 5000 | $5.0748 \times 10^{-5}$ |

Since these results are mutually exclusive, the expected return on a $1.25 ticket is found by multiplying each of the eight payoffs in Table 3.88 by its corresponding probability, adding the products, and subtracting the ticket price. We find that $E = -\$.33$, and the HA is 26.37%. The HA on a standard Pick 6 ticket, using page 6 as the pay table, is 25.04%, so playing for the Golden Ball bonus on a Pick 6 ticket entails a sacrifice of 1.33% more to the casino. ■

Repeating this calculation for Pick 1 through Pick 10 tickets shows that the advantage shifts between the two games as the number of spots chosen, $k$, changes. Figure 3.27 plots the value of the HA for Golden Ball Keno minus the HA for a standard game for $k = 1$–10.

While the average difference in HA is only .42%, with the edge to standard keno, several options give the advantage to Golden Ball Keno: for Pick 1, 2, 3, 5, and 7, the HA for Golden Ball is lower. The difference in HA begins to favor standard keno sharply for Pick 8, 9, and 10 games; this is in part because the aggregate limit begins to limit how much can be awarded for the Golden Ball Keno payoffs when all of the numbers chosen are caught, including the Golden Ball.

## Exacta Keno

Years after keno quinela bets came to Macau, *Exacta Keno* returned more of the language of the race track to keno when Gaming Arts introduced it as

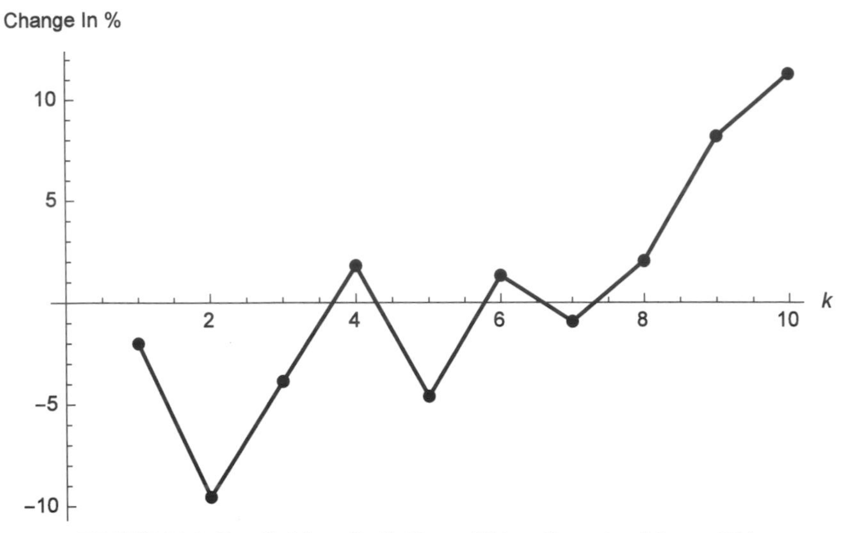


FIGURE 3.27: Golden Ball Keno HA – Standard keno HA.

a side bet. In horse racing, an *exacta* bet requires that the bettor pick the top two finishers in a given race, in the correct order. Exacta Keno is an optional extra bet that the player will catch the same number of marked numbers in two successive drawings for which he or she holds a ticket. The exact numbers matched need not be the same ones.

A minimum bet of \$1 on the main games is required to make the Exacta Keno bet, which costs an additional 25¢ on a \$1 main bet and rises in proportion, so a \$2 bet would require a 50¢ Exacta Keno side bet, for a total wager of \$4.50. The side bet is available on 1-spot through 10-spot games, and can pay off even if the number of numbers caught is not enough to qualify for a prize on the main ticket.

**Example 3.58.** The Pick 7 game at Jerry's Nugget pays off if 5 or more of the player's chosen numbers are drawn by the casino. The Exacta Keno side bet on a Pick 7 ticket pays off if three or more numbers are caught on both tickets—this must be the same number of catches on each ticket, of course. Catching 3 out of 7 wins you nothing on the first ticket, but there's the possibility of turning that loss into a win by catching three out of seven on the second ticket. ■

For a Pick 1 game, Exacta Keno is quite simple: a side bet, paying off at \$4 for a 25¢ wager, that you will win on two tickets in a row. This was a game variation we saw on page 124, as 2-Game Parlay Keno in Bellevue, Nebraska, where it carries a HA of 62.31%. Since the probability of winning a Pick 1 ticket is ¼, the probability of winning two straight games is $\frac{1}{16}$. With a standard \$3 payoff on the main wager, the expected value of a \$2.25 Exacta

Keno bet covering two Pick 1 tickets is

$$E = (10) \cdot \frac{1}{16} + 2 \cdot (3) \cdot \frac{1}{4} \cdot \frac{3}{4} - 2.25 = -\$0.50.$$

Dividing this expectation by the cost of the ticket yields a house edge of 22.22%—far better than in Bellevue.

On a Pick 3 game, the pay table at Jerry's Nugget looks like this:

| Catch | 1st Game | 2nd Game | Exacta |
|---|---|---|---|
| 2/3 | 0 | 0 | 2.25 |
| 3/3 | 54 | 54 | 1072 |

In the play of the game, there are nine different outcomes: the first ticket can catch 3, 2, or fewer than 2 numbers, and similarly for the second ticket. (Catching one number has the same effect as catching zero for both the standard bet and the Exacta Keno bet, so we consider those outcomes as a single case.)

| 1st Game | 2nd Game | Total win |
|---|---|---|
| Catch 0-1 | Catch 0-1 | 0 |
| Catch 0-1 | Catch 2 | 0 |
| Catch 0-1 | Catch 3 | \$54 |
| Catch 2 | Catch 0-1 | 0 |
| Catch 2 | Catch 2 | \$2.25 |
| Catch 2 | Catch 3 | \$54 |
| Catch 3 | Catch 0-1 | \$54 |
| Catch 3 | Catch 2 | \$54 |
| Catch 3 | Catch 3 | \$1180 |

To assess this wager, we define the following probabilities;

- $p_1$ = The probability of catching either 1 number or 0 numbers, $\approx .8473$.
- $p_2$ = The probability of catching 2 numbers, $\approx .1388$.
- $p_3$ = The probability of catching 3 numbers, $\approx .0139$.

The expected value of a \$2.25 Exacta Keno wager would then be

$$E = 54p_3(2p_2 + 2p_1) + 2.25p_2^2 + 1180p_3^2 - 2.25.$$

Substituting the numerical values given above yields $E \approx -\$.4983$, and so the HA is approximately 22.15%.

**Example 3.59.** At the Orleans, the Pick 3 Exacta Keno pay table is different, with higher payoffs when catching 2 of 3 and lower payoffs for catching all 3 numbers:

| Catch | 1st Game | 2nd Game | Exacta |
|---|---|---|---|
| 2/3 | 1 | 1 | 3 |
| 3/3 | 45 | 45 | 910 |

Since the probabilities $p_1$–$p_3$ are unchanged, the expectation of the Orleans wager can be evaluated by simply incorporating the new payoffs with the previous probabilities. We find that

$$E = 1(2p_1p_2) + 45(2p_1p_3) + 46(2p_2p_3) + 5p_2^2 + 1000p_3^2 - 2.25.$$

The expected return is about a penny higher at the Orleans, -\$0.4878, and the corresponding house edge is 21.68%. ■

## Power Keno and Super Keno

*Power Keno* is not a side bet *per se*, but a game variation often seen on video keno machines that is based on the principle of a keno quinela bet. Power Keno is so named because it amplifies the effect of the 20th number drawn: if that number appears on a player's winning ticket, the payoff is quadrupled. *Super Keno* offers the same quadruple bonus if the first ball drawn is part of a player's winning combination. Since the two games are similar, we shall explore the mathematics of Power Keno here and examine Super Keno in the exercises.

As one might reasonably expect, the pay table for Power Keno is slightly different from that for standard keno; the biggest payoffs are in general smaller due to the possibility of a quadrupled win. Table 3.89 compares standard and Power Keno payoffs for Pick 6 games.

TABLE 3.89: Standard and Power Keno Pick 6 pay tables (for 1)

| Catch | Standard | Power Keno |
|---|---|---|
| 3 | 1 | 1 |
| 4 | 4 | 5 |
| 5 | 80 | 65 |
| 6 | 2000 | 500 |

Suppose that $n$ numbers are chosen. For $k > 0$, what is the probability of catching $k$ of them, including one on the last ball drawn?

Computing this probability is similar to finding the probability of winning a bingo game in a fixed number of draws. Here's what has to happen:

- Exactly $k-1$ of $n$ numbers must appear among the first 19 numbers drawn.
- The 20th ball, chosen from the 61 that remain, must bear one of the $n-(k-1)=n-k+1$ remaining numbers.

Let this probability be denoted by $p(k \mid n)$. We have

$$p(k \mid n) = \frac{{}_nC_{k-1} \cdot {}_{80-n}C_{20-k}}{{}_{80}C_{19}} \cdot \frac{n-k+1}{61}.$$

**Example 3.60.** If $n=6$, we then have

$$p(k \mid 6) = \frac{{}_6C_{k-1} \cdot {}_{74}C_{20-k}}{{}_{80}C_{19}} \cdot \frac{7-k}{61},$$

which is tabulated for winning values of $k$ in Table 3.90.

TABLE 3.90: Power Keno $p(k \mid 6)$

| Catches ($k$) | $p(k \mid 6)$ |
|---|---|
| 3 | .0195 |
| 4 | .0057 |
| 5 | $7.739 \times 10^{-4}$ |
| 6 | $3.870 \times 10^{-5}$ |

Combining this table with the probabilities for matching $k$ of 6 numbers without regard for the last ball gives the probability distribution in Table 3.91.

TABLE 3.91: Power Keno Pick 6 probability distribution

| Catch | Payoff $X$ | $P(X)$ |
|---|---|---|
| 3 | 1 | .1104 |
| | 4 | .0195 |
| 4 | 5 | .0228 |
| | 20 | .0057 |
| 5 | 65 | .0023 |
| | 260 | $7.739 \times 10^{-4}$ |
| 6 | 500 | $9.029 \times 10^{-5}$ |
| | 2000 | $3.870 \times 10^{-5}$ |

Using the values in Table 3.91 shows that the expected return on a \$1 6-spot wager is -\$.1088, and the corresponding house advantage is 10.88%—substantially lower than the 25.04% HA on a standard 6-spot ticket. ■

By letting $n$ vary in Example 3.60, and incorporating the values in the Power Keno pay table [140], we can compute the house edge of any Power Keno ticket. The HA is remarkably consistent as $n$ ranges from 2–10, reaching a minimum of 10.75% for a 10-spot ticket; these house edges are shown in Table 3.92.

TABLE 3.92: House advantage of an $n$-spot Power Keno ticket (%)

| $n$ | 2 | 3 | 4 | 5 | 6 | 7 | 8 | 9 | 10 |
|---|---|---|---|---|---|---|---|---|---|
| **HA** | 14.02 | 11.61 | 11.10 | 11.01 | 10.88 | 11.03 | 10.94 | 11.19 | 10.75 |

## Cleopatra Keno

*Cleopatra Keno* is a video keno variation that incorporates a bonus round rather than an increased payoff for catching the last number drawn. The game is available in Pick 3 through Pick 10 options; one pay table is shown in Table 3.93.

TABLE 3.93: Cleopatra Keno pay table

| | **Numbers marked** | | | | | | | |
|---|---|---|---|---|---|---|---|---|
| **Catch** | **3** | **4** | **5** | **6** | **7** | **8** | **9** | **10** |
| 2 | 3 | 1 | | | | | | |
| 3 | 17 | 5 | 3 | 2 | 1 | | | |
| 4 | | 35 | 22 | 5 | 3 | 2 | 2 | 1 |
| 5 | | | 220 | 34 | 6 | 10 | 5 | 3 |
| 6 | | | | 375 | 95 | 56 | 10 | 5 |
| 7 | | | | | 500 | 180 | 80 | 25 |
| 8 | | | | | | 1000 | 200 | 150 |
| 9 | | | | | | | 1000 | 1000 |
| 10 | | | | | | | | 2000 |

If the player catches the last number drawn by the machine, a bonus of 12 free games is awarded. While these free games pay off at double the values in the pay table, they do not accumulate further free games if one of them also catches the last number.

**Example 3.61.** The pay table shown here carries a house advantage of about 10% on all options [138]. Cleopatra Keno does not offer a Pick 1 game, but what would the payoff for 1 need to be to return 90% of all money wagered and fall in line with Table 3.93?

The probability of catching the only marked number on the last number drawn is $\frac{1}{80}$; the probability of catching it earlier is $\frac{19}{80}$. In the bonus round, the number of wins is a binomial random variable with $n = 12$ and $p = \frac{1}{4}$; hence the expected number of wins is $np = 3$. Letting $X$ represent the payoff to be determined, we have

$$E = (X) \cdot \frac{19}{80} + (X + 3 \cdot 2X) \cdot \frac{1}{80} - 1 = -.10,$$

or

$$\frac{26X}{80} = .90.$$

Solving this equation for $X$ shows that paying 2.77 for 1 would yield an HA of approximately 10%. While this might be feasible on a video keno machine, which can pay off in pennies through the ticket in/ticket out option, it would require a minimum wager of \$1 if fractions of cents are to be avoided.

A standard keno pay table pays 3 for 1 when choosing and catching 1 number; the HA for Cleopatra Keno with that payoff would be only 2.5%. Paying 2 for 1, as in state lotteries, would give the casino a 35% edge. ■

## Million-Dollar Keno

With the demise of keno aggregate limits, a million-dollar payoff was only restrained by the casino's willingness to assume the risk of a large player win. In 2004, 5-spot keno players at the Peppermill Casino in Reno, Nevada were offered an optional 25¢ "eXtra Million" side bet. In addition to the standard pay table for a 5-spot ticket, eXtra Million paid \$1 million, in one lump sum rather than the 20-year annuity frequently offered to big lottery winners, if the player's 5 numbers were the *first* 5 numbers drawn by the casino, with smaller payoffs for matching two or more of the first 5 numbers drawn [132].

**Example 3.62.** Find the probability of winning the million dollars.

Since we're only looking at the first five numbers drawn, the denominator in our calculations will be ${}_{80}C_5$ rather than the usual ${}_{80}C_{20}$. The probability of hitting the million-dollar jackpot is

$$\frac{{}_5C_5}{{}_{80}C_5} = \frac{1}{24,040,016}.$$

■

Of course, if you win this prize, you also qualify for the prize for catching 5 of 5 numbers—there's an additional \$750 for you. That would make a good tip for the keno attendant or runner who handled your ticket.

The complete pay table for the eXtra Million bet, with all catches occurring within the first 5 numbers drawn, is shown in Table 3.94 [132].

TABLE 3.94: eXtra Million pay table

| Catch | Payoff |
|---|---|
| 2 | \$0.25 |
| 3 | \$5 |
| 4 | \$1000 |
| 5 | \$1,000,000 |

Since eXtra Million is a side bet, we expect a high house advantage, and we are not disappointed. The expected value of this bet—not including the standard Pick 5 payoffs—is –18¢, and the HA is 72%. We can use Theorem 2.9 to incorporate the expected value of a \$1 basic Pick 5 bet and find the expectation for the entire wager. Consider a \$1 Pick 5 wager made against Table 3.95 and backed up with a 25¢ eXtra Million side bet.

TABLE 3.95: Peppermill Casino Pick 5 pay table

| Catch | Payoff |
|---|---|
| 3 | 1 |
| 4 | 10 |
| 5 | 750 |

If $X_1$ is a random variable measuring the payoff of the basic bet and $X_2$ measures the payoff on the eXtra Million bet, we have

$$E(X_1 + X_2) = E(X_1) + E(X_2) = -\$0.3114 + -\$0.1800 = -\$0.4914,$$

nearly a 50¢ loss. The house advantage for the combined ticket is then 39.31%.

*Keno Millions* is a side bet currently offered in a number of casinos that brings back some of the features of eXtra Million. A player making a regular keno wager on four or more numbers is eligible for this bet. To play this side bet, the gambler designates four of their numbers as active and pays an additional amount. The side bet returns some money if any of the four specified numbers are caught among the first four numbers drawn.

Two versions of this wager are available. Keno Millions costs \$2, in addition to the initial wager, while Keno Millions Mini costs only an additional 25¢. Table 3.96 serves as the pay table for these bets.

The probability of winning something—even a break-even payment for catching 1 number in Keno Millions Mini or a net loss for the same catch in Keno Millions—is easily calculated. Only four numbers matter, and so this wager can be regarded as a Pick 4 game with 4 numbers chosen by the casino.

$$P(\text{Win}) = 1 - P(0|4) = 1 - \frac{{}_{76}C_4}{{}_{80}C_4} \approx .1888.$$

TABLE 3.96: Keno Millions pay table

| Catch | Keno Millions | Keno Millions Mini |
|---|---|---|
| 1 | $1 | 25¢ |
| 2 | $10 | $1 |
| 3 | $500 | $25 |
| 4 | $1,000,000 | $100,000 |

The house advantages on these two side bets are slightly better than for eXtra Million: 49.28% for Keno Millions and 50.68% for Keno Millions Mini. These HAs are again on a par with those offered by state lotteries.

## Keno Multipliers

The Oregon Lottery's video keno game includes an optional bonus bet, the Keno Multiplier, that can multiply a player's winnings for an additional $1 per dollar wagered—this becomes an offer to double one's bet for the possibility of a payoff that can be more than doubled. The multiplier is implemented through a virtual wheel that is spun before each drawing, which stops to reveal a multiplier of 1 to 10 times the standard payoff. The frequency of each multiplier is shown in Table 3.97.

TABLE 3.97: Probabilities for Oregon Lottery's Keno Multiplier [120]

| Multiplier | Probability |
|---|---|
| 1× | .450 |
| 2× | .283 |
| 3× | .160 |
| 5× | .097 |
| 10× | .010 |

This multiplier is a sound bet for the long-term keno player provided that the average value of the multiplier exceeds 2, since the player must double his or her bet in order to activate it. The expected value of the multiplier is

$$1 \cdot .450 + 2 \cdot .283 + 3 \cdot .160 + 5 \cdot .097 + 10 \cdot .010 = 2.081.$$

Since this is greater than 2, the multiplier returns more—provided that the initial keno bet wins—than it takes in, and thus is a rare bet where some advantage rests with the gambler [72]. The advantage here lies in doubling the wager for an average payoff that is more than doubled.

**Example 3.63.** For a direct comparison, consider a simple $1 Pick 2 ticket, which pays off at 11 for 1 if both numbers are caught. Without the multiplier, the expected value of this ticket is –\$.3386, and the HA is 33.86%. If the multiplier is active, then the expected return is

$$E = (2.081 \cdot 11) \cdot \frac{19}{316} - 2 = -\$.6236,$$

for a house edge of 31.18%.

While we must take note of the fact that buying in to the multiplier means putting an additional dollar at risk, making this side bet does mean decreasing the lottery's advantage by about 2.6%. On the other hand, that advantage still exceeds 30%. ■

In Ohio, this wager is called the Keno Booster, and its probabilities are listed in Table 3.98.

TABLE 3.98: Probabilities for Ohio Lottery's Keno Booster

| Multiplier | Probability |
|---|---|
| 1× | 1 in 2.3 |
| 2× | 1 in 2.5 |
| 3× | 1 in 16 |
| 4× | 1 in 16 |
| 5× | 1 in 27 |
| 10× | 1 in 80 |

The expected value of the Keno Booster is only 1.9825. Since a player must double his or her bet to activate this option, this bet does not improve the player's position, and so it should not be made.

## Betting on the Total

Several online casinos, including 12macau.com, offer a number of keno wagers that are based on the sum of the 20 numbers drawn. Since the average keno number is 40.5, it follows that the mean of all sums is $20 \cdot 40.5 = 810$, and the Big and Small bets are based on this number. The *Big* bet pays off if the sum of the drawn numbers is greater than 810, and the *Small* bet wins if that sum is less than 810. Both bets push if the sum is exactly 810. The online casinos charge a 5% commission on winning bets, which turns an even-money proposition into a 1.95 for 1 win. The *810* bet pays off at 108 for 1 if the sum is exactly 810 [66].

What is the probability of winning an 810 bet? Computing the exact probability of the sum totalling 810 would require summing each of the ${}_{80}C_{20}$ possible combinations, which is entirely too much work. An acceptably accurate approximation of this probability can be computed by a random sample. For this experiment, a total of 4,000,000 keno combinations were generated by a Java computer program. After summing, 17,737 of these simulated draws resulted in a total of 810. $P(\text{Sum} = 810)$ is then approximately .0044; this means that the expectation of \$1 wagered on 810 is

$$E = (108) \cdot .0044 - 1 \approx -\$.5211,$$

giving a house advantage of over 52%.

We can also use the result of this simulation to assess the Big and Small bets. Since each combination totalling more than 810 is precisely balanced by a combination totalling less than 810 (simply replace each drawn number $x$ by $81 - x$; the new combination has sum 1620 - the old sum), these two bets offer the same chance of winning. Denoting this common chance by $p$, we find that

$$p \approx \frac{1 - .0044}{2} = .4978.$$

The expected return on either bet is then

$$E = (1.95) \cdot p + (-1) \cdot (1 - .0044) \approx -\$.0243,$$

and the HA is a much more reasonable 2.43%. Without the commission, the payoff would be 2 for 1, and the game would then be fair with $E = 0$. If Big and Small bets lost on a total of 810, then a \$1 wager with commission would return

$$(1.95) \cdot p - 1 \approx -\$.0292,$$

and without commission,

$$(2) \cdot p - 1 \approx -\$.0043.$$

**Example 3.64.** The *Five Ranges* bet is a second wager based on the sum of the draw. Five Ranges divides the range of the sum, 210–1410, into 5 intervals and accepts bets on any of them [66]:

- **Gold**: 210–695, pays 9.2 for 1.
- **Wood**: 696–763, pays 4.6 for 1.
- **Water**: 764–855, pays 2.4 for 1.
- **Fire**: 856–923, pays 4.6 for 1.
- **Earth**: 924–1410, pays 9.2 for 1.

Since this bet is found in online casinos, decimal payoffs as listed are easy to accommodate because players' accounts are handled on a computer and there is no need to handle cash or chips.

Once again, some simulation is the quickest route to an accurate assessment of these bets, using experimental probabilities. Using the data from the 4,000,000-draw simulation on page 184 used to estimate the probability of the numbers summing to 810 gives the approximate probability distribution and house advantages shown in Table 3.99.

■

TABLE 3.99: Experimental probabilities of Five Ranges intervals

| Wager | Interval | Probability | HA |
|---|---|---|---|
| Gold | 210–695 | .1025 | 5.72% |
| Wood | 696–763 | .2013 | 7.42% |
| Water | 764–855 | .3883 | 6.82% |
| Fire | 856–923 | .2035 | 6.38% |
| Earth | 924–1410 | .1045 | 3.87% |

Since these intervals are approximately symmetric about 810 and span nearly identical ranges, it is not unreasonable to think that the Gold and Earth bets, as well as the Wood and Fire bets, should have the same probabilities of winning and house edges. If we approximate those probabilities by taking the means of the probabilities determined in our experiment, we have $P(\text{Win}) \approx .1035$ for Gold and Earth, giving a house advantage of 4.80%. Wood and Fire have a common winning probability of .2024 and common HA of 6.90%.

---

## 3.6 Keno Strategies: Do They Work?

Where there is gambling, there will surely be gambling systems, sold by entrepreneurs who claim to have a way to get the advantage on negative-expectation games. Keno is no exception. However, careful application of mathematics will show the truth of the statement "There is no betting system which can lead a gambler, with finite capital, to any greater expectations when he plays, for a large number of times, any game with fixed probabilities of winning.", which, ironically, appears in a book called *Win More At Keno* [12].

Simply stated: you cannot beat the house advantage. In this section, we'll look at some schemes that have been put forward to beat the odds at keno, and see why they fail.

### Hot and Cold Numbers Considered

At the Angel of the Winds Casino, the keno brochure offers players the option of betting on "hot" (frequently drawn) or "cold" (infrequently drawn) numbers by asking for the "What's Hot?" or "What's Cold?" ticket at the keno counter. Players may specify how many numbers they want to select and how many games they wish to play, and the keno computer takes care of selecting numbers and printing tickets.

Since successive keno drawings are independent, it would appear that taking these options does nothing to increase your chances of winning. The only possible exception would be at a casino that still uses ping-pong balls in keno, and then only if some balls were so defective as to decrease or increase their chance of being chosen. For example, a ball that is slightly heavier than the others would tend to remain near the bottom of the container, and thus would be less likely to be drawn when balls are picked from the top.

But...maybe there's a way to take advantage of this, if you're in a casino where the keno numbers are drawn mechanically rather than by a computer?

There are two problems with this optimistic line of thinking:

1. Developing a statistically significant sample requires observing the numbers across a large number of drawings. This might, of course, be something the casino is doing for you by offering these wagers. At Harvey's Casino in South Lake Tahoe, Nevada, the casino made it easy for players to identify cold numbers by posting a frequency board for its Gold Game, displaying how many games had passed since each of the 80 numbers had been drawn

2. The casinos have a vested interest in the outcome of each drawing being random—remember that casinos make their profits from the way the pay tables are set, not by cashing in on deviations from expected behavior. It follows that any defective ball would be identified, tested, and replaced so as to ensure completely random selections. The casino is certainly not collecting information on hot and cold numbers purely to sell it back to players.

Humans want to identify patterns, even when none are present. It is virtually certain that any string of keno drawings will turn up a few numbers that appear either more or less often than chance would predict. Figure 3.28 shows the cumulative results of 300 keno drawings conducted on April 15, 1991 at Bob Cashell's Horseshoe Club in Reno. Each spot on this keno ticket bears a number showing how many times it was drawn that day.

| | | | | | | | | | |
|---|---|---|---|---|---|---|---|---|---|
| 64 | 78 | 74 | 71 | 74 | 82 | 88 | 81 | 74 | 81 |
| 75 | 75 | 82 | 68 | 73 | 69 | 59 | 91 | 76 | 80 |
| 84 | 75 | 66 | 90 | 63 | 64 | 66 | 83 | 65 | 78 |
| 76 | 78 | 87 | 69 | 62 | 75 | 77 | 77 | 64 | 70 |
| 71 | 80 | 81 | 68 | 71 | 79 | 83 | 92 | 90 | 72 |
| 59 | 85 | 78 | 70 | 81 | 82 | 68 | 78 | 73 | 58 |
| 79 | 75 | 72 | 81 | 76 | 58 | 93 | 81 | 81 | 69 |
| 76 | 67 | 71 | 73 | 64 | 81 | 68 | 89 | 71 | 72 |

FIGURE 3.28: Cumulative keno results from Bob Cashell's Horseshoe Club, April 15, 1991 [52]. "Hot" numbers are shaded light gray; "cold" numbers are shaded dark gray.

With 300 drawings, we would expect each number to be drawn exactly 75 times, and indeed five numbers—11, 12, 22, 36, and 62—were drawn that often. Based on this chart, we would be inclined to identify 60, 66, 17 and 51 as "cold" and 67, 48, 18, 24, and 49 as "hot", for their frequencies are farthest from 75. However, to assess whether or not there is meaningful deviation from a random distribution, it is necessary to consider all 80 data points. There are statistical tests that can assess a collection of data and determine how likely it is to have arisen from a random process; one is called the $\chi^2$ *goodness of fit test*, where $\chi$ is the Greek letter chi, pronounced "ky". We can apply this test to calculate the probability of achieving a distribution with at least this much total deviation from 80 75s, under the assumption that the selection process was completely random. As described in Chapter 2, if that probability is less than .05, we are justified in suspecting that something other than randomness is at work in the selection of these numbers.

The $\chi^2$ test on these data returns a probability of .6987, meaning that we can be confident that the numbers chosen that day in this casino were randomly selected. Tests for randomness are easily conducted on any collection of keno draw results; any responsible keno outlet will certainly run this kind of test on its numbers from time to time.

**Example 3.65.** What is the probability that, over the course of 4 consecutive independent drawings, every number will be drawn exactly once?

- There are ${}_{80}C_{20}$ ways to choose the first 20 numbers. The probability of getting 20 new numbers on the first drawing is

$$p_1 = \frac{{}_{80}C_{20}}{{}_{80}C_{20}} = 1,$$

since no numbers have yet been drawn.

- There are then 60 numbers to choose from in the second drawing. The probability that two consecutive drawings have no numbers in common is then

$$p_2 = \frac{{}_{60}C_{20}}{{}_{80}C_{20}} \approx 1.186 \times 10^{-3},$$

already a very small probability. Approximately one keno draw in 843 will repeat no numbers from the previous drawing.

- If we have beaten these long odds and had two disjoint drawings, the probability that the third drawing then uses none of the 40 drawn numbers is

$$p_3 = \frac{{}_{40}C_{20}}{{}_{80}C_{20}} \approx 3.899 \times 10^{-8}.$$

- Finally, the conditional probability of drawing the last 20 numbers, given three previous mutually disjoint drawings, is

$$p_4 = \frac{{}_{20}C_{20}}{{}_{80}C_{20}} = \frac{1}{{}_{80}C_{20}} \approx 2.828 \times 10^{-19}.$$

The probability, then, of four mutually disjoint keno drawings is

$$p = p_1 \cdot p_2 \cdot p_3 \cdot p_4 \approx 1.308 \times 10^{-29}.$$

■

The probability of one or more numbers being repeated (and one or more other numbers being omitted) is $1 - p$, which is a decimal number starting with a string of 28 consecutive 9s. For all practical purposes, it's not really worth speculating about four straight keno drawings without any repeated numbers—indeed, two straight keno drawings are highly likely to repeat at least one number; that probability is

$$1 - p_1 \cdot p_2 \approx .9988.$$

While the Law of Large Numbers guarantees that the proportion of any number drawn over a large number of drawings will approach $\frac{1}{80}$, this does not mean that all numbers will appear equally often. The probability of uniform distribution gets much smaller if more drawings are considered.

Knowing which numbers have not been drawn in a while, or which numbers have been hitting more frequently than average, does not improve a gambler's chances of correctly choosing the numbers that will come up on the next drawing.

## Kirby Keno System

An overreliance on hot and cold numbers was a flaw at the heart of a 1978 keno betting system. In that year, *Gambling Times* published an article that laid out a purported strategy for winning at keno. The *Kirby Keno System* took as one of its premises that "Keno is a foolish game to play for nickels and dimes," suggesting that the game was only worth playing in a quest for large prizes [70]. The system was simple: it called for betting three 4-spot ways combined into three 8-spot tickets and one 12-spot ticket. At 70¢ per way, this required investment of only $2.80, and carried a chance of winning a five-figure prize on any of its four virtual tickets.

The Kirby system gave no prescription for choosing numbers, suggesting that players pick their favorite numbers or follow the game for hours before playing, looking for trends that might be ascribed to defective ping-pong balls and effectively exploited. The system's misinterpretation of the Law of Large Numbers was evident in its final recommendation: that players "count on the fact that eventually all the numbers must even out in coming up on the keno board". As we know, though, the fraction of games in which each number is drawn will reliably tend to ¼ as the number of drawings gets large, this does not mean that all of the numbers will appear equally often, even in the very long run.

A second major problem with this system was the aggregate limit, which

TABLE 3.100: Pay table for 70¢ ticket; $25,000 aggregate limit

| Pick 8 | | Pick 12 | |
|---|---|---|---|
| Catch | Payoff | Catch | Payoff |
| 5 | 6.30 | 6 | 4.20 |
| 6 | 63.00 | 7 | 19.60 |
| 7 | 1155.00 | 8 | 140.00 |
| 8 | 12,600.00 | 9 | 595.00 |
| | | 10 | 1680.00 |
| | | 11 | 9100.00 |
| | | 12 | 25,000.00 |

at the time stood at $25,000 in Nevada casinos. Table 3.100 shows the pay table for a 70¢ ticket subject to an aggregate limit of $25,000.

In the admittedly unlikely event that the ticket catches all 12 of its numbers, the aggregate limit kicks in. Instead of winning $62,800 for catching 3 8-spots and a 12-spot, the player would only win $25,000—a virtual "loss" of $37,800. Catching 11 numbers also gives a $25,000 payoff capped by the limit.

If you truly want to play keno for big money, it's best to choose a ticket where your big paydays aren't cut short by the rules of the game. The local aggregate limit needs to be taken into consideration when setting up a ticket.

## Chicken Keno: Enter the Martingale

The *martingale* is a betting system often promoted as a way to guarantee victory while making a sequence of even-money bets, such as a hand of blackjack or a bet on Red at roulette. To employ the martingale, begin by betting 1 unit. If it wins, collect your 1-unit profit and begin again with another 1-unit bet. If it loses, double your bet for the next round. If it wins, you have bet 3 units and won 4, a 1-unit profit. If it loses, double again to 4 units—and keep going. Eventually, you will win, it is said, and you'll then have a 1-unit profit.

Martingales fail because casino table games have maximum bet limits. If you run into a long losing streak, continued doubling will eventually bring you to a point when your next bet would exceed that limit. For example, if the limit is $1000 and your unit is $10, you can only make 7 bets before reaching the limit: your next scheduled bet would be $1280. Your losses to that point total $1270, and you can't recoup it all with one winning bet.

Lotto Nebraska runs local keno operations in dozens of western Nebraska cities, and recommends a version of the martingale to its patrons in the guise of a game called Chicken Keno. The Lotto Nebraska paybook describes the game like this:

> You can win big if you don't CHICKEN OUT! Pick a number and stick with it game after game until it wins. Double your

bet each game. *Remember the limit of $25,000 in aggregate non-progressive winnings per game.*

Recall that the aggregate limit in keno serves the same function as the bet limit at table games: reducing the amount of risk assumed by the casino in accepting bets. Lotto Nebraska's aggregate limit is $25,000 per game; since their Pick 1 ticket pays off at 3 for 1, this game is somewhat more lucrative than a standard martingale on a bet paying 1–1, which only wins one unit after all of the losing bets are deducted from the final payoff.

That doesn't make Chicken Keno a good gamble, though. Suppose you begin a bet sequence with a $1 wager on your chosen number. The chance of winning $3 is then $\frac{1}{4}$, and the chance of losing is $\frac{3}{4}$. The probability of winning for the first time on the $2 second wager is $\frac{3}{4} \cdot \frac{1}{4}$, which carries a payoff of $6. The probability of losing both of the first two bets is $\left(\frac{3}{4}\right)^2 = \frac{9}{16}$—still greater than 50%.

But the admonition in the paybook is to keep going, doubling your wager each time, until you win. The probability that the first win in this sequence comes on the $n$th bet, where the wager is $\$2^{n-1}$, is

$$P(n) = \left(\frac{3}{4}\right)^{n-1} \cdot \left(\frac{1}{4}\right) = \frac{3^{n-1}}{4^n}.$$

The corresponding payoff would be $3 \cdot 2^{n-1}$, provided that this amount does not exceed $25,000. This means that the maximum bet before the aggregate limit affects the payoff is $\$2^{13} = \$8192$: you have 14 bets to win one before that happens. A hypothetical 15th bet would be for $16,384; while this could be made, the payoff if it wins would only be $25,000, not the $49,152 that a 3–1 payoff would call for. You would still incur a net loss of $7767 even if the 15th bet wins.

The flaw in any martingale is that, while the wins are frequent and small, the infrequent losses are catastrophic. The probability of losing 14 straight 1-spot keno bets is

$$\left(\frac{3}{4}\right)^{14} \approx .0178,$$

or about 1.78%. At that point, your total losses come to

$$\$1 + 2 + 4 + \cdots + 8192 = \$16,383.$$

Prior to that though, Chicken Keno played this way is not a bad game—as is the case with any martingale. If you win a bet before crashing into the aggregate limit, you can walk away with a nice profit. Table 3.101 displays the net winnings if this scheme pays off within the first 14 draws.

If you do not make the 15th bet, and absorb the loss, the expected value of this game is

$$E = \sum_{n=0}^{13} \left[3 \cdot 2^{n-1} \cdot \frac{3^{n-1}}{4^n} - 2^{n-1}\right] - 16,383 \cdot \left(\frac{3}{4}\right)^{14} \approx -\$145.46.$$

TABLE 3.101: Chicken Keno profits, before the aggregate limit

| Drawing | Wager | Total Wagered | Payoff | Profit |
|---|---|---|---|---|
| 1 | 1 | 1 | 3 | 2 |
| 2 | 2 | 3 | 6 | 3 |
| 3 | 4 | 7 | 12 | 5 |
| 4 | 8 | 15 | 24 | 9 |
| 5 | 16 | 31 | 48 | 17 |
| 6 | 32 | 63 | 96 | 33 |
| 7 | 64 | 127 | 192 | 65 |
| 8 | 128 | 255 | 384 | 129 |
| 9 | 256 | 511 | 768 | 257 |
| 10 | 512 | 1023 | 1536 | 513 |
| 11 | 1024 | 2047 | 3072 | 1025 |
| 12 | 2048 | 4095 | 6144 | 2049 |
| 13 | 4096 | 8191 | 12,288 | 4097 |
| 14 | 8192 | 16,383 | 24,576 | 8193 |
| $n$ | $2^{n-1}$ | $2^n - 1$ | $3 \cdot 2^{n-1}$ | $2^{n-1} + 1$ |

While this is negative, the probability of losing money is very small and the payoffs along the way to ruin keep rising, which might draw in players. The house advantage of the entire sequence of bets is only .89%, the result of averaging out multiple small wins with one unlikely but catastrophic loss.

How does a Lotto Nebraska outlet make money from this game? After all, less than 2% of players will follow this bet sequence all the way to the end and lose a fortune. The secret lies in "all the way to the end". Some players might start out making the first few bets, but run out of money before scoring a win or hitting the 14-bet limit.

---

## 3.7 Exercises

Solutions begin on page 303.

**3.1.** Find a whole-number payoff for catching one chosen number on both tickets at Bellevue, Nebraska's 2-Game Parlay Keno (page 124) that gives a HA as close as possible to the 25% offered on a single Pick-1 ticket.

**3.2.** Barton's 93 Club in Jackpot, Nevada offered a multigame bonus payoff on its Pick 1 game. A winning 1-spot ticket paid off at 3.25 for 1 rather than 3 for 1, but this rate was obtained only by playing 5 or more consecutive games at a time. Find the probability that a gambler playing 5 games makes a profit on this game.

**3.3.** In July 2017, the monthly Big Red Keno special in Blair and Valley, Nebraska was a firecracker-themed Pick 3 Catch All game called *321 Boom*. For a $5 minimum wager, this ticket paid off at 321 for 5 (64.2 for 1) if the player caught all three chosen numbers. Where does the 321 Boom house edge rank among the wagers listed in Table 2.3 (page 31)?

**3.4.** The "Mark 10 Spots" keno game at the FireKeepers Casino pays off $1 on a $1 wager if either 0 or 5 of the player's 10 selected numbers are drawn by the casino. Which of these two events has the higher probability?

**3.5.** Spotlight Keno's (page 60) Pick 2 game uses Table 3.102 for catching both numbers on a $1.25 ticket.

TABLE 3.102: Spotlight Keno Pick 2 pay table

| **Color** | **Payoff** |
|---|---|
| Red/Mixed | $13 |
| White | $50 |
| Blue | $150 |

a. Find the probability that both of a player's chosen numbers are among the 8 blue numbers when they are first identified before the draw.

b. Find the probability of winning $150 by catching two blue numbers.

c. Find the probability of winning $50 by catching two white numbers.

**3.6.** Along with the quinela, a single-number keno side bet was also occasionally seen in Macau casinos. Players could invest an additional wager, from $2 up to ten times their ticket bet, and designate one number as a "Star" number. If the Star number was either the first or last number drawn by the casino, the bet paid off at 35–1. Find the house advantage for this bet.

**3.7.** Suppose that you buy into the 100 Dimes keno game described on page 132 for the minimum: 100 tickets at 10¢ each. Find the probability that you will make a profit on this $10 investment.

**3.8.** Confirm that the house advantage on a Nevada Numbers ticket is approximately 71.44% if the progressive jackpot is $5,000,000.

**3.9.** The Just 7777s game described on page 78 is said to have a house advantage exceeding 50%. Confirm this assertion by computing the exact HA.

**3.10.** By changing the payoff for catching 6 of 6 numbers in Shortcake Keno (see Table 3.70), find a whole-number payoff for which the revised game offers the same HA as the standard Pick 6 game at Binion's Horseshoe Club.

**3.11.** Another 70-number Israeli lottery game based on keno is called *Pais 777*. Pais 777 is also a Pick 7 game with Systematic 8- and 9-number options. Seventeen numbers are drawn twice a day for this game, seven days a week, and Table 3.103 is used as the pay table [98]. Pais 777 tickets cost NIS 7.

TABLE 3.103: Israel Lottery Pais 777 pay table (NIS 7 ticket) [98]

| Catch | Payoff |
|---|---|
| 0 | 5 |
| 3 | 5 |
| 4 | 20 |
| 5 | 50 |
| 6 | 500 |
| 7 | 70,000 |

For a NIS 7 ticket, which game, Pais 777 or Israeli Keno, is a better game for the player?

**3.12.** The probability of losing everything on a $1.90 Penny Keno "K" ticket is tricky to calculate, because there are many ways not to fill a complete 3-, 4-, or 5-spot block, from catching none of your 50 numbers (Example 3.38) to catching 20 numbers that are inconveniently spread across the ticket. Find the probability of losing $1.90 when the drawn numbers fall into each of the following arrangements:

a. Only one of the K numbers is caught.

b. Two numbers are caught from each of the 8 blocks of 3, and the remaining four numbers fall outside the K completely.

c. Exactly 11 of the 20 drawn numbers are catches: 9 from the 13 blocks of 2 and 2 from one of the 8 blocks of 3.

d. All 20 drawn numbers are inside the K: one number each from 20 of the 21 blocks. (Hint: There are two subcases to be considered here.)

**3.13.** Figure 3.29 shows the Penny Keno "F" configuration with the active numbers shaded and the 20 blocks labeled A-T.

a. Show that this set of blocks can be used to form a 170-way ticket consisting of 3-spot, 4-spot, and 5-spot ways.

b. At a penny per way, this ticket costs $1.70. With the following pay table in use:

| Catch | Payoff |
|---|---|
| 3 of 3 | $0.55 |
| 4 of 4 | $2.50 |
| 5 of 5 | $11.50 |

| A | E | I | M | M | M | N | N | N | 10 |
|---|---|---|---|---|---|---|---|---|---|
| A | E | I | O | O | O | P | P | P | 20 |
| B | F | J | 24 | 25 | 26 | 27 | 28 | 29 | 30 |
| B | F | J | Q | Q | Q | R | R | R | 40 |
| C | G | K | S | S | S | T | T | T | 50 |
| C | G | K | 54 | 55 | 56 | 57 | 58 | 59 | 60 |
| D | H | L | 64 | 65 | 66 | 67 | 68 | 69 | 70 |
| D | H | L | 74 | 75 | 76 | 77 | 78 | 79 | 80 |

FIGURE 3.29: Penny Keno bet slip with "F" bet groups labeled.

find the profit function $F$ for this ticket.

c. What configuration of drawn numbers maximizes the profit on the F ticket?

**3.14.** Figure 3.30 shows the Penny Keno "H" configuration with the active numbers shaded and the 21 blocks labeled A-U. The blocks include 15 3-spot blocks, 5 2-spot blocks, and a single king (J: number 58).

| A | A | A | 4 | 5 | 6 | 7 | P | R | R |
|---|---|---|---|---|---|---|---|---|---|
| B | B | B | 14 | 15 | 16 | 17 | P | Q | R |
| C | C | C | 24 | 25 | 26 | 27 | M | Q | O |
| D | D | D | S | S | S | U | M | Q | O |
| E | E | E | T | T | T | U | M | N | O |
| F | F | F | 54 | 55 | 56 | 57 | J | N | L |
| G | G | G | 64 | 65 | 66 | 67 | I | K | L |
| H | H | H | 74 | 75 | 76 | 77 | I | K | L |

FIGURE 3.30: Penny Keno bet slip with shaded "H" bet.

a. Show that this set of blocks can be used to form a 110-way ticket consisting of 4-spot and 5-spot ways.

b. At a penny per way, this ticket costs \$1.10. If matching all four numbers on a 4-way ticket pays \$2.50 and fully matching a 5-spot way pays \$11.50, find the profit function $F$ for this ticket.

c. What configuration of drawn numbers maximizes the profit on the H ticket?

d. Intuition tells us that it is better to catch the king number than not, and while this is not quite universally true, the maximum profit found in part c. does arise from a configuration including 58. What is the maximum profit when the king is *not* drawn by the casino?

**3.15.** The *Bank Shot* keno wager, offered at the Western Village Casino in Sparks, Nevada, is a very simple bet: Pick 20 numbers, and you win $50,000 if 12 or more are drawn by the casino, and nothing otherwise [91].

a. Find the house advantage for this bet if the ticket price is $7.

b. Advertising for Bank Shot claimed that this game was 23 times as easy to hit as a solid 8-spot ticket (an 8-spot ticket catching all of its numbers), which also paid off at $50,000 for 1. Is this a valid claim?

**3.16.** As mentioned on page 178, *Super Keno* is a variation on Power Keno that quadruples the payoff if the *first* ball is part of a player's winning combination. While this results in a game with some similarities to Power Keno, the games are not mathematically identical.

a. On an $n$-spot ticket, show that the probability of matching $k$ numbers, including the first number drawn, is

$$p(k \mid n) = \frac{k}{80} \cdot \frac{{}_{n-1}C_{k-1} \cdot {}_{80-n}C_{20-k}}{{}_{79}C_{19}}.$$

b. If $n = 6$, show that the *ratio* of the Power Keno probability $p(k \mid n)$ to its Super Keno counterpart is

$$\frac{6k + 330}{74k}.$$

c. Find the values of $k$ for which a 6-spot Power Keno ticket has a higher winning probability than a 6-spot Super Keno ticket.

**3.17.** In April 2016, Big Red Keno in Beatrice, Nebraska saluted income tax season with its "1040 Tax Special" Pick 5 game. The pay table for this $1 game is compared to Beatrice's standard Pick 5 game in Table 3.104.

TABLE 3.104: Big Red Keno: Standard and 1040 Tax Special pay tables

| Catch | Standard | 1040 Tax Special |
|---|---|---|
| 3 | $0.50 | $0.50 |
| 4 | $12 | $10 |
| 5 | $800 | $1040 |

Find the change in the house edge with the 1040 Tax Special ticket. Is this better than the standard game?

**3.18.** Eight Las Vegas-area casinos operated by Red Rock Resorts run a progressive keno game across all of their keno rooms. Jumbo Keno Progressive offers a progressive jackpot on Pick 6, Pick 7, Pick 8, and Pick 9 games, and the progressive jackpot in each game is paid when a ticket matches all of the chosen numbers. Each game starts its progressive jackpot at a fixed amount which increases by a percentage of the total amount wagered on the progressive games; the first verified ticket across the 8-casino network collects the jackpot, which then resets. Like many keno games, the aggregate payout on any one game of Jumbo Keno Progressive is capped, at $100,000, but only the initial amount and not the accumulated extra value is subject to this cap.

For the $1 Pick 8 game, the progressive jackpot starts at $50,000, and the following pay table is used:

| Catch | Payoff |
|---|---|
| 5 | $5 |
| 6 | $75 |
| 7 | $1500 |
| 8 | Progressive |

a. Find the expected value of a $1 ticket with the initial progressive jackpot amount.

b. As the jackpot grows, the house edge decreases. Find the minimum progressive jackpot (to the nearest dollar) that gives the player a positive expectation.

**3.19.** Table 3.105 shows pay tables and ticket prices for several Big Red Keno special Pick 4 games run in May 2016.

TABLE 3.105: Big Red Keno: May 2016 special Pick 4 pay tables

| **Location** | Beatrice | Blair | Hadar | Lincoln | Omaha | Valley |
|---|---|---|---|---|---|---|
| **Price** | $1.35 | $0.25 | $1.35 | $0.25 | $0.25 | $0.25 |
| **Catch 3** | 2 | 1 | 3 | 0.50 | 0.50 | 1 |
| **Catch 4** | 350 | 50 | 340 | 55 | 50 | 55 |

a. An alternative to computing house advantages when comparing keno games is a careful comparison of pay tables. By inspection, determine which Nebraska city offers the best 25¢ game.

b. The situation for the $1.35 games listed in this table is not so clear-cut, since each city is better than the other at one of the two pay levels. Which game is better for the player?

c. Convert the two games identified in parts a. and b. to equivalent payoffs for a $1 ticket. Which game offers the better Catch 3 payoff? Which game offers the better Catch 4 payoff? Which game is better for the player?

**3.20.** Find the probability of making a profit on a single Pick-7 ticket at Double-Up Keno.

**3.21.** At Jerzes Sports Bar, the *King Me Eights* game invites players to choose 8 kings, which are grouped into 56 3-spot tickets at 10¢ each and a single 8-spot ticket priced at 40¢, for a total investment of $6. Catching 3 numbers on any 3-spot way pays $6.25; the brochure for the game simply states "Catch All Numbers And Win OVER $10,000.00". If the pay table for Jerzes' 8-spot game pays off at 25,000 for 1 for catching all 8 numbers, find the exact amount of that payoff and the net profit when all 8 numbers are caught.

**3.22.** The regular Pick 5 pay table for a $2.50 wager at La Vista Keno uses the pay table in Table 3.106.

TABLE 3.106: La Vista Keno: Pick 5 pay table

| Catch | Payoff |
|---|---|
| 3 | $2.50 |
| 4 | $37.50 |
| 5 | $1875.00 |

Find the amount of the progressive jackpot at La Vista for which the expected value of a progressive ticket (see page 116) is equal to the expected value of a standard Pick 5 ticket.

**3.23.** A $10 wager called *Line Drive Keno*, available at the Western Village, involved all 80 numbers [91]. The bet offered a 5 for 1 payoff if any five of the casino's selected numbers formed either a horizontal or vertical line on the keno ticket, with higher payoffs (Table 3.107) for longer horizontal or vertical strings of drawn numbers.

TABLE 3.107: Western Village Line Drive Keno pay table

| Catch | Payoff |
|---|---|
| 5 in a row | 50 |
| 6 | 325 |
| 7 | 1500 |
| 8 | 9275 |
| 9 | 17,000 |
| 10 | 25,000 |

a. How many winning lines of 5 numbers are there on a keno bet slip?

b. Find the probability of catching 5 numbers in a horizontal or vertical line.

**3.24.** In Harrah's Pick 10 Stimulus ticket (page 88), what payoff for catching all 10 numbers would give a more reasonable 30% house advantage?

**3.25.** A discounted keno coupon need not always be a lucrative deal for the player. A coupon issued by the Monte Carlo Casino in Reno (now Diamond's Casino) gave players a shot at a \$40 payoff for catching 4 out of 4 numbers, for 25¢ plus the coupon. This represented a 50% discount on the 50¢ ticket.

Show that the house advantage on this coupon deal exceeds 50%.

**3.26.** The California Hotel and Casino in downtown Las Vegas is the favored casino of many Hawaiian visitors, and the resort features several island-related shops and restaurants. The theme continues into the keno lounge, where 80¢ Island Special tickets compete with the standard pay tables for gamblers' attention. Table 3.108 compares the Pick 4 and Pick 5 games with the California's indicated minimum ticket prices.

TABLE 3.108: California Casino standard and Island Special pay tables

| | **Pick 4** | | **Pick 5** | |
|---|---|---|---|---|
| | **Standard** | **Island Special** | **Standard** | **Island Special** |
| **Catch** | \$1 rate | 80¢ rate | \$1 rate | 80¢ rate |
| 2 | 0.50 | 0.80 | 0 | 0 |
| 3 | 1 | 2 | Free ticket | 0.80 |
| 4 | 200 | 120 | 4 | 6 |
| 5 | — | — | 1000 | 800 |

Determine which game has the lower house advantage at the Pick 4 and Pick 5 levels.

**3.27.** The Foxwoods Casino offers *Ticket To Ride*, a promotional game with free tickets awarded as prizes to bingo players. Ticket To Ride is a Pick 6 game with Table 3.109 as its pay table.

TABLE 3.109: Ticket To Ride pay table

| **Catch** | **Payoff** |
|---|---|
| 4 | \$2 |
| 5 | \$5 |
| 6 | \$500 |

As a free ticket, this has a positive expectation. Is it worth selling a Ticket To Ride ticket for 15¢ prior to the drawing?

**3.28.** The Lottoland web site (www.lottoland.com), based in Gibraltar and Australia, serves as a central clearinghouse for a number of international lottery games. Lottoland patrons do not purchase tickets for the lotteries, but may bet on the outcomes of a collection of games and receive the same payoffs awarded by those games if they win—this is called a *synthetic lottery*. A local game operated by Lottoland is *KeNow*, a variation on keno that involves choosing from 70 rather than 80 numbers [87]. KeNow tickets have a minimum cost of €1 and a maximum cost of €10, and drawings are conducted every four minutes from 5:34 AM to 10:54 PM GMT.

The KeNow pay table is shown in Table 3.110.

TABLE 3.110: KeNow pay table

| | Numbers marked | | | | | | | | | |
|---|---|---|---|---|---|---|---|---|---|---|
| **Catch** | **1** | **2** | **3** | **4** | **5** | **6** | **7** | **8** | **9** | **10** |
| 0 | | | | | | | | 1 | 2 | 2 |
| 1 | 1.50 | | | | | | | | | |
| 2 | | 5 | 1 | 1 | | | | | | |
| 3 | | | 15 | 2 | 2 | 1 | | | | |
| 4 | | | | 20 | 5 | 2 | 1 | 1 | | |
| 5 | | | | | 100 | 10 | 10 | 2 | 2 | 2 |
| 6 | | | | | | 500 | 100 | 10 | 5 | 5 |
| 7 | | | | | | | 1000 | 100 | 10 | 10 |
| 8 | | | | | | | | 10,000 | 1000 | 100 |
| 9 | | | | | | | | | 50,000 | 1000 |
| 10 | | | | | | | | | | 100,000 |

a. Show that the house advantage on a Pick 2 ticket exceeds 60%.

b. Of the four catches listed that pay off at 100 for 1, which has the highest probability?

c. (This question is best solved on a spreadsheet.) Are there any games where KeNow offers a better game to the gambler than the standard 80-ball game using the pay table on page 6?

**3.29.** *Crossroads Keno* is a Pick 19 game designed for play against a Pick 20 pay table. Players choose one row and one column of the keno bet slip. Since these must intersect, any choice determines 19 distinct numbers. The number common to both the row and column is called the *intersection number*, and counts as two catches if it is drawn. For example, in Figure 3.31, the intersection number is 64. Any of the 80 numbers on a bet slip may be chosen as the intersection number.

a. Find the probability of scoring zero catches.

b. Find the probability of scoring exactly 19 catches.

c. For $2 \leqslant k \leqslant 18$, the probability $P(k)$ of scoring $k$ catches can be written as the sum of two terms: one in which the intersection number is caught and one in which it is not. Derive the formula for $P(k)$.

| 1 | 2 | 3 | 4 | 5 | 6 | 7 | 8 | 9 | 10 |
|---|---|---|---|---|---|---|---|---|---|
| 11 | 12 | 13 | 14 | 15 | 16 | 17 | 18 | 19 | 20 |
| 21 | 22 | 23 | 24 | 25 | 26 | 27 | 28 | 29 | 30 |
| 31 | 32 | 33 | 34 | 35 | 36 | 37 | 38 | 39 | 40 |
| 41 | 42 | 43 | 44 | 45 | 46 | 47 | 48 | 49 | 50 |
| 51 | 52 | 53 | 54 | 55 | 56 | 57 | 58 | 59 | 60 |
| 61 | 62 | 63 | 64 | 65 | 66 | 67 | 68 | 69 | 70 |
| 71 | 72 | 73 | 74 | 75 | 76 | 77 | 78 | 79 | 80 |

FIGURE 3.31: Crossroads Keno bet slip with intersection number 64.

d. For practical reasons, it may be desirable for Crossroads Keno to use a $5 ticket and a pay table such as Table 3.46 (page 109) that pays on all catches, so as to avoid the situation where catching the intersection number turns a money-winning combination into a $5 loser. Using a spreadsheet, find the house advantage of a $5 Crossroads Keno ticket using that pay table.

# Chapter 4

# Lotteries

## 4.1 Return of the Numbers Game

Recall Example 1.2, about the New York City's "You Pick 'Em Treasury Ticket", in which players chose a five-digit number and were paid off in accordance with the last five digits of the published U.S. Treasury balance each day. If a player matched all five digits, the payoff was 300 for 1; matching only the last four paid 30 for 1 and a match of the last three paid 3 for 1.

There is 1 number that is an exact match, 9 numbers that match only the last four digits, and 99 that match only the last three. The expected value of a $1 Treasury ticket is then

$$E = (300) \cdot \frac{1}{100,000} + (30) \cdot \frac{9}{100,000} + (3) \cdot \frac{99}{100,000} - 1 = -\frac{99,133}{100,000},$$

or –99.13¢—a 99.13% house advantage.

As high as the player's disadvantage may be in a game of keno or a state lottery, it's far better than it was in the days of the Numbers Game—only the special Catch All Pick 10 tickets described on page 88 come close to this HA. Despite the long odds, the Numbers Game was quite popular. In Detroit, an employee at the *Free Press* once clipped the two lines containing the day's Treasury balance from the business pages so that the stock market tables would all fit. When the newspaper hit the streets the next morning, the paper's telephone lines were swamped with calls wanting to know where the numbers were [37].

### New Jersey Pick-It

New Jersey introduced its own legal version of the Numbers Game in 1975, with *Pick-It*, the first modern state lottery allowing players to pick their own numbers. Pick-It called for players to choose their own three-digit number and match it against the state's drawn number. This was a *pari-mutuel* game, one whose payoff varies based on the number of tickets sold. A portion of each ticket price is placed into a pool from which the prizes are paid. Pari-mutuel jackpots are common in multi-million dollar lotteries such as Powerball, where the top prize rises with each week's drawing until someone wins it, at which point the jackpot resets to $40,000,000. The Pick-It pari-mutuel pool was divided in half: 50% went to the state and 50% comprised the prize pool. The approximate payoff to daily winners averaged about 575–1, somewhat better than the odds offered in the Numbers Game [76].

Pick-It lives on today in New Jersey's daily Pick 3 lottery.

## Massachusetts Lottery Numbers Game

In Massachusetts, the Numbers Game was launched in 1976 in an effort to compete with illegal numbers games. A four-digit number is drawn twice daily, in the middle of the day and in the evening, and the pay table includes payoffs for matching the digits in order or out of order. Players may choose their number and the amount they wish to wager when buying a ticket.

The Numbers Game, like Pick-It, is also a pari-mutuel game. The purchase price of each Numbers Game ticket is placed into a prize fund and 63% of the total handle is allocated to prizes, leaving 37% as the state's profit. Matching all four digits drawn in order, with probability 1/10,000, carries a variable payoff which depends on the total amount wagered and on the number of tickets sold bearing the winning number or portion of it. In December 2015, the 62 drawings resulted in payoffs for a $1 wager ranging from $2491 on the evening of December 17 (with winning number 1231) to $7058 midday on December 19 (with winning number 9920) [114].

The pari-mutuel nature of the Numbers Game suggests that, as with games like Powerball (page 277), there is value in choosing a four-digit number that is not likely to be chosen by many other players. In December 2015, the two lowest payoffs for matching all four digits in order were on the numbers 1231 and 1213, which suggests a bias in favor of the sequence 1-2-3. The four highest payoffs all included a zero among their digits, which may mean that players avoid choosing 0s when picking numbers [114].

## Daily Numbers Games

Many state and national lotteries offer their own games which mimic the Numbers Game by inviting players to select a favorite 2–6 digit number and compare it to a number drawn each afternoon or evening. While these games have the considerable advantage of being legal, the house advantage tends to run around 50%, making these games slightly less favorable than the best illegal games.

### Daily 2

In Pennsylvania, *Pick 2* is the simplest of the daily numbers games, asking players to select a two-digit number for drawings held twice a day. The simplest bet is a "straight" bet, where the gambler picks a two-digit number and wagers $1. If the player's number matches the number drawn by the state, the payoff is 50 for 1. Since the pool of two-digit numbers includes numbers with a leading zero such as 00 or 09, the probability of winning on a $1 straight bet is $\frac{1}{100}$. It follows immediately that the state enjoys a 50% advantage over the Pick 2 player.

To improve the chance of winning, at the cost of a lower payoff, the player may choose a *box* bet, which pays off if the player's two digits are drawn in either order, so if the number chosen is 29, both 29 and 92 are winning numbers. The cost of a ticket remains $1, and the chance of winning is doubled to .02, but the payoff odds drop to 25 for 1. Boxing a number is is not an option, nor is it desirable, if the player is betting on a number with two identical digits such as 55, since there would still be only one way to win, and the payoff would be cut in half.

**Example 4.1.** Pick 2 also offers *Front Digit* and *Back Digit* wagers, which pay off at 5 for 1 if the selected digit is matched in the state's drawing. These bets are mathematically identical; the expected value of a $1 digit bet is

$$E = (5) \cdot \frac{10}{100} - 1 = -\$.50,$$

and the house advantage is 50%. ■

## Daily 3

The Michigan Lottery offers a *Daily 3* game, for which players pick a three-digit number. Numbers are drawn by the state twice a day, seven days a week. As with Pennsylvania's Pick 2, the simplest bet is a straight bet, that the state's number exactly matches the player's. Since there are 1000 possible three-digit numbers, from 000 through 999, the probability of winning this lottery is $\frac{1}{1000} = .001$. The expected value of a $1 bet is

$$E = (500) \cdot \frac{1}{1000} - 1 = -\frac{500}{1000} = -\$.50$$

—and so the state takes half of every dollar wagered on a Daily 3 ticket.

**Example 4.2.** Box bets are also an option in Daily 3. A player who "boxes" a three-digit number wins if the three digits comprising that number come up, in any order, in the state's chosen number. For example, if you make a box bet on the number 269, you win if the state's number is 269, 296, 629, 692, 926, or 962. The payoff structure depends on whether the player's number contains two or three different digits; we shall first consider the case where the three digits are all distinct. This leads to what is called a "six-way boxed bet," since there are ${}_3P_3 = 6$ different permutations of a three-digit number consisting of three different digits. This bet pays off at 83 for 1 if any of the six permutations is drawn, and so the expected value is

$$E = (83) \cdot \frac{6}{1000} - 1 = -\frac{502}{1000} = -\$.502,$$

slightly less than the expectation for a \$1 straight bet. This difference in expectation arises due to rounding the prize on a box bet down to the nearest whole number. If a winning ticket paid off at $83\frac{1}{3}$ for 1, then the HA would be exactly 50%. ■

A box bet on a 3-digit number with two copies of a single digit, for example 313, is called a "three-way boxed bet", since there are three ways to rearrange the digits. The payoff is 166 for 1, and the HA is the same as for a six-way boxed bet: 50.2%.

In Missouri, the state lottery offers a Pick 3 and Pick 4 game with a twist. *1-Off* advertising aims to appeal to the frustrated lottery player who would have won, except that his or her number was "one off" the drawn winning number. The bet is a winner if all three of the digits in the state's number either match the player's digits or are within 1, either above or below. For the purposes of this bet, the digit 0 is deemed to be one above the digit 9.

**Example 4.3.** If the player has bet on the number 517, the following numbers are winners:

- **Straight Number**: 517 pays 300 for 1. This should be compared to the 600 for 1 payoff in the Missouri Pick 3 game without the 1-Off option.
- **1-Off One Digit**: 417, 617, 507, 527, 516, and 518 pay 29 for 1.
- **1-Off Two Digits**: 506, 508, 526, 528, 416, 418, 616, 618, 407, 427, 607, and 609 pay 4 for 1.
- **1-Off Three Digits**: 406, 408, 606, 608, 426, 428, 626, and 628 pay 9 for 1.

By a simple counting argument, winning numbers have one of three digits in each of three positions, hence there are $3^3 = 27$ winning numbers, and 973 losers. Similarly, a 1-Off Pick 4 ticket has $3^4 = 81$ winning numbers and 9919 losers.

With a \$1 ticket, the expected value of a 1-Off Pick 3 ticket is

$$E = (300) \cdot \frac{1}{1000} + (29) \cdot \frac{6}{1000} + (4) \cdot \frac{12}{1000} + (9) \cdot \frac{8}{1000} - 1 = -\$\frac{406}{1000},$$

and so Missouri enjoys a 40.6% advantage on each ticket, which is approximately the 40% HA that the state holds on its straight Pick 3 ticket. ■

### Daily 4

Michigan also offers a *Daily 4* lottery game. Like the Daily 3, players pick a four-digit number for one of two drawings per day. Bets on this number may be made for 50¢ or \$1.

- For a straight bet, the player wins if his or her number is exactly the one matched by the state. This pays off at 5000 for 1.

- Another option is the *wheel* bet, which, for \$24, allows a straight bet on every permutation of a four-digit number with four different digits. In effect, this combines 24 separate tickets into one and is thus easier for both the gambler and the clerk who prints the ticket. If any one of the 24 permutations is drawn, the ticket pays \$5000, or 5000 for 24 ($208\frac{1}{3}$ for 1).

- There are several varieties of boxed bets available.

  - A *24-way* boxed bet is made on a four-digit number with four different digits, such as 8427. If any combination of the four digits is chosen, the bet pays off at 208 for 1.
  - A *12-way* boxed bet covers a number which has two copies of one digit and two nonmatching digits, for example, 0226. If any of the 12 rearrangements of the number hits, the payoff is 416 for 1.
  - The *6-way* boxed bet is for a four-digit number with two pairs of identical digits, as in 3113. The payoff on a 6-way box is 833 for 1.
  - Finally, a *4-way* boxed bet can be made on a number where one of the two different digits is repeated three times, like in 7707. This payoff is 1250 for 1.

**Example 4.4.** Compute the HA for the straight and wheel bets. How do they compare to the HA for the straight bets in the Daily 3 game?

For the straight bet, the expectation is

$$E = (5000) \cdot \frac{1}{10,000} - 1 = -\frac{5000}{10,000} = -\$.50,$$

and the HA is therefore 50%.

For a wheel bet, we have

$$E = (5000) \cdot \frac{24}{10,000} - 24 = -\frac{12,000}{10,000} = -\$12,$$

or half the amount wagered. As was the case in the Daily 3 straight bet, the HA is 50%. ■

**Example 4.5.** How do the HAs for the various Daily 4 boxed bets compare to one another and to the HA for the straight bet on the same four-digit number?

If you make an $n$-way boxed bet, there are $n$ numbers that win and $10,000 - n$ that lose. Let the payoff for a specified boxed bet be given as $x$ for 1; your expectation for any of these bets is then

$$E = (x) \cdot \left(\frac{n}{10,000}\right) - 1 = \frac{xn - 10,000}{10,000}.$$

Using the values for $x$ given on page 206 yields the expectations and HAs given in Table 4.1.

TABLE 4.1: Expected return for Michigan Daily 4 box tickets

| $n$ | $x$ | $E$ | HA |
|---|---|---|---|
| 24 | 208 | -\$.5008 | 50.08% |
| 12 | 416 | -\$.5008 | 50.08% |
| 6 | 833 | -\$.5002 | 50.02% |
| 4 | 1250 | -\$.5000 | 50.00% |

All four bets have a house edge close to 50%, as we expect from a lottery. There's a slight increase in the expectation as you bet fewer combinations, but it's scarcely enough to recommend one bet over another. ■

In Indiana, the Hoosier Lottery offers the *Superball* option with its Daily 3 and Daily 4 tickets. For an additional wager equal to the amount wagered on the straight four-digit number (which can range from 50¢ to $10), the player gains access to the Superball, an extra digit drawn after the winning 4-digit number is selected. The Superball digit can be substituted in any position of the drawn number; if any of the four new numbers so formed matches the player's number, the payoff is 1360 for 1.

**Example 4.6.** If the winning number is 1963 and the Superball is 5, the four winning Superball numbers are 5963, 1563, 1953, and 1965. ■

The standard Hoosier Lottery Pick 4 payoff is 5000 for 1, so its HA is 50%. A $1 Superball wager has expectation

$$E = (1360) \cdot \frac{4}{10,000} - 1 = \frac{5440}{10,000} - 1 = -\frac{4560}{10,000} = -\$0.456,$$

so the HA on that wager is lower: 45.6%. Since a Superball wager must be made together with a standard Pick 4 bet, Theorem 2.9 states that the expectations of the two bets can be added to find the total expectation:

$$E = -\$0.50 + -\$0.456 = -\$0.956,$$

which, when divided by the $2 wagered, reveals an overall house advantage of 47.8%.

Superball bets can also be placed in conjunction with various boxed Pick 4 tickets; for example, adding the Superball option to a 24-way boxed Pick 4 ticket pays off $58 if transposing the Superball number into a player's number gives a winning combination. The expectations continue to be additive and the house advantages remain near 50%.

### Daily 5

The District of Columbia Lottery runs *DC5*, a twice-daily 5-digit number game with a wide assortment of play options and payoffs. Tickets may be purchased for either 50¢ or \$1. The lottery advertises "120 ways to win", which counts the number of possible permutations of a 5-digit number with all digits different, as one might choose by boxing that number [30].

The simplest DC5 wager is a straight ticket, which pays 50,000 for 1 if the lottery's number matches yours exactly. Since there are 100,000 possible 5-digit numbers, it is easy to see that the house advantage on this wager by itself is 50%.

DC5 pays off on partial matches as well as on exact matches, and a player may choose any or all of six partial match prizes by wagering separately on each possibility. When any of these options are chosen, the ticket wins a prize if the first or last 2–4 digits of a player's chosen number match the drawn number exactly. The combinations are called "Front" and "Back", depending on the location of the matched digits.

**Example 4.7.** If the winning number is 15861, the following numbers would be winners, where $x$ denotes any digit except the matching one and $y$ denotes any digit.

| **Front** | **Back** |
|---|---|
| $1586x$ | $x5861$ |
| $158xy$ | $yx861$ |
| $15xyy$ | $yyx61$ |

A ticket numbered $15x61$, with $x \neq 8$, could win both the Front Match 2 and Back Match 2 prizes if both options are selected and paid for. ■

To count the number of winning tickets at each match level, note that there are 9 choices for $x$, since this digit cannot match the winning number. Each digit $y$, if any, can be filled in in 10 ways, leading to the values in Table 4.2. The last line of this table includes the 9 two-way Match 2 tickets.

TABLE 4.2: DC5 match ticket payoffs [30]

| **Match** | **Count** | **Payoff** |
|---|---|---|
| 4 | 9 | 5000 |
| 3 | 90 | 500 |
| 2 | 900 | 50 |

The expected return of a \$1 wager on any one of these 6 tickets can be computed with a single equation. Suppose that the ticket matches exactly $n$ numbers at either end. The payoff for such a win is $\$5 \times 10^{n-1} = \$\frac{10^n}{2}$, and the probability of winning is

$$\frac{9 \cdot 10^{4-n}}{100,000} = \frac{9}{10^{n+1}}.$$

The expectation is then

$$E = \binom{10^n}{2} \cdot \frac{9}{10^{n+1}} - 1 = \frac{9}{20} - 1 = -\$\frac{11}{20} = -\$.55,$$

independent of $n$, and the lottery commission holds a 55% advantage.

A variety of boxed bets (Table 4.3) is also available, with payoffs depending on the number of repeated digits in the original number. These can be classified using poker hand terminology.

TABLE 4.3: DC5 box ticket options [30]

| Ticket | Numbers | Example | Payoff |
|---|---|---|---|
| 5 way | 4 of a kind | 00700 | 10,000 for 1 |
| 10 way | Full house | 90909 | 5000 for 1 |
| 20 way | 3 of a kind | 96939 | 2500 for 1 |
| 30 way | 2 pairs | 19009 | 1700 for 1 |
| 60 way | 1 pair | 48101 | 850 for 1 |
| 120 way | No pairs | 52143 | 425 for 1 |

**Example 4.8.** The combinations corresponding to a 30-way ticket can be counted as follows, with the digits denoted $xxyyz$:

- There are ${}_5C_2 = 10$ ways to place the two $x$s.
- Once the $x$s are in position, there are then ${}_3C_2 = 3$ ways to choose where the $y$s go.
- This leaves only 1 choice for $z$. Multiplying gives $10 \cdot 3 \cdot 1 = 30$ ways.

■

These six options have a HA of either 50%, for 5–20 ways, or 49%, for 30–120 ways.

### Daily 6 Bullseye

In New Zealand, the national lottery offers a daily game in which players choose a 6-digit number. The winning 6-digit number drawn each day is called the Bullseye number, and payoffs are made according to how close to the Bullseye a player's number is, with a prize awarded to any ticket bearing a number within 50,000. Table 4.4 shows the pay table, with the top prize starting at NZ$100,000 and rising with each drawing that has no winner. If the Bullseye jackpot rises to $400,000, the next drawing features a "Bullseye Must Be Won" drawing. If the top prize is not won, $400,000 is added to the Division 2 prize pool and evenly divided among all winning tickets at that level. If there are no Division 2 winners, the jackpot continues to roll down

TABLE 4.4: New Zealand Bullseye pay table [111]

| Division | Prize (NZ$) | Winning Range |
|---|---|---|
| 1 | Jackpot | Exact match |
| 2 | 10,000 | Within $\pm 5$ |
| 3 | 500 | $\pm 6$ to $\pm 50$ |
| 4 | 100 | $\pm 51$ to $\pm 500$ |
| 5 | 25 | $\pm 501$ to $\pm 5000$ |
| 6 | Free ticket | $\pm 5001$ to $\pm 50,000$ |

until it reaches a division with winning tickets. Table 4.4 shows the value of Division 2–6 prizes assuming no amount rolls down from Division 1.

The numbers are arranged in a circle for the purposes of determining winners; thus 999999 is adjacent to 000000, and if 999999 is the Bullseye, the Division 2 prize is paid on the numbers 999994–000004. There are 100,001 winning numbers in any drawing, making the probability of winning a prize very slightly more than 10%.

**Example 4.9.** If the Bullseye is 386114, all numbers from 336114 through 436114 are winners, divided among the six prize divisions as follows:

| Division | Winning Range |
|---|---|
| 1 | 386114 |
| 2 | 386109–386113 and 386115–386119 |
| 3 | 386064–386108 and 386120–386164 |
| 4 | 385614–386063 and 386165–386614 |
| 5 | 381114–385613 and 386615–391114 |
| 6 | 336114–381113 and 391115–436114 |

■

The probability of winning a free ticket at the Division 6 level is found by taking the 100,001 prize winning numbers and removing the tickets that win bigger prizes:

$$\frac{100,001 - 10,001}{1,000,000} = \frac{90,000}{1,000,000} = 9\% \approx \frac{1}{11}.$$

The expected value of a NZ$2 ticket with the average jackpot of NZ$180,000 is –NZ$1.30. However, in a "Bullseye Must Be Won" situation, the expected value turns positive.

### Daily 7: Encore

*Encore*, a wagering option offered in Ontario, Canada, is not a game in its own right, but an optional $1 addition to many of the Ontario Lottery's other draw games. Players choosing the Encore option on a ticket receive a random

7-digit number, and the lottery agency draws its number twice a day. There are 22 different ways to win on an Encore ticket, and the payoffs mimic DC5 as they award cash prizes for matching various combinations of the 7 digits. The top prize, for matching all 7 digits, is $1,000,000; the smallest prize, $2, is paid when only the last digit is matched [117].

**Example 4.10.** Matching 6 out of 7 digits is an event with a low probability. If the position of the unmatched digit is specified, then there are 9 choices for that digit and 1 for each other digit, making the probability

$$\frac{9 \cdot 1^6}{10^7} = \frac{9}{10,000,000},$$

but *which* 6 digits you match is critically important in Encore, where there are several prizes for matching 6 of the 7 digits that vary based on which digits are matched.

For this example, assume that the winning number is 8675309 and let $x$ denote any of the 9 digits that does not match the digit in the indicated position. Six combinations are winners.

- $\mathbf{867530x}$ matches the first six digits and pays $10,000.
- $\mathbf{86753x9}$, a match of the first 5 digits and the last digit, pays $502.
- $\mathbf{8675x09}$, matching the first 4 and last 2 digits, pays $55.
- $\mathbf{867x309}$, matching the first 3 and last 3 digits, pays $20.
- $\mathbf{86x5309}$, a match of the first 2 and last 4 digits, pays $105.
- $\mathbf{x675309}$ matches the last six digits and pays $100,000.

Note that $\mathbf{8x75309}$, which matches the first digit and the last 5 with the same $9 \times 10^{-7}$ probability as the winning combinations, pays nothing. ■

**Example 4.11.** Continuing Example 4.10, consider the winning combination $\mathbf{86xyz09}$, which matches the first two and last two digits, but not the middle three. Finding the probability of this event, which pays $10, requires careful work with the rest of the pay table to rule out other winning combinations where the letters $x, y$, and $z$ represent winning digits and thus pay off differently.

- $x$ cannot equal 7, because matching the first three digits together with the last 1–3 digits is a separate winner, but otherwise has no other restrictions. There are 9 choices for $x$.
- There are 10 choices for $y$, because there is no additional payoff for a number of the form $86x5z09$.
- As with $x$, there are 9 choices for $z$. Matching the first 2 and last 3 digits (if $z = 3$) pays $15, nor $10.

Multiplying gives 810 winning combinations of the form $86xyz09$, and a corresponding probability of $\frac{810}{10,000,000}$. ■

The probability of winning something in an Encore drawing is .1090, or 1 chance in 9.17—remember that Encore is purchased along with another lottery ticket, which presumably has a slightly higher likelihood of winning a prize [117]. The expected value of the Encore option is –48.6¢.

---

## 4.2 Passive Lottery Tickets

Unlike the Numbers Game, early state lotteries sold tickets preprinted with numbers rather than allowing patrons to choose their own numbers. These are called *passive* tickets by collectors. Most passive games would later be replaced by daily games or multi-state affairs allowing player choice.

### Virgin Islands Lottery Traditional Game

The U.S. Virgin Islands are home to the longest continuously-running lottery on American territory, which was founded in 1937 [77, p. 2]. The local lottery offers its Traditional Ordinary and Traditional Extra Ordinary games, with passive tickets, as well as participating in Powerball. Tickets are priced at $1.50 and are offered for sale in sheets of 20 for the Ordinary game and 25 for Extra Ordinary, though patrons may purchase any number of tickets they desire and are not restricted to buying full sheets. All tickets on a given sheet bear the same number, so opting for a full sheet (Figure 4.1) means not having to share a prize; 34,000 sheets of tickets, for a total of 680,000 individual tickets, are printed for each drawing of the Ordinary game, and 40,000 sheets (1 million tickets) for Extra Ordinary [154].

Drawings are conducted twice monthly, except for a single drawing in December. A variety of payoffs are on offer in the Ordinary and Extra Ordinary games [158]. As with the DC5 lottery, it is possible to win a prize while matching only some digits of a winning number. However, each ticket is eligible for all partial match prizes with a single payment—it is not necessary to make separate bets on matching some digits as is the case with DC5.

- The **Perfect Match** prize is awarded to tickets bearing any one of the five chosen winning numbers.
  - First prize: $150,000
  - Second prize: $65,000
  - Third prize: $40,000

FIGURE 4.1: Sheet of Virgin Islands Lottery tickets for the 12 March 1992 drawing.

 - Fourth prize: $30,000
 - Fifth prize: $20,000

These prizes are divided among all players holding the drawn number, so if all of the tickets bearing a winning number were sold, each one wins $\frac{1}{20}$ (Ordinary) or $\frac{1}{25}$ (Extra Ordinary) of the total prize. This is where buying a full sheet of tickets eliminates the chance of sharing a prize.

Variations on the five prize numbers can win other prizes.

- **Terminal** prizes are awarded when the last digit on a ticket matches the last digit of a prize number, so if the winning number is 07683, all tickets ending in 3 win a terminal prize. This is a simple refund of the $1.50 cost of the ticket, multiplied by the number of tickets held, so a full sheet in the Ordinary game winning a Terminal prize wins $30.

- **Triple Terminal** prizes are paid on tickets matching the last three digits of any of the five prize numbers; in our example above, these would be tickets ending in 683. The payoff is $500, $300, $200, $120, or $100, depending on which of the prize numbers is a partial match [155]. There are 165 winning Triple Terminal numbers for each drawing.

- **Approximation** prizes are won by tickets within 1 of any winning number. In the Ordinary game, for example, if a prize number is 07683, the numbers 07682 and 07684 pay off: $2500 for approximations to the first-prize number, $1200 for the second-prize number, $900 for third, $600 for fourth, and $300 for fifth [155].

- The **Series** prize is paid to ticket holders with tickets within 50 of a prize number: anything from 07633–07733 in our earlier example. The prize is $100 in the Ordinary game and $200 in the Extra Ordinary game. Approximation prize winners are also eligible for a Series prize.

**Example 4.12.** Advertising for the Virgin Islands lottery claims that an individual ticket has 1 chance in 8 of winning a prize. How accurate is that figure?

We shall consider a simple scenario where none of the five prize numbers generate any overlapping numbers for the smaller prizes. Suppose that the five Perfect Match winning numbers are the winning numbers from the 11 February 2016 drawing:

- First prize: 30338
- Second prize: 33362
- Third prize: 17614
- Fourth prize: 08947
- Fifth prize: 25425.

The numbers in this lottery were in the range 00000–33999. The number of winning tickets may be counted by running through the list of prizes.

- 5 numbers win a Perfect Match prize.
- 3399 numbers—all numbers ending in 8 except for 30338—win a Terminal prize.
- $33 \cdot 5 = 165$ numbers win a Triple Terminal prize.
- 10 numbers (30337, 30339, 33361, 33363, and so on) win an Approximation prize.
- $97 \cdot 5 = 475$ numbers win the Series prize. The ten numbers winning an Approximation prize are removed from the count, making the first factor 97 rather than 99.

This gives a total of 4054 winning numbers. Dividing by 34,000 shows that the probability of winning is approximately .1192, so the chance of winning is about 1 in 8.39—acceptably close to 1 in 8.

■

As with the Newbury Lottery (page 11), the HA in the Virgin Islands can be calculated by comparing the total ticket income to the prize pool. For the Ordinary game, the prize structure above results in $704,300 in prizes, paid on \$780,000 in ticket sales if all tickets printed for a given drawing are purchased [155]. The Ordinary game thus returns 90.3% of ticket proceeds to players, for an effective house advantage of only 9.7%. In the Extra Ordinary game, \$1,342,800 is prizes is paid out based on \$1,500,000 in sales; a payoff percentage of 89.5% and a HA of 10.5%[153].

## American Lotteries Return: New Hampshire Sweepstakes

The first modern state lottery in the USA was run by New Hampshire, beginning in 1964. The New Hampshire Sweepstakes was tied to a $1\frac{3}{16}$-mile horse race of the same name held at Rockingham Park in Salem, New Hampshire. Tickets were sold for \$3, and a drawing was held when \$1 million in sales was recorded—after every 333,334 tickets were sold. For the first race, 332 three-year-old thoroughbreds were nominated, and so 332 tickets were drawn, each one assigned to one of the nominated horses [38]. Each of these ticket holders won a cash prize even if his or her horse did not run in the Sweepstakes; the largest prize was reserved for the ticket corresponding to the winning horse.

The probability of winning any prize at all, assuming that the bettor bought only one ticket, can be found with the Complement Rule. The probability of *losing* is simply the chance that all 332 winners are chosen from the 333,333 other tickets sold; this is

$$p = \frac{{}_{333,333}C_{332}}{{}_{333,334}C_{332}} = \frac{166,501}{166,667};$$

and so the probability of winning is

$$1 - p = \frac{166}{166,667} \approx \frac{1}{1004}.$$

**Example 4.13.** New York resident Martin Zayacher had two tickets drawn in the first drawing [38]. Find the probability of this event, assuming that he purchased only two tickets.

The probability $p$ that two tickets are drawn is

$$p = 1 - P(\text{Neither ticket is drawn}) - 2 \cdot P(\text{Only one ticket is drawn}),$$

where the second term accounts for the cases that either ticket A or B is drawn alone. Rewrite this equation as $p = 1 - p_0 - 2 \cdot p_1$.

We have

$$p_0 = \frac{{}_{333,332}C_{332}}{{}_{333,334}C_{332}} = \frac{55,444,999,501}{55,555,611,111}$$

and

$$p_1 = \frac{{}_{333,332}C_{331}}{{}_{333,334}C_{332}} = \frac{55,278,332}{55,555,611,111};$$

consequently,

$$p = \frac{54,946}{55,555,611,111} \approx \frac{1}{1,011,094}.$$

If we assume that the large number of tickets makes the probability of a single ticket being drawn at any time constant, the probability that two tickets are both drawn is approximately

$$\left(\frac{1}{1004}\right)^2 = \frac{1}{1,008,016}.$$

This value is within .3% of the true probability—acceptably close, and far easier to compute. ■

The volume of tickets sold in the first year of the New Hampshire Sweepstakes led to six drawings for the first race, with six winners assigned to each of the 332 nominated horses. Eleven horses eventually ran the race; Roman Brother won by half a length, and six bettors won $100,000 each [38].

While sweepstakes organizers hoped that the New Hampshire Sweepstakes would become a fourth jewel in thoroughbred racing's Triple Crown, the novelty soon wore off, and competition from newly-legalized lotteries in other states led to the end of the Sweepstakes, with the last race running in 1972. The New Hampshire Lottery continued, but the link to live horse racing ended.

## Michigan Green Ticket Game

The first state lottery drawing in Michigan, the "Green Ticket" game, was held on 24 November 1972. The 50¢ passive tickets (Figure 4.2) bore two different three-digit numbers, and that first drawing produced the winning numbers 130 and 544.

FIGURE 4.2: Michigan Lottery's "Green ticket".

Tickets matching one of the two numbers won $25, while ticket holders matching both numbers were entered into a superdrawing with a top prize of $200,000, three $10,000 prizes, and one or more $50,000 prizes—a $50,000 prize was added for each additional qualifying ticket past five [93].

Twelve people matched both numbers, resulting in eight $50,000 prizes in the first drawing.The average return from the drawing was simply the mean of the 12 prizes on offer, or \$52,500. The expected return on the ticket was then

$$E = (25) \cdot 2 \cdot \frac{2}{1000} \cdot \frac{998}{999} + (52,500) \cdot \frac{2}{1000} \cdot \frac{1}{999} - .50 \approx -\$.295.$$

The state's edge on this ticket was approximately 59%.

An additional feature of the Green Ticket game was a million-dollar drawing, conducted from all \$25 winning tickets after 30,000,000 tickets had been sold; 120 prizes were available in this drawing [93]:

| Prize | Number |
|---|---|
| \$1,000,000 | 1 |
| \$100,000 | 1 |
| \$50,000 | 1 |
| \$5000 | 7 |
| \$1000 | 110 |

With 30,000,000 tickets sold, lottery officials expected

$$30,000,000 \cdot P(\text{Match 1 number}) = 30,000,000 \cdot \left(\frac{3992}{999,000}\right) \approx 119,880,$$

or approximately 120,000, tickets to qualify for the million-dollar drawing, so roughly one out of 1000 tickets would win an additional prize.

## Buckeye 300

In the 1970s, many states starting lotteries included prize drawings for which holders of winning tickets could qualify, as in Michigan's Green Ticket game. Ohio's first lottery game debuted in 1974 with two bonus drawings. A *Buckeye 300* ticket gave players five 3-digit numbers, divided into 3 numbers printed in blue boxes and 2 numbers in green boxes. In Figure 4.3, the blue numbers are in the first, third, and fifth boxes (886, 803, and 724), and the green numbers in the second and fourth (197 and 915).

FIGURE 4.3: Ohio Lottery's Buckeye 300 ticket. The "Commemorative Issue" imprint was used on tickets for the first drawing, on August 22, 1974.

Each week, the state drew three numbers:

- The first number drawn was the Single Number. A ticket matching this number in any of its 5 positions won \$20 and an entry into the Millionaire Pool drawing. The probability of matching the Single Number was $\frac{5}{1000} = \frac{1}{200}$; it was estimated that up to 25,000 people each week would win a \$20 prize [116].

  Several \$20-winning tickets each week were selected for a further Millionaire Pool prize; the prize drawing was held when 100 winners accumulated. The 100 Millionaire Pool prizes are listed in Table 4.5.

  TABLE 4.5: Buckeye 300 Millionaire Pool prizes [116]

  | **Prize** | **Count** |
  |---|---|
  | \$1,000,000 | 1 |
  | \$100,000 | 1 |
  | \$50,000 | 1 |
  | \$10,000 | 7 |
  | \$2000 | 10 |
  | \$1000 | 80 |

  In the first drawing, on August 22, 1974, the Single Number was 178 [36].

- The next numbers were the Double Numbers. Tickets matching both numbers of the pair were paid according to the color of the numbers:

  - If one number was blue and the other green, the prize was \$500. The probability of a blue-green pair was

    $$\frac{3 \cdot 2}{{}_{1000}C_2} = \frac{1}{83,250}.$$

    In the numerator, we count 3 ways to pick the blue box and 2 ways to pick the green box.

  - If both numbers were blue, the prize was \$1000. The chance of matching two blue numbers was

    $$\frac{{}_3C_2}{{}_{1000}C_2} = \frac{1}{166,500}.$$

  - If both numbers were green, an event with probability

    $$\frac{1}{{}_{1000}C_2} = \frac{1}{499,500},$$

    the ticket qualified for entry into a bonus drawing. Up to 12 tickets

> each week qualified for this weekly drawing, which offered a top prize of $300,000 and additional prizes of $60,000, $30,000, and as many $15,000 prizes as needed to offer a prize to each ticket.
>
> Since Buckeye 300 was a passive game, lottery officials and contractors could control the numbers printed on the tickets to limit the number of tickets in the bonus drawing. Imposing a maximum of 12 qualifying tickets meant that each of the 499,500 two-number combinations could appear in the green boxes on 12 tickets. The total number of possible tickets each week was then $12 \cdot 499,500 = 5,994,000$, a number surely adequate for the gambling population of Ohio during the duration of Buckeye 300.

The first pair of Double Numbers was 002 and 264 [36].

Two bonus drawings make computing the expected value of this 50¢ ticket challenging. In the bonus drawing for tickets matching both Double Numbers in green, the average prize, given $n$ eligible tickets, is

$$\frac{300,000 + 60,000 + 30,000 + 15,000 \cdot (n-3)}{n} = \frac{345,000 + 15,000n}{n},$$

provided that $3 \leqslant n \leqslant 12$. If you hold one of the $n$ tickets in this drawing, your expected return is

$$E = \frac{1}{n} \cdot \left(\frac{345,000 + 15,000n}{n}\right),$$

which is a maximum of $130,000 at $n = 3$ and decreases to $43,750 with the maximum number of 12 tickets.

The probability of a single ticket qualifying for this drawing, is $\frac{1}{499,500}$—making that rather a longshot, as one might expect. Multiplying the payoffs by this probability shows that the bonus drawing contributes anywhere from 26¢ (at $n = 3$) down to 8.76¢ ($n = 12$) to the final expectation.

## Superplay

In 1981, the Michigan Lottery launched its Superplay game, with three numbers per ticket. Each $1 ticket bore a red one-digit number, a white two-digit number, and a blue three-digit number (Figure 4.4).

The one-digit numbers repeated every ten tickets, which meant that buying 10 or more tickets at once would guarantee at least one win.

The prizes for Superplay were based on the numbers matched and are listed in Table 4.6. Matching all three numbers on a single ticket gained entry into a televised prize drawing, with prizes ranging from $10,000 to $1,000,000. As with 23 For A Dollar, it was possible to win more than one prize on a single ticket.

FIGURE 4.4: Michigan Lottery's Superplay ticket

TABLE 4.6: Prize table for Michigan Lottery's Superplay game

| **Numbers Matched** | **Prize** |
|---:|---|
| Red | Free ticket |
| White | \$10 |
| Blue | \$100 |
| Red & White | \$10 + free ticket |
| Red & Blue | \$500 |
| White & Blue | \$5000 |
| Red, White, & Blue | Prize Drawing Finals entry |

**Example 4.14.** Construct a probability distribution for Superplay.

The probability of each of the eight outcomes (seven winning and one losing) is the product of three probabilities, since we must rule "win" or "lose" on each of three independent drawings. For example:

$$P(\text{Free ticket}) = \frac{1}{10} \cdot \frac{99}{100} \cdot \frac{999}{1000} = \frac{98,901}{1,000,000} = .098901.$$

Similarly,

$$P(\$500) = \frac{1}{10} \cdot \frac{99}{100} \cdot \frac{1}{1000} = \frac{99}{1,000,000} = .000099.$$

The complete probability distribution is shown in Table 4.7.

■

## 23 For A Dollar

From 1979 until 1982, the Pennsylvania Lottery ran the passive game *23 For A Dollar*. As the name suggests, this \$1 ticket, shown in Figure 4.5,

TABLE 4.7: Probability distribution for Superplay

| **Prize** | **Probability** |
|---|---|
| Free ticket | .098901 |
| $10 | .008991 |
| $100 | .000891 |
| $10 + free ticket | .000999 |
| $500 | .000099 |
| $5000 | .000009 |
| Prize Drawing Finals entry | .000001 |
| Nothing | .890109 |

offered 23 chances to win. A single ticket included 7 different 3-digit numbers, 8 4-digit numbers, and 8 5-digit numbers.

FIGURE 4.5: Pennsylvania Lottery's 23 For A Dollar ticket.

Once a week, the state drew one number of each of the three lengths. Matching the 3-digit number won $50, matching the 4-digit number won $100, and matching the 5-digit number won $500 [1].

It was possible to win more than one prize on a single 23 For A Dollar ticket. While the probability of matching the 3-digit number is a simple .007, the probability of winning $50 is slightly less, as we must account for losing the other two prizes. The probability of winning exactly $50 is

$$\frac{7}{1000} \cdot \frac{9992}{10,000} \cdot \frac{99,992}{100,000} \approx .006994.$$

This is approximately .08% less than .007.

The probability that a 23 For A Dollar ticket wins no prizes is quite high:

$$\frac{993}{1000} \cdot \frac{9992}{10,000} \cdot \frac{99,992}{100,000} \approx .9921,$$

or over 99%. The expected value of a 23 For A Dollar ticket is –52¢, entirely in line with state lottery expectations.

## Minnesota Millionaire Raffle

*Raffles* are a familiar form of lottery frequently used as charitable fundraisers, where the probability of winning depends on the number of tickets sold. In Minnesota, the Millionaire Raffle is a passive game drawn once a year. Seven hundred thousand consecutively-numbered tickets are printed for each year's drawing, and the winning numbers are drawn when all tickets have been sold, or on New Year's Day if fewer than 700,000 tickets are sold. The lottery offers 4164 prizes; 42 of these are designated "Bonus Prizes" and award merchandise prizes valued at $4000 or more, including new cars, instead of cash. Tickets are priced at $10 each, and $3,560,331.28 is allocated to prizes, leaving approximately half the proceeds from ticket sales as profit for the state [100]. Two million dollars are allocated to the top two prizes of $1,000,000 each.

The probability of winning any specified prize in the Millionaire Raffle is simply $1/N$, where $N$ is the number of tickets sold, which will not exceed 700,000. Assuming this maximum number of tickets sold, the chance of winning something is the probability of your ticket being drawn among the 4164 winners; once again, we approach this question by using the Complement Rule:

$$P(\text{Win}) = 1 - P(\text{Lose}) = 1 - \frac{{}_{699,999}C_{4164}}{{}_{700,000}C_{4164}} = 1 - \frac{\frac{699,999!}{695,835!\cdot 4164!}}{\frac{700,000!}{695,836!\cdot 4164!}}.$$

Though the numerator and denominator of this fraction are both on the order of $10^{11,068}$, the factorials mean that the expression can be greatly simplified:

$$P(\text{Win}) = 1 - \frac{173,959}{175,000} = \frac{4164}{700,000}$$

—which is the value found by simply dividing the number of prizes by the number of tickets.

This probability also emerges from a different approach to the problem. Once again, we consider the probability of not winning any prizes. The probability of losing the first prize is easily found to be

$$p_1 = \frac{699,999}{700,000}.$$

We now let conditional probability take over. The probability of losing the second prize, given that the first prize was lost, is

$$p = P(\text{Lose second prize} \mid \text{Lose first prize}) = \frac{699,998}{699,999},$$

and so the probability of losing *both* of the first two prizes is

$$p_2 = p_1 \cdot p = \frac{699,998}{700,000}.$$

Continuing in this vein gives

$$p_3 = p_2 \cdot \frac{699,997}{699,998} = \frac{699,997}{700,000}$$

for the probability of losing the first three prizes, and we see a pattern developing: The probability of losing the first $n$ prizes is then easily shown to be

$$p_n = \frac{700,000 - n}{700,000}.$$

Setting $n = 4164$ gives the probability of losing all of the prizes:

$$p_{4164} = \frac{700,000 - 4164}{700,000} = \frac{695,836}{700,000},$$

from which it follows that the probability of winning a prize is

$$1 - p_{4164} = \frac{4164}{700,000}.$$

No matter how you calculate it, of course, this probability is still quite small: approximately .0059, or roughly .6%.

## Premium Bonds

One of the world's largest raffles is conducted monthly in Great Britain. *Premium Bonds* are issued by the British government, in a fund-raising practice common to many governments. Instead of taking the traditional step of paying interest on its bonds, Great Britain takes the money that would have been allocated to interest payments and distributes it in a monthly drawing from all Bonds held by purchasers. Bonds cost £1 apiece, and must be purchased in lots of at least 100; patrons are limited to £50,000 in bond holdings. At any time, a Bond may be cashed in for its purchase price, which led British Chancellor of the Exchequer Harold MacMillan to declare, on introducing Premium Bonds in 1956, "This is not gambling, for the subscriber cannot lose." [50] Of course, inflation eats away at the value of a bond held for any great length of time; the chance of winning a prize between £25 and £1,000,000 is deemed by many to be sufficient compensation for the lack of regular interest payments and value lost to inflation. An additional benefit to Premium Bonds is that prizes won in the monthly drawings are tax-free.

The parameters of the monthly drawing are based on a fixed probability of any one bond winning a prize; this is currently set at 1 in 30,000. Bonds are eligible for only one prize per month, but may win prizes in multiple drawings until they are cashed in. Over 2 million prizes, including two top prizes of £1,000,000, are awarded monthly, which means that the number of outstanding bonds is over 60 billion. The current interest rate paid into the Premium Bonds prize fund is 1.25% annually, or roughly .1% per month, so approximately £60 million in prizes are distributed each month [102]. The vast majority of these prizes are £25; Table 4.8 shows the distribution of prizes for December 2016, when 2,114,608 prizes totalling £66 million were awarded.

TABLE 4.8: Premium Bond prizes: December 2016 [103]

| **Prize** | **Count** |
|---|---|
| £1,000,000 | 2 |
| £100,000 | 3 |
| £50,000 | 4 |
| £25,000 | 11 |
| £10,000 | 27 |
| £5000 | 52 |
| £1000 | 1321 |
| £500 | 3963 |
| £100 | 67,433 |
| £50 | 67,433 |
| £25 | 1,974,359 |

Winning Bond numbers are chosen by a computer dubbed ERNIE: **E**lectronic **R**andom **N**umber **I**ndicator **E**quipment. Through four generations of computers, ERNIE has always provided truly random numbers, rather than the pseudorandom numbers generated by most computers. Rather than selecting winning numbers randomly from a list of all valid Bond numbers, ERNIE uses thermal noise generation in a small electrical circuit—a genuinely random signal—to generate a list of 10-digit numbers, some of which correspond to Bonds [50]. Figure 4.6 is a British envelope commemorating the 50th anniversary of the first Premium Bonds drawing on June 1, 1957, and shows the first ERNIE.

FIGURE 4.6: Envelope commemorating the first Premium Bonds drawing.

**Example 4.15.** A Bond owner with the maximum permissible holding of £50,000 is, in effect, paying £52.08 per month in forgone interest for 50,000 shots at the array of prizes. Since the number of Bonds is so large compared to the owner's holdings, we may build a simple yet acceptably accurate model by assuming the the chance of any Bond winning is independent of the others. With a fixed probability of 1/30,000 of winning a prize with any one Bond, the probability of scoring no wins in a given month is

$$\left(\frac{29,999}{30,000}\right)^{50,000} \approx .1889,$$

or about 19%.

Someone with the minimum holding of 100 Bonds has a probability of

$$\left(\frac{29,999}{30,000}\right)^{100} \approx .9967$$

of winning no prizes in each monthly drawing; their chance of winning one or more prizes is about .33%. ■

**Example 4.16.** Any single given Bond has a 1/30,000 chance of winning a prize each month. How many months, on the average, will it take for that Bond to have a 50% chance of having won something?

We seek to solve for $n$ in the equation

$$1 - \left(\frac{29,999}{30,000}\right)^{n} = .5,$$

or

$$\left(\frac{29,999}{30,000}\right)^{n} = .5.$$

The solution is

$$n = \frac{\ln .5}{\ln\left(\frac{29,999}{30,000}\right)} \approx 20,794.07 \text{ months},$$

corresponding to 1732.84 years' worth of drawings. ■

## Instant Lottery Tickets

The raffle model can also be used to examine instant lottery tickets, which are sometimes called "scratchers". These tickets are designed for immediate play; the player scratches off a thin latex coating to reveal numbers, dollar amounts, or special symbols, certain combinations of which correspond to prizes. Instant tickets were first sold in the USA in Massachusetts, beginning in 1974.

With instant tickets, the odds of winning depend on the number of tickets available and the number of prizes not yet won, which can be deduced from publicly-available information such as state lottery Web sites. These values are set at the start of the game, but will fluctuate over the game's lifetime. Depending on when the prizewinning tickets are sold, the odds of winning a given prize may change significantly, possibly dropping to 0 if all of the prizes are sold early on.

**Example 4.17.** The Michigan Lottery's Holly Jolly Jackpot instant game (Figure 4.7) is a Christmas-themed \$1 ticket issued in the fall during the runup to the winter holidays.

FIGURE 4.7: Michigan Lottery's Holly Jolly Jackpot instant ticket, 2015.

Table 4.9 shows the prizes offered in the 2015 game and the number of each prize available.

TABLE 4.9: Prize and ticket information for 2015 Holly Jolly Jackpot instant lottery tickets [94]

| Prize | Count |
|---|---|
| \$5000 | 5 |
| \$150 | 96 |
| \$90 | 479 |
| \$50 | 1896 |
| \$30 | 6168 |
| \$20 | 42,749 |
| \$15 | 21,352 |
| \$10 | 142,341 |
| \$5 | 313,349 |
| \$3 | 769,289 |
| \$1 | 1,716,098 |

Holly Jolly Jackpot thus offered a total of 3,013,822 prizes. The lottery advertised that the overall odds of winning were 1 in 4.73 [94]. Multiplying

3,013,822 by 4.73 indicates that approximately 14,255,378 tickets were printed. It follows that the probability of any one ticket winning the $5000 grand prize was $\frac{5}{14,255,378}$, or approximately 1 in 2,851,076. ■

On April 19, 2016, the lottery Web site noted that only 1 $5000 ticket remained unsold. On that date, 27.96% of all prizes remained; assuming an even distribution of winning tickets would mean that 3,985,115 tickets were still available. The chance of winning the jackpot in April had fallen to 1 in 3,985,115, roughly 71.5% of the probability when the game was launched in October 2015.

## Change Game and Change Play

Other than instant lottery tickets, passive tickets declined in popularity with the rise of games that allowed players to choose their own numbers. Some passive games which remained past the launch of such games as the Massachusetts Numbers Game and Powerball were tied to an unusual game element: something more than just picking numbers. For a brief time in the early 2000s, the Georgia and Michigan lotteries offered their versions of a raffle-like game with a variable ticket price, called *Change Game* in Georgia from 2002–2004 and *Change Play* in Michigan from 2002–2003. Based on the finding that the average American handled $1.65 in loose change per day, these games were designed as an impulse purchase for patrons with change on their hands from other purchases [44]. Tickets could be had for any amount from 25–99¢ that the player wished; the ticket amounts were intended to facilitate ticket sales at lottery outlets by allowing patrons to roll the coins from a non-lottery purchase into a passive ticket.

Both Change Game, which was the first to launch, and Change Play were pari-mutuel games. Tickets in both lotteries contained a unique combination of two letters and four digits, making for 6,760,000 different possible tickets. Change Game chose three winners per day, and Change Play picked six, from among all of the day's tickets. The winners divided the prize pool in proportion to the amount they had spent on their tickets. In Michigan, the prize pool was 64% of the day's wagers [2]; in Georgia, Change Game paid out 45% of wagers as prizes [43, p. 17].

**Example 4.18.** For the Georgia game, suppose that the three winners had paid 36, 93, and 95 cents for their tickets, and that the total prize pool came to $11,200. The winners would divide the prize in proportions of their wager divided by the total sum of the winning wagers: $2.24. The three prizes would be, respectively, $1800, $4650, and $4750. ■

Both games fell prey to lack of interest among players, with declining sales leading to smaller prizes, which worked to depress sales further. Change Play lasted less than a year, while Change Game only ran for two years.

## Poker-based Lottery Games

A number of state lotteries offer games that are based on 5-card poker hands. In Arizona, *5 Card Cash* gives each player five cards from a standard deck and offers two chances to win. Tickets win immediately if the poker hand on the card matches a standard poker hand of a pair of jacks or better, as listed in Table 4.10. For this reason, games such as 5 Card Cash are necessarily passive games, with no room for the player to choose his or her cards.

TABLE 4.10: Five-card poker hands

| **Hand** | **Description** |
|---|---|
| Royal flush | AKQJ10 of the same suit |
| Straight flush | Five cards of the same suit and in sequence |
| Four of a kind | Four cards of the same rank |
| Full house | Three of a kind and a pair |
| Flush | Five cards of the same suit, not in sequence |
| Straight | Five cards in sequence, not all the same suit |
| Three of a kind | Three cards of the same rank |
| Two pairs | Two pairs of cards of different ranks |
| One pair | Two cards of the same rank |
| High card | Five cards of different ranks, not all the same suit |

The probability distribution for five-card poker hands and the payoffs for 5 Card Cash are shown in Table 4.11. A "losing hand" here is defined as a pair of 10s or lower, or a high-card hand.

TABLE 4.11: Arizona 5 Card Cash probability distribution [6]

| **Hand** | **Payoff** | **Probability** |
|---|---|---|
| Royal flush | \$5000 | $1.539 \times 10^{-6}$ |
| Straight flush | \$500 | $1.385 \times 10^{-5}$ |
| Four of a kind | \$100 | $2.401 \times 10^{-4}$ |
| Full house | \$75 | $1.441 \times 10^{-3}$ |
| Flush | \$40 | $1.965 \times 10^{-3}$ |
| Straight | \$20 | $3.925 \times 10^{-3}$ |
| Three of a kind | \$5 | .0211 |
| Two pairs | \$4 | .0475 |
| Pair, jacks or better | \$2 | .1300 |
| Losing hand | \$0 | .7943 |

Additionally, five cards are chosen at random by lottery officials each evening, and players may win by matching 2–5 of them. The probability distribution for this game may be computed by generalizing from keno probabilities: instead of catching $k$ numbers from 20 drawn from 1–80, we look for the number of cards matched from 5 drawn from a standard 52-card deck.

$$P(\text{Match } k \text{ cards}) = \frac{{}_5C_k \cdot {}_{47}C_{5-k}}{{}_{52}C_5}.$$

Matching all 5 cards, an event with probability $\frac{1}{2,598,960}$, wins $100,000. Matching 4, 3, or 2 cards wins $500, $25, and $2, respectively [6].

Since each $2 ticket covers two independent games, the expected value of a single ticket can be computed by finding the expected win from each game without concern for the ticket price, adding these numbers, and then subtracting $2 at the end. The expected win from the instant game is 85.93¢, and the expected win from the drawing is 31.25¢. The expected value of a ticket is then −82.8¢, and so the ticket as a whole carries a 41.41% house advantage.

Several other states and provinces operate games similar to 5 Card Cash, with varying payoff amounts and consequently, varying house advantages. The game is called Poker Lotto in Indiana and Michigan, Poker Pick in Oklahoma, Lotto Poker in Quebec, and is also called 5 Card Cash in Kentucky and Maryland.

Table 4.12 compares the pay tables of the various games. Two states, Maryland and Oklahoma, offer a free ticket as the prize for a pair of jacks or better. This amounts to an equivalent prize for a high pair of $1.01 and $0.84, respectively–a lesser prize on the most probable win than the break-even payoff offered in the other games.

**Example 4.19.** In Kentucky, 5 Card Cash uses a slightly different pay table, Table 4.13, which awards a free ticket on a pair of 9s or 10s as well as a pair of jacks or better. This apparent generosity is more than balanced by decreased payoffs elsewhere in the pay table.

Tickets for this lottery cost $3, which leads to a HA of 54.42%—the highest of all of these games. ■

## Lotto :D

A Canadian passive game played with computer-simulated dice is *Lotto :D*, which is operated in Quebec [81]. Lotto :D combines an instant dice game, the Quick Play, with an evening drawing. This drawing is not a standard lottery where players pick numbers; it is also based on dice.

For the Quick Play, nine dice are rolled. The first of these is red and is known as “Dé rouge” (red die). The remaining eight dice, dubbed “Vos dés” (your dice) are white and are the player’s dice. The player wins a fixed amount, as shown in Table 4.14 (page 232) for a $2 bet, if two or more of the white dice match the red die.

The probabilities of each of these outcomes can be calculated using the

TABLE 4.12: Poker lottery pay tables

| | Location | | | | | |
|---|---|---|---|---|---|---|
| **Hand** | AZ | IN | MD | MI | OK | QC |
| Royal flush | 5000 | 5000 | 10,000 | 5000 | 10,000 | 5000 |
| Straight flush | 500 | 500 | 1000 | 1000 | 1000 | 500 |
| 4 of a kind | 100 | 100 | 150 | 150 | 100 | 100 |
| Full house | 75 | 75 | 75 | 75 | 50 | 50 |
| Flush | 40 | 40 | 50 | 50 | 30 | 30 |
| Straight | 20 | 20 | 25 | 20 | 10 | 20 |
| 3 of a kind | 5 | 5 | 5 | 5 | 5 | 10 |
| 2 pairs | 4 | 4 | 3 | 3 | 3 | 4 |
| Pair, JQKA | 2 | 2 | Ticket | 2 | Ticket | 2 |
| **Match** | AZ | IN | MD | MI | OK | QC |
| 5 | 100,000 | 250,000 | 100,000 | 100,000 | 100,000 | 100,000 |
| 4 | 500 | 500 | 1000 | 500 | 1000 | 500 |
| 3 | 25 | 20 | 30 | 25 | 20 | 20 |
| 2 | 2 | 2 | 3 | 3 | 4 | 2 |
| **Price** | 2 | 2 | 2 | 2 | 2 | 2 |
| **Expectation** | –$.83 | –$.79 | –$.81 | –$.77 | –$.96 | –$.80 |
| **HA** | 41.41% | 39.57% | 40.51% | 38.74% | 47.88% | 39.96% |
| **Source:** | [6] | [57] | [89] | [95] | [115] | [82] |

TABLE 4.13: Kentucky 5 Card Cash pay table [68]

| **Hand** | **Payoff** |
|---|---|
| Royal flush | 5000 |
| Straight flush | 500 |
| Four of a kind | 200 |
| Full house | 75 |
| Flush | 50 |
| Straight | 30 |
| Three of a kind | 5 |
| Two pairs | 4 |
| Pair, 9s or better | Ticket |
| **Match** | **Payoff** |
| 5 | 100,000 |
| 4 | 500 |
| 3 | 25 |
| 2 | 4 |

TABLE 4.14: Pay table for Lotto :D's Quick Play instant game [81]

| Match | Prize |
|---|---|
| 8 | \$2000 |
| 7 | \$200 |
| 6 | \$100 |
| 5 | \$10 |
| 4 | \$6 |
| 3 | \$3 |
| 2 | \$2 |

binomial formula with $n = 8$ and success probability $p = \frac{1}{6}$:

$$P(\text{Match } k \text{ dice}) = {}_8C_k \cdot \left(\frac{1}{6}\right)^k \cdot \left(\frac{5}{6}\right)^{8-k}.$$

The expected number of matching dice is seen to be $np = \frac{4}{3}$, making this bet, on average, a losing proposition.

**Example 4.20.** Find the probability of losing or breaking even at Quick Play.

This event corresponds to 0, 1, or 2 of the white dice matching the red die. This probability is

$$\left(\frac{5}{6}\right)^8 + 8 \cdot \left(\frac{1}{6}\right) \cdot \left(\frac{5}{6}\right)^7 + 28 \cdot \left(\frac{1}{6}\right)^2 \cdot \left(\frac{5}{6}\right)^6 \approx .7256.$$

■

The probability of winning at Quick Play is then about $1 - .7256 = .2744$.

In the evening, a player's dice are converted into an 8-digit number with digits in the range 1–6, and the lottery draws its own 8-digit number from $6^8 = 1,679,616$ possibilities. Matching all 8 digits wins a \$25,000 payoff for a \$2 wager or \$100,000 for a bet of \$5. Smaller payoffs are won if two or more digits of the player's number, beginning at either end, match the drawn number.

**Example 4.21.** If the winning number is 21143626, the following numbers are also winners, where $x$ denotes any number from 1–6 except for the matching digit and $y$ any digit from 1–6:

| **Front end match** | **Back end match** |
|---|---|
| 2114362$x$ | $x$1143626 |
| 211436$xy$ | $yx$143626 |
| 21143$xyy$ | $yyx$43626 |
| 2114$xyyy$ | $yyyx$3626 |
| 211$xyyyy$ | $yyyyx$626 |
| 21$xyyyyy$ | $yyyyyx$26 |

TABLE 4.15: Pay table for Lotto :D's Quick Play evening drawing [81]

| **Match** | **Payoff** | |
|---|---|---|
| **(first or last)** | **$2 ticket** | **$5 ticket** |
| 7 | $1000 | $2000 |
| 6 | $100 | $150 |
| 5 | $20 | $50 |
| 4 | $10 | $15 |
| 3 | $5 | $10 |
| 2 | $2 | $5 |

How many winning numbers are there?

For the Match 7 payoffs, there are $6-1=5$ choices for $x$—any number but the one in the 8-digit winning number. See Table 4.15. Multiplying by 2 accounts for the front end and back end matches, giving 10 tickets. For Match 6, of the 36 possible combinations for the two $x$s, $5\cdot 6=30$ are valid Match 6 winners, since the $x$ cannot match the winning number and the $y$ can be any digit from 1–6. Again, we multiply by 2, to get 60 winning Match 6 tickets.

In general, in a ticket with an $x$ and $k$ $y$s, the $x$ can be chosen from 5 possibilities, while any of the 6 numbers can be used to fill in each of the $y$s. Including the 1 Match 8 winning number and the factor of 2 gives a total of

$$1+2\cdot\left(5+5\cdot 6+5\cdot 6^2+5\cdot 6^3+5\cdot 6^4+5\cdot 6^5\right)=1+10\cdot\left(\frac{6^6-1}{5}\right)=93,311.$$

We conclude that approximately 5.56% of all possible tickets bear winning numbers. ■

## 4.3 Lotto Games

"Lotto" is the name given to a class of lottery games that resemble keno in that the player may choose his or her numbers—or cities, dates, hockey teams, playing cards, or race horses, as the games call for—from a range specified by the lottery agency. Some of these draw directly from keno, in that players choose a set of $r$ numbers in the range from 1 to $s$; these are called $r/s$ *lotteries.* The player goal is to match as many of the numbers drawn by the lottery commission as possible.

**Example 4.22.** In New Mexico, Roadrunner Cash calls for players to choose 5 numbers in the range 1–37, making this a 5/37 game. ■

**Example 4.23.** A number of lotto games require players to pick from two separate lists of numbers. Powerball (page 277) currently asks players to pick 5 numbers from 1–69 and one number from 1–26. We denote this as a $5/69 + 1/26$ lottery. ■

The forerunner of the modern $r/s$ lottery is a 6/90 game first played in Genoa, Italy in the 16th century. This game took hold throughout Italy and has evolved further into *Lo Giuoco del Lotto*, which offers eleven different 6/90 drawings thrice weekly. Ten of the drawings, called "wheels", bear the names of prominent Italian cities: Bari, Cagliari, Florence, Genoa, Milan, Naples, Palermo, Rome, Turin, and Venice; the eleventh is the national wheel, or "Ruota Nazionale" [79]. Players must specify which wheel they are playing when buying a ticket, though it is certainly possible to bet on more than one wheel in a given drawing.

Most modern $r/s$ lotteries, except for games that are directly based on keno, give the player usually no choice of the value of $r$: the player must choose the same specified number of numbers that the state draws. Beginning with its origins in Genoa, Lo Giuoco del Lotto is less restrictive, and allows bets on 1–5 numbers as the various wheels each pick 5 numbers. The following wagers are available; they pay off only if all of a player's chosen numbers appear on the specified wheel [137]:

- **Ambata**, a bet on one number, which pays 11.23 for 1.
- **Ambo**, a two-number bet paying 250 for 1.
- **Terno**, three numbers, with a 4250 for 1 payoff.
- **Quaterno**, four numbers paying 80,000 for 1.
- **Cinquina**, five numbers offering the top prize: 1,000,000 for 1.

It was unusual for mathematicians of the 16th–18th centuries to weigh in on the mathematics behind the lotteries of the day: players were attracted by high payoffs without concern for the long odds against them, and lottery operators weren't interested in doing the math so long as the money kept rolling in [29]. An exception was when mathematician Leonhard Euler turned his talents toward the original Genoese lottery, calculating the various success probabilities in somewhat more generality than the lottery required [15, 33]. While the various wagers listed above pay off only if all of the numbers selected are drawn, Euler's work, in modern notation, showed that the probability of matching $k$ out of $r$ numbers selected in an $r/s$ lottery is

$$\boxed{p(k \mid r, s) = \frac{{}_rC_k \cdot {}_{s-r}C_{r-k}}{{}_sC_r}}$$

This is, of course, the model we have developed for keno with the numbers 20 and 80 replaced by $r$ and $s$, respectively. As with keno, we must account for every number selected by the lottery agency in computing probabilities.

Euler also devoted some attention to the question of house advantage and its effect on game design, calculating the appropriate prizes for wagers in the Genoese lottery so that the game would be fair. Recognizing that a fair game would be disadvantageous to the lottery operator who incurred expenses in the course of running the game and might be ruined if someone matched 5 numbers on a Cinquina bet, he extended his work to suggest a payoff scheme guaranteed to generate a given profit margin; suggesting 10% as a reasonable house edge [33].

**Example 4.24.** For the Ambo wager, Euler's analysis results in the probabilities shown in Table 4.16.

TABLE 4.16: Probability distribution for Ambo wager (Euler) [33]

| $k$ | $p(k \mid 2, 90)$ |
|---|---|
| 0 | $\frac{7140}{8010}$ |
| 1 | $\frac{850}{8010}$ |
| 2 | $\frac{20}{8010}$ |

Under the reasonable assumption that matching no numbers should win nothing, deciding on a fair payoff structure for a 1-unit bet amounts to solving the equation

$$(x) \cdot \frac{20}{8010} + (y) \cdot \frac{850}{8010} - 1 = 0$$

for the two payoff amounts $x$ and $y$. This equation, of course, admits infinitely many solutions, even if we require that both prizes be positive. Euler chose to leave this matter open by dividing the 1-unit payoff between the two cases in the proportion $\alpha : \beta$, so that

$$\alpha + \beta = 1$$

and we then have

$$x = \frac{8010\alpha}{20}, \quad y = \frac{8010\beta}{850}.$$

For example, setting $\alpha = 1$ and $\beta = 0$ corresponds to a zero payoff for matching only 1 number, consistent with the rules for Lo Giuoco del Lotto, and yields $x = 400.50$. Setting $\alpha = \beta = \frac{1}{2}$ gives $x = 200.25$ and $y \approx 4.71$. ■

If the lottery operator wishes to ensure a 10% profit in the long run, Euler suggested the simple solution of multiplying the prizes determined from the above calculations by .9; this generalizes to other desired house edges. Ambo's 250 for 1 payoff for matching two numbers results in a 37.6% HA.

**Example 4.25.** *Pick 5*, a $1 game run by the Nebraska Lottery, is a 5/38 game, which means that there are ${}_{38}C_5 = 501,942$ different combinations available to players. The probability function for this game is

$$p(k) = p(k \mid 5, 38) = \frac{{}_5C_k \cdot {}_{33}C_{5-k}}{{}_{38}C_5},$$

and the corresponding probability distribution and prize structure are in Table 4.17. The progressive jackpot for matching all 5 numbers starts at $50,000 and rises until it is won.

TABLE 4.17: Probability distribution for Nebraska Pick 5

| $k$ | **Probability** | **Prize** |
|---|---|---|
| 0 | .4728 | 0 |
| 1 | .4076 | 0 |
| 2 | .1087 | Free ticket |
| 3 | .0105 | $9 |
| 4 | $3.2872 \times 10^{-4}$ | $450 |
| 5 | $1.9923 \times 10^{-6}$ | Progressive |

The free ticket prize for matching 2 out of 5 numbers is effectively a cash prize of $E + 1$: the expected value of a single ticket without subtracting the ticket price. Using the minimum value for the progressive jackpot leads to a linear equation in the expectation:

$$E = (E + 1) \cdot p(2) + (9) \cdot p(3) + (450) \cdot p(4) + (50,000) \cdot p(5) - 1.$$

The solution to this equation, and hence the minimum expected return, is $E = -.5059$. This corresponds to a maximum HA of 50.59%. If the jackpot doubles to $100,000, the new HA is 49.10%. ■

**Example 4.26.** South Carolina's Education Lottery offers *Palmetto Cash 5*, which is also a 5/38 game. The pay table, Table 4.18, differs at every prize level from Nebraska's 5/38 game; in particular, the top prize is fixed at $100,000 and is not a progressive jackpot.

TABLE 4.18: Pay table for South Carolina's Palmetto Cash 5

| **Match** | **Prize** |
|---|---|
| 2 | $1 |
| 3 | $5 |
| 4 | $300 |
| 5 | $100,000 |

With a fixed \$100,000 jackpot, Palmetto Cash 5 carries a HA of 54.09%, right in line with Nebraska's lottery, but there's an additional option that changes the game. Much like keno administered by state lotteries in Ohio and Oregon (p. 183), Palmetto Cash 5 offers a chance to double one's wager and multiply the prize—by 2, 3, 4, or 5—with the *Power-Up* multiplier. The multiplier is determined by a separate drawing from 16 ping-pong balls: 8 labeled "2", 5 labeled "3", 2 labeled "4", and 1 labeled "5". The expected value of the Power-Up multiplier is $2\frac{3}{4}$, so as with the Oregon Keno Multiplier, it is in a player's best interest to make this side bet.

How does the multiplier affect the HA? On the average, payoffs are multiplied by 2.75 while the ticket price rises to \$2; the new HA can be found to be 36.87%. The Nebraska jackpot would need to rise to \$160,680 to match this house advantage. ■

Table 4.19 (page 238) shows lotto games from across the United States and Canada. American entries are restricted to "standard" $r/s$ lotteries, with no separate bonus number drawing such as is seen in Powerball, and games for which tickets are only sold in one state except for Tri-State Gimme 5, which is run by a three-state coalition consisting of Maine, New Hampshire, and Vermont. None of these three states sponsors its own statewide lotto game. In Canada, several provinces and multi-province alliances in the east and west manage lotto games local to their region as well as the national lotto games. Provincial or regional games typically include a bonus number along with the numbers listed.

Table 4.20 (page 239) contains a list of international lotto games. Many international games include a bonus ball drawn from the same pool as the other numbers, which can add to certain prizes for matching fewer than all of the "ordinary" winning numbers. In these games, the bonus ball typically does not augment the prize when all of the main numbers are matched. This does not change the total number of possible tickets, which remains ${}_sC_r$. For an analysis of this type of game, see Example 4.42 (page 260).

**Example 4.27.** In Jamaica, the local *Lotto* game is a 6/39 offering with a seventh bonus ball drawn from the same pool of numbers that increases Match 5 and Match 4 payoffs if it is also matched by the player. An unusual betting option is a *combination ticket*, which is similar to way tickets in keno [148]. A player may select 4–10 numbers, and his or her ticket will be read as a multi-way ticket. For 4 numbers, the player's choice will be combined with all pairs of the remaining numbers to generate ${}_{35}C_2 = 595$ tickets. At a minimum wager of 100 Jamaican dollars (about 78.5¢ US) per ticket, this 4-spot ticket would cost at least J\$59,500 (approximately 486 US dollars). A 5-spot selection would generate 34 tickets.

If a player chooses 7–10 numbers, all possible 6-number subsets are formed from their selection, leading to 7, 28, 84, or 210 different tickets. ■

**Example 4.28.** Australia offers a lotto game called *The Pools*: a 6/38 game

TABLE 4.19: State and provincial lotto games

| State/Province | Lottery Name | $r$ | $s$ | ${}_sC_r$ |
|---|---|---|---|---|
| Arizona | The Pick | 6 | 44 | 7,059,052 |
| Arkansas | Natural State Jackpot | 5 | 39 | 575,757 |
| California | Fantasy 5 | 5 | 39 | 575,757 |
| Colorado | Lotto | 6 | 42 | 5,245,786 |
| Connecticut | Lotto | 6 | 44 | 7,059,052 |
| Florida | Florida Lotto | 6 | 53 | 22,957,480 |
| Georgia | Jumbo Bucks Lotto | 6 | 47 | 10,737,573 |
| Idaho | Weekly Grand | 5 | 32 | 201,376 |
| Illinois | Lucky Day Lotto | 5 | 45 | 1,221,759 |
| Indiana | Hoosier Lotto | 6 | 46 | 9,366,819 |
| Louisiana | Lotto | 6 | 40 | 3,838,380 |
| Maine/NH/VT | Tri-State Gimme 5 | 5 | 39 | 575,757 |
| Massachusetts | Mass Cash | 5 | 35 | 324,632 |
| Michigan | Lotto 47 | 6 | 47 | 10,737,573 |
| Minnesota | Gopher 5 | 5 | 47 | 1,533,939 |
| Missouri | Show Me Cash | 5 | 39 | 575,757 |
| Montana | Montana Cash | 5 | 45 | 1,221,759 |
| Nebraska | Pick 5 | 5 | 38 | 501,942 |
| New Mexico | Roadrunner Cash | 5 | 37 | 435,897 |
| New York | Take 5 | 5 | 39 | 575,757 |
| North Carolina | Carolina Cash 5 | 5 | 41 | 749,398 |
| Ohio | Classic Lotto | 6 | 49 | 13,983,816 |
| Oklahoma | Cash 5 | 5 | 36 | 376,992 |
| Oregon | Oregon's Game Megabucks | 6 | 48 | 12,271,512 |
| Pennsylvania | Cash 5 | 5 | 43 | 962,598 |
| South Carolina | Palmetto Cash 5 | 5 | 38 | 501,942 |
| South Dakota | Dakota Cash | 5 | 35 | 324,632 |
| Texas | Lotto Texas | 6 | 54 | 25,827,165 |
| Virginia | Cash 5 | 5 | 34 | 278,256 |
| Washington | Lotto | 6 | 49 | 13,983,816 |
| West Virginia | Cash 25 | 6 | 25 | 177,100 |
| Wisconsin | Badger 5 | 5 | 31 | 169,911 |
| Wyoming | Cowboy Draw | 5 | 45 | 1,221,759 |
| | | | | |
| Atlantic Canada | Atlantic 49 | 6 | 49 | 13,983,816 |
| British Columbia | BC/49 | 6 | 49 | 13,983,816 |
| Ontario | Lottario | 6 | 45 | 8,145,060 |
| Quebec | Québec Max | 7 | 49 | 85,900,584 |
| Western Canada | Western Max | 7 | 49 | 85,900,584 |

TABLE 4.20: A sample of international lotto games

| **Location** | **Lottery Name** | $r$ | $s$ | ${}_sC_r$ |
|---|---|---|---|---|
| Albania | Loto 6 out of 39 | 6 | 39 | 3,262,623 |
| Argentina | Loto 5 | 5 | 37 | 435,897 |
| Aruba | Lotto di Dia | 5 | 30 | 142,506 |
| Australia | Oz Lotto | 7 | 45 | 45,379,620 |
| Barbados | Mega 6 | 6 | 33 | 1,107,568 |
| Bénin | Loto Star | 5 | 90 | 43,949,268 |
| Brazil | Mega-Sena | 6 | 60 | 50,063,860 |
| Canada | Lotto Max | 7 | 47 | 62,891,499 |
| Costa Rica | Lotto | 5 | 35 | 324,632 |
| Dominican Republic | Loto Pool | 5 | 27 | 80,730 |
| Fiji | Pick 6 | 6 | 36 | 1,947,792 |
| Germany | Lotto | 6 | 49 | 13,983,816 |
| Grenada | Lotto | 5 | 34 | 278,256 |
| Guyana | Daily Millions | 5 | 26 | 65,780 |
| Hong Kong | Mark Six | 6 | 49 | 13,983,816 |
| Iceland | Lottó | 5 | 40 | 658,008 |
| India | Jaldi 5 | 5 | 36 | 376,992 |
| Ireland | Lotto | 6 | 47 | 10,737,573 |
| Israel | Lotto | 6 | 37 | 2,324,784 |
| Italy | SuperEnalotto | 6 | 90 | 622,614,630 |
| Jamaica | Lotto | 6 | 39 | 3,262,623 |
| Malta | Super 5 | 5 | 45 | 1,221,759 |
| Mexico | Melate | 6 | 56 | 32,468,436 |
| Mongolia | 6/42 Jackpot Lotto | 6 | 42 | 5,245,786 |
| New Zealand | Lotto | 6 | 40 | 3,838,380 |
| Nigeria | Lotto Nigeria | 6 | 49 | 13,983,816 |
| Philippines | Grandlotto | 6 | 55 | 28,989,675 |
| Poland | MINI Lotto | 5 | 42 | 850,668 |
| Russia | GosLoto | 5 | 36 | 376,992 |
| St. Lucia | Double Daily Grand | 4 | 22 | 7315 |
| Samoa | Samoa National Lotto | 6 | 35 | 1,623,160 |
| South Korea | Lotto | 6 | 45 | 8,145,060 |
| Spain | Daily 6/49 | 6 | 49 | 13,983,816 |
| Sweden | Lotto | 7 | 35 | 6,724,520 |
| Switzerland | Swiss Lotto | 6 | 42 | 5,245,786 |
| Tanzania | Lotto | 6 | 49 | 13,983,816 |
| Trinidad & Tobago | Cash Pot | 5 | 20 | 15,504 |
| Turkey | Süper Loto | 6 | 54 | 25,827,165 |
| Uganda | Billion Lotto | 6 | 49 | 13,983,816 |
| Vietnam | Mega | 6 | 45 | 8,145,060 |
| Windward Islands | Super 6 | 6 | 28 | 376,740 |

where the numbers are keyed to Australian and European soccer matches. The goal for players is to choose numbers corresponding to drawn games and high-scoring games. Instead of using a random drawing, the winning numbers are chosen by choosing the top 6 of the 38 games, ranked by the following algorithm [8]:

1. Score draws (draws other than at 0–0) rank highest. In the case of multiple drawn games, games with more goals scored rank higher.

2. 0–0 draws rank second. If there are more than 6 drawn games, the scoreless draws with higher assigned match numbers are selected first.

3. Wins for the visiting side rank next. Ties are broken by looking first to smaller margins of victory—3–2 outranks 2–0, for example—and then to total goals scored, where 5–4 ranks higher than 2–1.

4. Home team wins rank last. The tiebreaker used in step 2 is also applied here.

A seventh game is chosen as a supplemental winning number [124].

Barring unusual player insight into soccer, which cannot be totally ruled out, this is effectively a 6/38 game with random selection of winning numbers, and many players pick their numbers without regard for the games to which they're attached. There are five "divisions" of prizes, which are listed in Table 4.21. All prizes are pari-mutuel, based on the total amount wagered at A$0.75 per ticket; the top prize starts at A$75,000 and grows until someone wins it. ■

TABLE 4.21: Australian Pools prize divisions [124]

| Division | Match |
|---|---|
| 1 | 6 |
| 2 | 5 + bonus |
| 3 | 5 |
| 4 | 4 |
| 5 | 3 + bonus |

## Texas Triple Chance

*Texas Triple Chance* is an exception to the rule that $r/s$ lotteries call for the state to choose $r$ numbers as players pick $r$. From the player's perspective, this is a 7/55 game, with the unusual bonus that a single $2 ticket offers three opportunities—dubbed "Chances"—to win: the player's original set of 7 numbers and 2 randomly-generated Quick Pick tickets. The Texas Lottery, however, draws 10 numbers rather than only 7. Numbers are drawn six days per week, omitting only Sunday, and payoffs start when any one of the player's three chances matches at least 3 of the 10 drawn numbers; a single ticket can win on all three chances.

The probability of matching $k$ numbers, $0 \leqslant k \leqslant 7$, when 10 are chosen is

$$P(k \mid 10) = \frac{{}_7C_k \cdot {}_{48}C_{10-k}}{{}_{55}C_{10}},$$

while the probability of matching $k$ numbers when only 7 are drawn in a standard 7/55 lottery is

$$P(k \mid 7) = \frac{{}_7C_k \cdot {}_{48}C_{7-k}}{{}_{55}C_7}.$$

The number of possible choices when the state picks 10 numbers is bigger, by a factor of about 144, so the denominator goes up in picking 10. However, there are a lot more ways to draw into a winning combination when there are 3 extra winning numbers, so the numerator also rises. How do these factors balance—which of these probability functions gives the more favorable distribution? Table 4.22 compares the two functions.

TABLE 4.22: Probability distribution functions: $P(k \mid 10)$ vs. $P(k \mid 7)$

| $k$ | $P(k \mid 10)$ | $P(k \mid 7)$ |
|---|---|---|
| 0 | .2236 | .3628 |
| 1 | .4014 | .4233 |
| 2 | .2709 | .1772 |
| 3 | .0881 | .0336 |
| 4 | .0147 | .0030 |
| 5 | .0012 | $1.1673 \times 10^{-4}$ |
| 6 | $4.6568 \times 10^{-5}$ | $1.6558 \times 10^{-6}$ |
| 7 | $5.9134 \times 10^{-7}$ | $4.9278 \times 10^{-9}$ |

- The probability of winning, found by summing the values for $k = 3$ to 7, goes up considerably, from .0367 to .1041, when the state chooses 10 numbers.
- The probability of winning at each individual prize level goes up.
- The chances of matching 0 or 1 numbers—both losing outcomes—both go down with 10 numbers drawn, but the probability of losing with 2 matched numbers increases.

A single chance on a Texas Triple Chance ticket has a price of $\$\frac{2}{3}$ and an expected value of –\$.3039. This expectation might seem encouraging by comparison with other games until we remember that tickets cost less than \$1. Dividing by the ticket price shows that the HA of 45.58% is in line with other lottery games.

## Super 7: A Game's Evolution

Much like Texas Triple Chance, the Pennsylvania Lottery's *Super 7* game, which was offered off and on from 1986 to 2010, drew more numbers than players were asked to pick. During the course of its two separate runs, a number of game parameters were adjusted as lottery officials sought to maintain interest in the game.

On its initial launch in 1986, a $1 Super 7 ticket called for players to choose 7 numbers from 1–80, while the state drew 11 weekly. The probability of matching $k$ numbers was then

$$P(k \mid 7) = \frac{{}_7C_k \cdot {}_{73}C_{11-k}}{{}_{80}C_{11}}.$$

In 1991, the game was modified. Players still picked 7 numbers, but the range was shrunk to 1–74 and the state's draw reduced from 11 to 10. These values were in force until the game was discontinued in 1995. The new PDF was

$$P(k \mid 7) = \frac{{}_7C_k \cdot {}_{67}C_{10-k}}{{}_{74}C_{10}}.$$

When Super 7 was revived in 2009, the rules were again changed, with the state back to choosing 11 numbers from 1–77. Players still picked 7 numbers, though tickets now cost $2. These rules prevailed for the remainder of the game's run, which ended in April 2010. The final probability distribution function was

$$P(k \mid 7) = \frac{{}_7C_k \cdot {}_{70}C_{11-k}}{{}_{77}C_{11}}.$$

Throughout all of the changes, the match probabilities changed very little with each successive version of the game. Table 4.23 shows the value of $P(k \mid 7)$ for each set of game parameters.

TABLE 4.23: Probability distribution functions for Pennsylvania's Super 7

| **Years** | **1986–1991** | **1991–1995** | **2009–2010** |
|---|---|---|---|
| **Range** | 1–80 | 1–74 | 1–77 |
| **State's draw** | 11 | 10 | 11 |
| $k$ | | $P(k \mid 7)$ | |
| 0 | .3396 | .3452 | .3238 |
| 1 | .4151 | .4166 | .4156 |
| 2 | .1946 | .1907 | .2044 |
| 3 | .0449 | .0424 | .0495 |
| 4 | .0054 | .0049 | .0063 |
| 5 | $3.412 \times 10^{-4}$ | $2.823 \times 10^{-4}$ | $4.121 \times 10^{-4}$ |
| 6 | $1.003 \times 10^{-5}$ | $7.468 \times 10^{-6}$ | $1.268 \times 10^{-5}$ |
| 7 | $1.039 \times 10^{-7}$ | $6.668 \times 10^{-8}$ | $1.372 \times 10^{-7}$ |

With the first two iterations of Super 7, players had to match at least 4 of their 7 numbers to win a prize. The Match 4 prize was fixed at $7 for the 1986–1991 game and $15 for the 1991–1995 version; the higher-level prizes were pari-mutuel, paid from fixed percentages of the prize pool remaining after the Match 4 prizes were paid. Under these rules, the probability of winning something was very low: 1 in 172.6 for the original game and 1 in 194.1 for the second version. The slightly lower probability of winning a prize may have been intentional, to increase the average jackpot size.

When Super 7 returned in 2009, a fixed $2 Match 3 prize was added, which refunded the ticket price, and the remaining prizes were pari-mutuel. The probability of winning rose considerably, to 1 in 17.8. Interest never developed, and the revived game lasted just over a year.

## Amigo

*Amigo*, a game operated by the French lottery, puts a twist on the idea of bonus numbers. Amigo is a 7/28 lottery with 12 numbers drawn in all. The 7 primary numbers are called "blue" numbers; the other 5 are yellow "bonus" numbers. Winning numbers are drawn every 5 minutes throughout the day, much like state lottery keno games in the USA. Payoffs (Table 4.24) are based on the total number of matches, beginning at 4, and the distribution of matched numbers between blue and bonus numbers [150].

To compute the probabilities of the outcomes in this table, we need to separate the blue numbers from the bonus numbers. Suppose that the ticket matches $b$ blue numbers and $k$ bonus numbers. The probability of matching exactly $b$ blue numbers is

$$p_1(b) = \frac{{}_7C_b \cdot {}_{21}C_{7-b}}{{}_{28}C_7}.$$

If $b$ blue numbers are matched, there are $7-b$ player numbers left and 21 numbers left from which the 5 bonus numbers will be drawn. There are thus ${}_{7-b}C_k$ ways to choose the $k$ bonus numbers, and $21-(7-b) = 14+b$ numbers remain that have not been chosen by the player or matched among the blue numbers, giving ${}_{14+b}C_{5-k}$ ways to pick the unmatched bonus numbers.

Putting all of the factors together shows that the probability of matching $k$ of the 5 bonus numbers is

$$p_2(k \mid b) = \frac{{}_{7-b}C_k \cdot {}_{14+b}C_{5-k}}{{}_{21}C_5}.$$

The probability of matching exactly $b$ blue numbers and $k$ bonus numbers is then

$$p(b,k) = p_1 \cdot p_2 = \frac{{}_7C_b \cdot {}_{21}C_{7-b}}{{}_{28}C_7} \cdot \frac{{}_{7-b}C_k \cdot {}_{14+b}C_{5-k}}{{}_{21}C_5}.$$

**Example 4.29.** The probabilities of matching 7 numbers in the various configurations listed in Table 4.24 are shown in Table 4.25.

TABLE 4.24: Amigo pay table (€2 ticket) [150]

| Matches | Blue | Bonus | Payoff (€) |
|---|---|---|---|
| 4 | 4 | 0 | 5 |
| | 3 | 1 | 2 |
| | 2 | 2 | 2 |
| | 1 | 3 | 2 |
| | 0 | 4 | 2 |
| 5 | 5 | 0 | 40 |
| | 4 | 1 | 8 |
| | 3 | 2 | 3 |
| | 2 | 3 | 3 |
| | 1 | 4 | 3 |
| | 0 | 5 | 3 |
| 6 | 6 | 0 | 500 |
| | 5 | 1 | 55 |
| | 4 | 2 | 20 |
| | 3 | 3 | 15 |
| | 2 | 4 | 15 |
| | 1 | 5 | 15 |
| 7 | 7 | 0 | 25,000 |
| | 6 | 1 | 600 |
| | 5 | 2 | 140 |
| | 4 | 3 | 105 |
| | 3 | 4 | 100 |
| | 2 | 5 | 100 |

TABLE 4.25: Amigo: Probabilities of Match 7 configurations

| Blue | Bonus | Probability |
|---|---|---|
| 7 | 0 | $8.4457 \times 10^{-7}$ |
| 6 | 1 | $2.9560 \times 10^{-5}$ |
| 5 | 2 | $1.7736 \times 10^{-4}$ |
| 4 | 3 | $2.9560 \times 10^{-4}$ |
| 3 | 4 | $1.4780 \times 10^{-4}$ |
| 2 | 5 | $1.7736 \times 10^{-5}$ |

The total probability of matching 7 of the 12 numbers is the sum of these values, or $6.6890 \times 10^{-4}$: 1 chance in 1495. ■

Applying the formula for $p(b,k)$ to a €2 ticket gives an expected value of −€.649. An occasional Amigo promotion doubles the prize of the jackpot to €50,000; this moves the expectation up, but only by about 2.2¢.

**Example 4.30.** Another occasional Amigo promotion offers a payoff when none of a player's 7 numbers appears in the lottery's 12 numbers. Find the probability of this event.

We are simply looking for the probability that the 12 winning numbers—regardless of whether they're blue or yellow—all fall among the 21 numbers not chosen by a player. That probability is

$$\frac{{}_{21}C_{12}}{{}_{28}C_{12}} = \frac{293,930}{30,421,755} \approx .0097$$

—just under 1%. ■

## Atlantic Bucko

*Atlantic Bucko* is a game related to keno that is played in the four maritime provinces of Canada. Unlike most other lotto games, Atlantic Bucko is a passive game. A \$1 ticket gives a player three sets of five numbers in the range 1–41 on a single ticket, as in Figure 4.8. Numbers may be repeated on multiple lines: for example, 39 appears in both the first and second lines in this figure.

| | | | | |
|---|---|---|---|---|
| 19 | 25 | 27 | 30 | 39 |
| 4 | 13 | 20 | 31 | 39 |
| 10 | 12 | 24 | 34 | 40 |

FIGURE 4.8: Atlantic Bucko ticket.

The lottery authority draws five numbers nightly, and a single ticket is effectively playing two games:

- **Single Line**: Each line of 5 numbers can be played as a separate 5/41 ticket, with Table 4.26 as the pay table.
- **Combo**: The ticket as a whole pays off if four or more, or none, of a player's 15 numbers are drawn. The payable for this wager is given in Table 4.27.

The Combo bet illustrates why Atlantic Bucko must be played as a passive game; if players were allowed to choose their own numbers and repetitions were

TABLE 4.26: Atlantic Bucko Single Line pay table [84]

| Catch | Payoff |
|---|---|
| 3 | \$3 |
| 4 | \$100 |
| 5 | \$20,000 |

TABLE 4.27: Atlantic Bucko Combo pay table [84]

| Catch | Payoff |
|---|---|
| 0 | Free ticket |
| 4 | \$1 |
| 5 | \$5 |
| 6 | \$20 |
| 7 | \$75 |
| 8 or more | \$1000 |

permitted (as they must be if the payoffs for other than 0 matches are to be meaningful), a smart player could net \$4959 every day by simply buying 41 tickets, each one containing a different single number repeated 15 times. Five of those tickets would win \$1000 apiece. Setting up a collection of rules to eliminate schemes like this would almost surely be more complicated than desirable, so making Atlantic Bucko a passive game was the right decision.

One intriguing question is how likely it is to score 8 matches on a 15-number ticket when there are only 5 winning numbers. Clearly, this cannot be done unless the ticket contains repeated numbers among its three lines. Counting the number of ways to match 8 numbers is equivalent to counting the number of ways to write 8 as the sum of five integers in the range 0–3:

$$8 = a + b + c + d + e,$$

where the integers $a - e$ represent the number of times each of the 5 numbers drawn by the lottery agency is present on a ticket. The following are valid sums, with order not considered:

$$\begin{array}{c} 3+3+2+0+0 \\ 3+3+1+1+0 \\ 3+2+2+1+0 \\ 3+2+1+1+1 \\ 2+2+2+2+0 \\ 2+2+2+1+1 \end{array}$$

By considering the possibilities for rearrangement, we see that each of these sums can be attained in several ways: for example, the combination

$2 + 2 + 2 + 2 + 0$, with four repeated numbers, can occur ${}_5C_1 = 5$ ways, depending on which of the five numbers is not present on a ticket. Adding up over all of the combinations above gives 155 ways to record 8 catches when 5 numbers are drawn.

However, this presumes that the ticket contains at least 3 repeated numbers, and that the repeated numbers are among the winning numbers.

**Example 4.31.** Given the importance of repeated numbers to the Combo payoffs, what is the probability of a ticket with no repeated numbers?

Consider the three lines as independent selections of 5 numbers from 41. The first line contains 5 numbers; the probability that the second line does not duplicate any of them is then

$$\frac{{}_{36}C_5}{{}_{41}C_5} = \frac{376,992}{749,398} \approx .5031$$

—approximately the same probability as a coin toss. Assuming that the first two lines have no numbers in common, the probability that the third line contains none of these 10 different numbers is then

$$\frac{{}_{31}C_5}{{}_{41}C_5} = \frac{169,911}{749,398} \approx .2267.$$

Multiplying these two probabilities together shows that the probability of drawing 15 different numbers is approximately .1141, so almost 89% of tickets will contain at least one repeated number. ∎

**Example 4.32.** Find the probability of a ticket with the minimum number of repeated numbers to qualify for the $1000 Combo prize: 3 repeated numbers, 2 appearing 3 times and the third appearing twice.

Denoting the repeated numbers by $a, b$, and $c$, the ticket must have the following configuration, where the blank cells can contain any number and the $c$'s can appear in any two of the three lines.

| | | | | |
|---|---|---|---|---|
| $a$ | $b$ | $c$ | | |
| $a$ | $b$ | $c$ | | |
| $a$ | $b$ | | | |

If we specify that none of the numbers in the blank cells are repeated, the number of ways to fill each cell is recorded here:

| | | | | |
|---|---|---|---|---|
| 41 | 40 | 39 | 38 | 37 |
| 1 | 1 | 1 | 36 | 35 |
| 1 | 1 | 34 | 33 | 32 |

Multiplying these 15 numbers gives a total of 4,068,245,531,750,400 tickets; this number must be multiplied by 3 to account for the ways to position the $c$'s in two of the three rows. There are $({}_{41}C_5)^3$ different possible tickets, making the probability

$$\frac{12,204,736,595,251,200}{420,859,940,190,832,792} \approx .0290,$$

or just under 3%. ■

There are several ways to win the Combo jackpot with this ticket by catching exactly 8 numbers.

- One possibility is when the numbers $a, b$, and $c$ are drawn among the 5 winning numbers; the probability of this happening is

$$\frac{{}_{31}C_2}{{}_{41}C_5} = \frac{465}{749,398}.$$

- A second option is when $a, b$, and two other numbers from the 7 appearing once on the ticket are drawn. This event occurs with probability

$$\frac{{}_7C_2 \cdot {}_{31}C_1}{{}_{41}C_5} = \frac{651}{749,398}.$$

- Finally, either $a$ or $b$ can be drawn together with $c$ and 3 other numbers on the ticket. This event's probability is

$$\frac{{}_7C_3}{{}_{41}C_5} = \frac{35}{{}_{41}C_5},$$

which must be multiplied by 2 to account for the choice of either $a$ or $b$.

Adding up these probabilities gives

$$P(\text{Catch } 8) = \frac{1186}{749,398} \approx .0016.$$

This ticket can also win the Combo top prize by catching 9 or 10 numbers. Nine numbers can be caught by catching $a, b, c$, and one other number, or by catching $a, b$, and any two other numbers. This gives the probability

$$P(\text{Catch } 9) = \frac{{}_7C_1 \cdot {}_{31}C_1}{{}_{41}C_5} + \frac{{}_7C_2 \cdot {}_{31}C_1}{{}_{41}C_5} = \frac{217 + 651}{749,398} = \frac{868}{749,398}.$$

There is only one way to catch 10 numbers: when the official draw includes $a, b, c$, and two of the other 7 numbers. The probability is

$$P(\text{Catch } 10) = \frac{{}_7C_2}{{}_{41}C_5} = \frac{21}{749,398}.$$

The total probability of winning the Combo jackpot with this ticket is then

$$\frac{1186 + 868 + 21}{749,398} \approx \frac{1}{361}.$$

Similar analysis of other tickets that can catch 8 or more numbers gives the probability of winning the $1000 prize as approximately $\frac{1}{29,515}$ [84].

This prize contributes about 3¢ to the expected value of an Atlantic Bucko ticket.

## Lotzee

In his 1793 essay *Idea Concerning A Lottery*, Alexander Hamilton asserted that a successful lottery had two characteristics: simplicity and low-priced tickets [51]. *Lotzee*, a game operated by the New Jersey Lottery from 1998–2003, was perhaps too complicated to be successful. The numerical parameters for Lotzee were changed over time; in this section, we shall consider the rules and payoffs in force in 2000.

Lotzee stood for "Lots of Ways to Win", and the ticket certainly provided that, for a $2 ticket price [99]. In 2000, Lotzee was a variation on a 4/100 game: players could choose 4 numbers in the range 00–99 or leave them to a Quick Pick option. Players then received 20 additional Quick Pick sets of 4 numbers in that same range, divided into 5 tiers and sorted by the value of the top prize available [109]:

- 1 set, the player's selection, on the $500,000 tier
- 2 sets on the $100,000 tier
- 4 sets on the $50,000 tier
- 6 sets on the $25,000 tier
- 8 sets on the $10,000 tier

The state's 4 Lotzee numbers were drawn weekly. Tickets matching 2 or more numbers in a single set were winners; the prize won depended both on the number of matches and the tier in which the set was placed. Table 4.28 shows the pay table.

TABLE 4.28: Lotzee pay table in 2000 [109]

| | Tier | | | | |
|---|---|---|---|---|---|
| **Match** | **$500,000** | **$100,000** | **$50,000** | **$25,000** | **$10,000** |
| 2 | $15 | $3 | $3 | $3 | $3 |
| 3 | $1000 | $50 | $50 | $50 | $50 |
| 4 | $500,000 | $100,000 | $50,000 | $25,000 | $10,000 |

**Example 4.33.** While the payoffs may differ among tiers, the probability $P(k)$ of matching $k$ numbers out of the 4 in a single set is constant from tier to tier; it is

$$P(k) = \frac{{}_4C_k \cdot {}_{96}C_{4-k}}{{}_{100}C_4},$$

which is tabulated in Table 4.29.

TABLE 4.29: Lotzee probability distribution function

| $k$ | $P(k)$ |
|---|---|
| 0 | .8472 |
| 1 | .1458 |
| 2 | .0070 |
| 3 | $9.793 \times 10^{-5}$ |
| 4 | $2.550 \times 10^{-7}$ |

■

The probability of winning anything on any line by matching 2 or more numbers can be seen to be less than 1%. The mean number of matches on a single line is .16. While successive lines are not independent, due to the possibility of duplicated numbers, a rough approximation of the total number of matches on a single 21-line ticket can be found by assuming independence: $.16 \cdot 21 = 3.36$.

The expected value of a single set of 4 Lotzee numbers depends, of course, on its tier. For the top tier, the expected win without taking the ticket price into account is 33.01¢. Adding up across all 21 lines and subtracting the $2 ticket price gives an expected return of –99.27¢, and so the house advantage is 49.63%.

Lotzee was discontinued, after several attempts to simplify the games and stimulate interest, in 2003. At the game's end, tickets contained only 7 lines of numbers, chosen from 1–77. This was the first New Jersey Lottery game discontinued since 1991 [99].

## Pozo Millonario

*Pozo Millonario* ("Millionaire Well") is a passive game played in Ecuador that has certain lotto-like features. While players do not get to choose their own numbers, Pozo Millonario functions otherwise like a 14/25 lotto game. This provides for ${}_{25}C_{14} = 4,457,400$ different possible number combinations, and since the game is passive, no combination will be duplicated, which means that a grand prize winner never has to split the progressive jackpot.

Fourteen numbers are drawn each week, which means that no ticket can match fewer than 3 numbers. The lottery pays off for matching 3, 4, 12, 13, or all 14 drawn numbers, and the pay table is shown in Table 4.30.

The top two prizes are subject to an automatic 14% tax, which reduces the prize for matching 13 numbers to $430. Lottery officials estimate that approximately 1 million $1 tickets are sold weekly, and lottery rules stipulate that 50% of the ticket income is added to the jackpot fund, which is known

TABLE 4.30: Pozo Millonario pay table [83]

| Match | Payoff |
|---|---|
| 3 | $100 |
| 4 | $10 |
| 12 | $20 |
| 13 | $500 |
| 14 | Jackpot |

as the "well". As a result, it is expected that the jackpot will be won about once every 5 weeks, and will typically run about $2–3 million [131].

Pozo Millonario includes a second drawing each week: every ticket is also emblazoned with one of 15 animals, the mascot of that ticket. One animal is drawn with each set of winning numbers, and a ticket matching the mascot wins a refund of the $1 ticket price.

Let the probability of matching exactly $k$ numbers be $P(k)$, so that

$$P(k) = P(k \mid 14) = \frac{{}_{14}C_k \cdot {}_{11}C_{14-k}}{{}_{25}C_{14}},$$

and denote the jackpot by $J$. Accounting for the tax and the $\frac{1}{15}$ chance of matching the mascot, the expected value of a $1 ticket is

$$E = (100)\cdot P(3) + (10)\cdot P(4) + (20)\cdot P(12) + (430)\cdot P(13) + (.86J)\cdot P(14) - \frac{14}{15}.$$

We then have $E = 0$ when $J = \$4,473,733$, an amount considered "exceptional" by lottery officials, which means that a positive expectation would be a very unlikely event [131].

## Wild Card 2

From 1998 until 2016, the Multi-State Lottery Association (MUSL) operated *Wild Card* and *Wild Card 2* in several western states. Wild Card 2 combined a standard 5/33 number lottery with an independent 1/16 lottery in which players chose 5 numbers in the range 1–33 and one face card or ace in the suit of their choice from a standard card deck. *Wild Card* was an earlier version that used 31 numbers instead of 33.

A twist to Wild Card 2 was that players received two tickets for their $1 entry fee, which was the minimum price—buying a single ticket was not an option. Any ticket that matched the drawn card won a prize; the prizes were bigger when more of the numbers were matched, as shown in Table 4.31. The progressive jackpot for matching all 6 drawn elements started at $200,000.

**Example 4.34.** The two-ticket rule in Wild Card 2 confounds simple mathematical analysis. To simplify the problem, we shall consider the case where

TABLE 4.31: Pay table for Wild Card 2 [145]

| Match | Prize |
|---|---|
| 5 + wild card | Jackpot |
| 5 | \$6000 |
| 4 + wild card | \$500 |
| 4 | \$30 |
| 3 + wild card | \$6 |
| 3 | \$2 |
| 2 + wild card | \$2 |
| 1 + wild card | \$1 |
| Wild card only | \$1 |

the two tickets have no numbers or cards in common; call these tickets *disjoint*. Find the probability that a \$1 investment in two disjoint tickets wins the jackpot.

The total number of possible tickets is

$$_{33}C_5 \cdot {}_{16}C_1 = 237,336 \cdot 16 = 3,797,376.$$

Since a pair of disjoint tickets offers two chances at the one winning combination, the probability of winning is

$$\frac{2}{3,797,376} = \frac{1}{1,898,688}.$$

■

**Example 4.35.** Suppose that the two tickets are not disjoint, but have three numbers in common, as, for example, 4, 10, 11, 18, 20, $J♣$ and 9, 10, 11, 18, 30, $A♣$.

Winning the jackpot now requires each of the following events to happen:

- $E_1$ = Draw all three of the common numbers. Both tickets are still viable jackpot winners at this point.
- $E_2$ = Draw the other two numbers from one of the tickets. Only one ticket can now win the jackpot.
- $E_3$ = Draw the card corresponding to that ticket.

We then have the following probabilities:

$$p_1 = P(E_1) = \frac{1}{_{33}C_3} = \frac{1}{5456}.$$

$$p_2 = P(E_2 \mid E_1) = \frac{2}{_{30}C_2} = \frac{2}{435}.$$

$$p_3 = P(E_3) = \frac{1}{16}.$$

Multiplying these three probabilities together gives a win probability of

$$p_1 \cdot p_2 \cdot p_3 = \frac{1}{18,986,880}.$$

Disjoint tickets represent a chance of winning that is 10 times greater than this. ■

## All or Nothing

In Arizona, the *All or Nothing* lotto game is a stripped-down version of keno where the state draws 10 numbers in the range 1–20. Players pay $2 for the chance to select ten numbers from 1–20 and are paid off if the number of catches is either close to 0 or close to 10. The ticket gets its name from the fact that the top jackpot of $25,000 is paid when the player catches either all 10 of his or her numbers among the state's set of 20, or none of them.

TABLE 4.32: Pay table for All or Nothing

| Catch | Prize |
|---|---|
| 0 or 10 | $25,000 |
| 1 or 9 | $250 |
| 2 or 8 | $10 |
| 3 or 7 | $2 |

The symmetry in Table 4.32 is possible because players are required to pick half of the possible numbers, and so the probability of catching $x$ numbers is the same as the probability of catching $10 - x$ numbers:

$$P(\text{Catch } x) = \frac{{}_{10}C_x \cdot {}_{10}C_{10-x}}{{}_{20}C_{10}} = \frac{{}_{10}C_{10-x} \cdot {}_{10}C_x}{{}_{20}C_{10}} = P(\text{Catch } 10 - x).$$

Dropping the pool of numbers from keno's 80 down to 20 means that the denominator in probability calculations for All or Nothing is smaller, but what does this do to the probabilities themselves? As we have done several times, we compute the probability of losing, here by catching 4, 5 or 6 out of 10 numbers.

$$P(\text{Lose}) = \frac{{}_{10}C_4 \cdot {}_{10}C_6 + {}_{10}C_5 \cdot {}_{10}C_5 + {}_{10}C_6 \cdot {}_{10}C_4}{{}_{20}C_{10}} = \frac{151,704}{184,756} \approx .8211.$$

Adding in the break-even probability of approximately .1559 reveals that the chance of winning anything on an All or Nothing ticket is a low 2.30%.

## 2by2

*2by2* is a game offered by the Multi-State Lottery Association, the organization also responsible for running Powerball, in Kansas, Nebraska, and North

Dakota. This is a double 2/26 lotto game: players pick two numbers, in the range 1 to 26, in each of two colors: red and white. Two numbers in each color are drawn, and a ticket wins if at least one of its four numbers matches them. The payoff table for 2by2 is shown in Table 4.33.

TABLE 4.33: 2by2 payoff table [3]

| **Matches** | | |
|---|---|---|
| **Red** | **White** | **Prize** |
| 2 | 2 | \$22,000 |
| 2 | 1 | \$100 |
| 1 | 2 | \$100 |
| 2 | 0 | \$3 |
| 0 | 2 | \$3 |
| 1 | 1 | \$3 |
| 1 | 0 | Free ticket |
| 0 | 1 | Free ticket |

Due to symmetry—the chance of matching 2 red balls and 1 white ball is the same as the chance of matching 1 red and 2 white balls, for example—there are five different winning probabilities. The probability of matching $m$ red numbers and $n$ white numbers, where $0 \leqslant m, n \leqslant 2$, is

$$p(m,n) = \frac{{}_2C_m \cdot {}_{24}C_{2-m}}{{}_{26}C_2} \cdot \frac{{}_2C_n \cdot {}_{24}C_{2-n}}{{}_{26}C_2}.$$

This gives the probability distribution shown in Table 4.34.

TABLE 4.34: 2by2 probability distribution

| $p(m,n)$ | | $n$ **(White)** 0 | 1 | 2 |
|---|---|---|---|---|
| $m$ **(Red)** | 0 | .7212 | .1254 | .0026 |
| | 1 | .1254 | .0218 | .0005 |
| | 2 | .0026 | .0005 | $9.4675 \times 10^{-6}$ |

Using Tables 4.33 and 4.34, we have the following equation for $E$:

$$\begin{aligned} E &= (22,000) \cdot 9.4675 \times 10^{-6} + 2 \cdot (100) \cdot .0005 \\ &\quad + 2 \cdot (3) \cdot .0026 + \cdot(3) \cdot .0218 + 2 \cdot (E+1) \cdot .1254 - 1, \end{aligned}$$

or

$$E = .2508E - .3689.$$

The solution to this equation is $E = -.4924$, and the HA is 49.24%.

**Example 4.36.** 2by2 offers a bonus game for players who buy a multi-draw ticket whose number of drawings is a multiple of 7. The *2by2 Tuesday* bonus doubles any amounts won on all Tuesday drawings over the course of the drawings. Find the house advantage of a $1 ticket on Tuesdays.

Denote the expected value of a Tuesday ticket by $E^*$. Since the free tickets won on Tuesday are for non-Tuesday drawings, $E^*$ depends on $E$, the expected value found above for a single ordinary 2by2 ticket. $E^*$ is given by the following equation:

$$E^* = (44,000) \cdot 9.4675 \times 10^{-6} + 2 \cdot (200) \cdot .0005 \\ + 2 \cdot (6) \cdot .0026 + (6) \cdot .0218 + 2 \cdot 2(E+1) \cdot .1254 - 1,$$

or

$$E^* = 4 \cdot .1254E + .2802,$$

from which we find that $E^* = -.0692$, and the HA is only 6.92%. ■

Since the prizes are doubled, but the cost of the ticket is not, this value is a fraction of the HA of a non-Tuesday 2by2 ticket.

## Nebraska MyDaY

*MyDaY* is a Nebraska Lottery lotto offering where players choose a date from the calendar instead of a number. The state draws a single date in a daily drawing, and prizes are paid according to how many components of the date are matched. The date is composed of three separate choices:

- The month, which is a number between 1 and 12.
- The day, which is a number between 1 and 31. Only valid calendar dates are accepted in MyDaY, so the computer system will block attempts to wager on nonexistent dates such as November 31. February 29 may be chosen only if the year is a bona fide leap year, including the year 2000. In the first 7½ years of MyDaY, beginning in October 2008, February 29 was drawn once: February 29, 1932 was the winning date in the January 18, 2016 drawing.
- The year, which is a two-digit number from 00 through 99. These digits represent the last two digits of the year, with the understanding that 00 represents 2000 (which was a leap year, permitting bets on 2/29/00) rather than 1900 (which was not a leap year) [105].

This effectively sets up a date pool consisting of all the days between January 1, 1901 and December 31, 2000—though if you wanted to bet on your child's birthday by interpreting 5/5/01 as May 5, 2001 instead of 1901, no one would object. Twenty-five of those years were leap years, making the denominator in MyDaY calculations $365 \cdot 100 + 25 = 36,525$. The probability of matching all three components of a MyDaY wager is thus somewhat less than the chance of winning on a Pick 4 lotto ticket, where the denominator is 10,000.

Table 4.35 contains the pay table for the MyDaY game, whose tickets cost $1.

TABLE 4.35: Nebraska MyDaY pay table [105]

| **Match** | **Prize** |
|---|---|
| Month, day, and year | $5,000 |
| Day and year | $365 |
| Month and year | $52 |
| Month and day | $12 |
| Year only | $7 |
| Day only | $4 |
| Month only | $1 |

The lottery imposes a cap on payouts by restricting betting to 40 tickets for each possible date; the state's maximum possible liability for the top prize is therefore $200,000.

**Example 4.37.** To calculate the probability of winning the $52 prize for matching the month and year, it is necessary to consider all three components and include the probability that the day is not also matched. This last probability depends on the month involved.

- If the month has 31 days, the probability is

$$\frac{1}{12} \cdot \frac{30}{31} \cdot \frac{1}{100} = \frac{30}{37,200}.$$

- If the month has 30 days, the probability is

$$\frac{1}{12} \cdot \frac{29}{30} \cdot \frac{1}{100} = \frac{29}{36,000}.$$

- If the month is February and the year is not a leap year, we have a probability of

$$\frac{1}{12} \cdot \frac{27}{28} \cdot \frac{1}{100} = \frac{27}{33,600}.$$

- If the month is February and the year is a leap year, then the probability is

$$\frac{1}{12} \cdot \frac{28}{29} \cdot \frac{1}{100} = \frac{28}{34,800}.$$

These four probabilities must be combined in a weighted sum that accounts for how the 36,525 days fall into these four groups. If $p$ is defined as the probability of winning $52, we have

$$p = \frac{21,700}{36,525} \cdot \frac{30}{37,200} + \frac{12,000}{36,525} \cdot \frac{29}{36,000} + \frac{2100}{36,525} \cdot \frac{27}{33,600} + \frac{725}{36,525} \cdot \frac{28}{34,800},$$

or

$$p = \frac{157}{194,800} \approx \frac{1}{1240.76}.$$

If you bought one ticket per day, with each ticket bearing a randomly-selected date, it would take slightly over $3\frac{1}{3}$ years, on the average, to match both the month and year but not the day. ■

## Quebec Astro

*Astro* is a $1 calendar-based game similar to MyDaY that is operated by the Quebec provincial lottery. The three date components of month, day, and year are joined by an astrological sign, adding a factor of 12 to the possible choices. Astro cares not for calendar accuracy, however; a bettor may select dates that do not exist such as February 30, and the chosen astrological sign need not be consistent with the block of days typically associated with that sign.

The number of possible choices is then $12 \cdot 31 \cdot 100 \cdot 12 = 446,400$. The pay table does not distinguish among the elements that are matched by the province's draw; only the number of matching components (out of 4) is counted. The pay table is shown in Table 4.36.

TABLE 4.36: Quebec Astro pay table [80]

| Match | Prize |
|---|---|
| 4 | $25,000 |
| 3 | $500 |
| 2 | $10 |
| 1 | Free ticket |

**Example 4.38.** The probability of matching 2 of the 4 components of the selected date is the sum of 6 terms, each one the probability of matching exactly 2 parts of a full Astro selection: month, day, year,and sign. Each probability is the product of 4 factors, one for each component. The probability of a winning component has the form $\frac{1}{x}$, while a losing component's factor looks like $\frac{x-1}{x}$.

- $P(\text{Match month and day only}) = \frac{1}{12} \cdot \frac{1}{31} \cdot \frac{99}{100} \cdot \frac{11}{12} = \frac{1089}{446,400}$.
- $P(\text{Match month and year only}) = \frac{1}{12} \cdot \frac{30}{31} \cdot \frac{1}{100} \cdot \frac{11}{12} = \frac{330}{446,400}$.
- $P(\text{Match month and sign only}) = \frac{1}{12} \cdot \frac{30}{31} \cdot \frac{99}{100} \cdot \frac{1}{12} = \frac{2970}{446,400}$.
- $P(\text{Match day and year only}) = \frac{11}{12} \cdot \frac{1}{31} \cdot \frac{1}{100} \cdot \frac{11}{12} = \frac{121}{446,400}$.
- $P(\text{Match day and sign only}) = \frac{11}{12} \cdot \frac{1}{31} \cdot \frac{99}{100} \cdot \frac{1}{12} = \frac{1089}{446,400}$.
- $P(\text{Match year and sign only}) = \frac{11}{12} \cdot \frac{30}{31} \cdot \frac{1}{100} \cdot \frac{1}{12} = \frac{330}{446,400}$.

Adding these six probabilities together gives

$$P(\text{Match 2}) = \frac{5929}{446,400} \approx .0133,$$

or approximately once every 75.29 tickets purchased.

■

The free ticket prize for matching one of the four components of the date means that computing the expected value of a \$1 ticket leads to a linear equation:

$$E = (25,000) \cdot \frac{1}{446,400} + (500) \cdot \frac{151}{446,400} + (10) \cdot \frac{5929}{446,400} + (E+1) \cdot \frac{80,949}{446,400} - 1,$$

or

$$E = \frac{80,949E - 205661}{446,400},$$

whose solution is $E = -\$.5628$. As we would expect from a lottery, the house edge, 56.28%, is high.

## Ontario Megadice Lotto

In 2014, the lottery commission in Ontario, Canada, launched *Megadice Lotto*, a \$2 passive lotto ticket with an additional instant video game played at the point of purchase. Like Quebec's Lotto :D, the instant game gives the player an electronically-rolled set of dice, seven in this game, and pays off in accordance with Table 4.37 if a particular combination is rolled.

The various probabilities for the dice game are easily calculated, since the dice are independent.

TABLE 4.37: Pay table for Megadice Lotto's instant dice game [118]

| **Die Roll** | **Prize** |
|---|---|
| 7 ones | \$7500 |
| 7 of a kind, 2s-6s | \$1500 |
| 6 of a kind | \$100 |
| 4 of a kind and 3 of a kind | \$25 |
| 5 of a kind | \$10 |
| 3 of a kind and 3 of a kind | \$5 |
| 3 of a kind and 2 pairs | \$4 |
| 6-number straight | \$3 |
| 4 of a kind | \$2 |

**Example 4.39.** The probability of winning the top prize by rolling seven ones is

$$\left(\frac{1}{6}\right)^7 = \frac{1}{279,936}$$

and the probability of rolling any other seven-of-a-kind combination is

$$5 \cdot \left(\frac{1}{6}\right)^7 = \frac{5}{279,936}.$$

■

While the probability goes up by a factor of 5 and is balanced by the payoff decreasing by the same factor, neither has an especially attractive expectation.

**Example 4.40.** The 6-number straight must necessarily contain a pair. We select the six dice comprising the straight in order: choosing the die (from first to seventh) with a ⚀, then the die with a ⚁, and so on. The number of ways to select the six dice in order is ${}_7P_6 = 5040$, and the probability of getting six different numbers on those six dice is

$$\frac{{}_7P_6}{6^6} = \frac{5040}{46,656}.$$

While the seventh die can be anything, we need to divide by 2 to eliminate double-counting—for example, the single winning combination

⚄ ⚀ ⚁ ⚄ ⚅ ⚂ ⚃

arises from the two choices 2,3,6,7,1,5 and 2,3,6,7,4,5 of the six dice in the straight. The final probability is

$$\frac{{}_7P_6}{2 \cdot 6^6} = \frac{5040}{93,312} \approx .0540.$$

■

**Example 4.41.** For four of a kind plus three of a kind, the one point to remember when computing the probability is to ensure that the four of a kind and the three of a kind are made up of different numbers—so that the roll is not seven of a kind. The probability is

$$\underbrace{6 \cdot {}_7C_4 \cdot \left(\frac{1}{6}\right)^4}_{\text{4 of a kind}} \times \underbrace{5 \cdot \left(\frac{1}{6}\right)^3}_{\text{3 of a kind}} = \frac{175}{6^6} \approx .0038,$$

where the factor of 6 counts the number of values for the four of a kind and the 5 counts the remaining options for the three of a kind. ■

The Megadice Lotto's lottery ticket is a standard 6/39 game drawn each night, with the added twist that a seventh bonus number is drawn from the 33 numbers remaining after the first 6 numbers are drawn, and matching that bonus number along with 5 of the 6 ordinary numbers wins a bonus payoff. Its pay table is shown in Table 4.38—note that the top prize does not accumulate if there are no winners.

TABLE 4.38: Pay table for Megadice Lotto's lottery drawing [118]

| Match | Prize | Probability |
|---|---|---|
| 6 | \$100,000 | $3.065 \times 10^{-7}$ |
| 5 + bonus | \$5000 | $1.839 \times 10^{-6}$ |
| 5 | \$250 | $5.885 \times 10^{-5}$ |
| 4 | \$15 | .0024 |
| 3 | \$3 | .0334 |

**Example 4.42.** A look at Table 4.38 might cause a casual lottery fan to wonder what the point of the bonus ball is: it's only used for one prize, and that prize is not the top prize. Suppose that the bonus ball was dropped from the game. What would be the equivalent payoff on all Match 5 payoffs?

The probability of matching exactly 5 of the 6 drawn numbers is

$$P(5) = \frac{{}_6C_5 \cdot {}_{33}C_1}{{}_{39}C_6} = \frac{198}{{}_{39}C_6}.$$

The bonus payoff is triggered when the sixth number, drawn from 33, is the bonus ball. The expected return on the Match 5 part of the ticket, omitting the cost of the ticket, is

$$E = P(5) \cdot \left(\frac{1}{33} \cdot 5000 + \frac{32}{33} \cdot 250\right) = P(5) \cdot \frac{13,000}{33}.$$

If we eliminate the bonus ball and pay a flat amount of $\$x$ for matching 5 of 6 numbers, the expectation is

$$E = P(5) \cdot x.$$

Setting these two expressions for $E$ equal to each other gives

$$P(5) \cdot \frac{13,000}{33} = P(5) \cdot x,$$

from which it follows that $x \approx \$393.94$. ■

A fixed payoff of $400 for matching 5 numbers would be a convenient amount and yield nearly the same profit for the lottery commission. Whether dropping the bonus ball would affect participation or interest is another matter entirely.

The total probability of winning a prize in the lotto drawing is

$$\frac{117,239}{3,262,623} \approx .0359;$$

the corresponding probability for the instant dice game is

$$\frac{66,786}{279,936} \approx .2386.$$

Taken together, the probability of winning something at Megadice Lotto is

$$1 - P(\text{Lose in both games}) \approx 1 - [(1 - .0359)\,(1 - .2386)] = .2659,$$

or slightly better than 1 chance in 4. This is close to the chance of winning a prize at Lotto :D.

## Ontario NHL Lotto

In Ontario, the *NHL Lotto* game, just like Megadice Lotto, combines an instant video game with a lotto drawing in a single $2 ticket. The video game features a National Hockey League player taking three shots at a goalie, and the drawing is a 5/30 game with an additional 1/30 component that calls for the player to choose one NHL franchise in an effort to match the franchise also chosen by the province in its nightly drawing. (This game predates the NHL awarding its 31st franchise to the Vegas Golden Knights in 2016.)

The goal-scoring game is effectively a variation on a video slot machine. The hockey player, who wears the uniform of the team selected by the gambler, takes three shots at targets placed behind a goalie wearing a different team's uniform. Each target bears a prize amount: $2, $5, $10, $100, and $5000. The number and value of any goals scored are determined by a computerized random number generator, as is done in video slots.

What is different about NHL Lotto is that the Ontario lottery commission has released the probabilities of winning each of the 13 prizes available in the instant game, which are shown in Table 4.39. Slot machine manufacturers compile these probabilities in Probability Accounting Report (PAR) sheets, which are seldom available to the public.

TABLE 4.39: NHL Lotto instant game probabilities [119]

| Prize ($) | Odds of winning |
|---|---|
| 2 | 1 in 3.6 |
| 4 | 1 in 42.8 |
| 5 | 1 in 85.5 |
| 7 | 1 in 143 |
| 10 | 1 in 178 |
| 12 | 1 in 214 |
| 20 | 1 in 611 |
| 22 | 1 in 2036 |
| 100 | 1 in 14,251 |
| 200 | 1 in 42,752 |
| 5000 | 1 in 610,740 |
| 10,000 | 1 in 1,425,600 |
| 15,000 | 1 in 4,275,180 |

Missing from Table 4.39 is the chance of winning nothing in the instant game. This can be easily calculated using the Complement Rule. By converting the figures given in Table 4.39 into fractional probabilities, adding the probabilities together, and subtracting the sum from 1, we find that the chance of winning $0 is approximately .6677—which is not readily apparent from a look at the payoff table. Roughly two tickets out of every three will score no goals, and thus win no money, in the instant game. In terms of odds as shown in Table 4.39, the chance of winning $0 would be stated as about 1 in 1.5.

The lotto drawing is conducted every evening and uses Table 4.40 as its pay table.

TABLE 4.40: NHL Lotto number draw pay table [119]

| Match | Payoff |
|---|---|
| 5 + NHL team | $100,000 |
| 5 without team | $1000 |
| 4 + NHL team | $100 |
| 4 without team | $12 |
| 3 + NHL team | $12 |
| 3 without team | $4 |
| 0–2 + NHL team | $3 |

The probability of winning the $100,000 jackpot is

$$\frac{1}{{}_{30}C_5} \cdot \frac{1}{30} = \frac{1}{142,506} \cdot \frac{1}{30} = \frac{1}{4,275,180}.$$

This is also the probability of winning the maximum prize of $15,000 in the instant game. This is surely no coincidence—since the instant game is computer-controlled, setting the probabilities is a simple exercise in programming and need not correspond to counting anything.

**Example 4.43.** Which of the two events paying \$12 has the higher probability: matching 4 numbers without matching the team or matching 3 of 5 numbers with the team?

The lotto draw and the team selection are independent events, so we can use the Multiplication Rule to combine their probabilities:

$$P(\text{Match 4 without team}) = \frac{{}_5C_4 \cdot {}_{25}C_1}{{}_{30}C_5} \cdot \frac{29}{30} = \frac{3625}{4,275,180}$$

and

$$P(\text{Match 3 with team}) = \frac{{}_5C_3 \cdot {}_{25}C_2}{{}_{30}C_5} \cdot \frac{1}{30} = \frac{3000}{4,275,180}.$$

It's almost 21% more likely to match 4 out of 5 numbers without matching the team. ■

## Max 5

*Max 5* was a nonstandard lotto game operated by the Hoosier Lottery in Indiana from 2002–2003 [23]. The game was simple in its design: a 2/10 + 2/10 lotto where players chose two numbers in the range 1–10 and two from 11–20, and the state drew one number in each subset in each drawing. A player won if his or her ticket matched both of the state's numbers, an event with probability

$$\frac{2}{10} \cdot \frac{2}{10} = \frac{4}{100}.$$

The twist came in the payoff formula. Each \$5 ticket was active for exactly 5 consecutive drawings, and the payoff on a winning ticket was based on how many of those drawings matched both numbers. Table 4.41 shows the payoffs.

TABLE 4.41: Hoosier Lottery Max 5 pay table [56]

| **Winning Draws** | **Prize** |
|---|---|
| 1 | \$5 + Free ticket |
| 2 | \$50 |
| 3 | \$500 |
| 4 | \$20,000 |
| 5 | \$1,000,000 |

Since successive drawings are independent, the binomial distribution with $n = 5$ and $p = .04$ is the appropriate model for $X$, the number of winning draws out of 5.

$$P(X = x) = {}_5C_x \cdot \left(\frac{4}{100}\right)^x \cdot \left(\frac{96}{100}\right)^{5-x}.$$

Setting $x = 5$ shows that the probability of winning in 5 straight drawings is

$$\left(\frac{4}{100}\right)^5 = \frac{1024}{10,000,000,000};$$

this means that the million-dollar prize contributes 10.24¢ to the expected value of a single Max 5 ticket. Despite this low probability of winning, Indiana lottery officials included a provision stating that in the unlikely event that more than 2 tickets were winners in all 5 drawings, the ticket holders would divide a $2,000,000 prize pool equally. Other prize levels had their total payout capped at $500,000 [56].

At the other extreme, the probability of losing five times is $.96^5 \approx 81.5\%$.

## 4.4 Games Where Order Matters

### 4 This Way

From 2004–2007, the New Mexico Lottery offered a twist on Pick 4 lotteries with its *4 This Way* game, which called for players to choose a 4-digit number and paid off according to how many digits, counting from right to left, were matched in their correct places in the state's nightly drawing. Table 4.42 shows the pay table.

TABLE 4.42: 4 This Way pay table [110]

| Digits matched | Payoff |
|---|---|
| 1 | $1 |
| 2 | $4 |
| 3 | $40 |
| 4 | $4000 |

**Example 4.44.** For a ticket bearing the number 8128, any drawn number ending in 8 would win one of the four prizes. As with other games based on matching digits of a number, probabilities for 4 This Way involve the chance of matching some digits and the chance of not matching other digits. The

probability of winning the $1 prize, working digit by digit from left to right, is

$$\frac{10}{10} \cdot \frac{10}{10} \cdot \frac{9}{10} \cdot \frac{1}{10} = \frac{900}{10,000} = .09.$$

Here, the first two digits of the drawn number can be anything, the third digit must *not* be the 2 that would match the ticket's third number, and the last digit must be 8. ∎

Similar analysis gives the probability distribution function shown in Table 4.43. The chance of winning some prize is found to be 10%.

TABLE 4.43: 4 This Way PDF

| Payoff | Probability |
|---|---|
| $1 | .0900 |
| $4 | .0090 |
| $40 | .0009 |
| $4000 | .0001 |
| –$1 | .9000 |

A standard Daily 4 game, which New Mexico does not offer, pays 5000 for 1 the player's number matches all 4 digits in order and nothing otherwise. How does 4 This Way compare to this game, with its HA of 50%?

The expectation of a $1 4 This Way ticket is

$$E = -\frac{4380}{10,000},$$

giving a HA of 43.8%. The smaller payoffs for matching some digits in order served to give the player a slightly better game.

## Mix & Match

Example 2.49 on page 39 described Mix & Match, the $2 Pennsylvania Lottery 5/19 game that paid off for matching 1–5 numbers in the same order as drawn ("Match") and 3–5 numbers in any order ("Mix"). Mix & Match's pay table is shown in Table 4.44.

**Example 4.45.** The probability of winning the $1000 Match prize by matching exactly 4 numbers in order is simply

$$\frac{{}_5C_4 \cdot {}_{14}C_1}{{}_{19}P_5} = \frac{70}{1,395,360},$$

or approximately 1 in 19,934. Note that combinations are used in the numerator, since we are simply counting which 4 of the 5 numbers are drawn in their correct places. ∎

TABLE 4.44: Mix & Match pay table ($2 ticket) [126]

| **Numbers matched** | **Mix prize (Any order)** | **Match prize (In order)** |
|---|---|---|
| 1 | — | Free ticket |
| 2 | — | 4 |
| 3 | 2 | 100 |
| 4 | 20 | 1000 |
| 5 | 2000 | Jackpot |

Mathematical analysis of Mix & Match is facilitated by the fact that tickets could win more than one prize, which meant that 17 different prize combinations were available. This stands in contrast to many other draw-based lotteries which state that only the highest prize is awarded to a winning ticket.

**Example 4.46.** Suppose that the state's ordered draw was $\langle 10, 3, 11, 8, 7\rangle$. A ticket carrying the numbers $\langle 10, 3, 7, 14, 4\rangle$ has two numbers in the right places and one additional winning number, and so would win $6: $4 for the 2 Match numbers and an additional $2 Mix prize for matching 3 of the 5 winning numbers. ■

The ticket in Example 4.45 cannot qualify for an additional Mix prize, since the fifth number, if drawn, would also have to be in its correct place.

## Daily Derby

*Daily Derby* is a horse racing-themed game operated by the California Lottery. Players choose 3 horses from a field of 12 and also guess the last 3 digits of the time, in minutes, seconds, and hundredths of seconds, that the hypothetical race will last. The time of the race is guaranteed to be between 1:40.00 and 1:49.99, so choosing the time amounts to picking a 3-digit number between 000 and 999 [20].

The field is listed in Table 4.45. The player's three horses must be selected in order; thus there are ${}_{12}P_3 = 1320$ different choices. Multiplying by the 1000 ways to fill out the time of the race gives 1,320,000 different possible selections.

There are effectively two distinct games on a Daily Derby ticket: one for the horse race and one for the time. The top prize for the race is won when hitting the trifecta: selecting the three winning horses in the correct order. Lesser prizes are available for the exacta—matching the top two horses in order—or for simply picking the winner. Correctly guessing the winning time wins a prize of its own, and combining a winning trifecta with a correct time wins the grand prize.

California law provides that Daily Derby is a pari-mutuel game, meaning that the prizes are based on fixed percentages of a pool of prize money. This pool is formed by taking any funds not won in the previous day's drawing and

TABLE 4.45: California Lottery Daily Derby field

| Number | Name |
|---|---|
| 01 | Gold Rush |
| 02 | Lucky Star |
| 03 | Hot Shot |
| 04 | Big Ben |
| 05 | California Classic |
| 06 | Whirl Win |
| 07 | Eureka |
| 08 | Gorgeous George |
| 09 | Winning Spirit |
| 10 | Solid Gold |
| 11 | Money Bags |
| 12 | Lucky Charms |

adding approximately 50% of the day's ticket sales. This pool is then divided into separate subpools for each prize, as indicated in Table 4.46.

TABLE 4.46: California Lottery Daily Derby prize pool allocation [21]

| Prize | Allocation |
|---|---|
| Grand Prize | 14% |
| Trifecta | 36% |
| Exacta | 16% |
| Win | 25% |
| Race Time | 5% |

The individual pool amounts are then divided by the number of winning tickets in each category to determine the final prize amounts. Prizes are rounded down to the nearest dollar, except when such rounding would take a prize below the $2 ticket price. These prizes are always paid at $2, with extra funds, if necessary, drawn from the 4% of the total pool that is held out of the prize pools and allocated to a prize reserve for this purpose [21]. Excess funds from rounding, as well as the money allocated to prizes where there were no winning tickets, are added to the pool for the next day's Grand Prize, which means that the Grand Prize is something of a progressive jackpot.

The probability of winning the Race Time prize, which typically pays out somewhere between $40 and $60, is slightly less than 1 in 1000, since we must exclude the chance of also hitting the trifecta and winning the grand prize. That probability is

$$P(\text{Pick the right time}) \cdot P(\text{Lose the trifecta prize}) = \frac{1}{1000} \cdot \frac{1319}{1320} \approx \frac{1}{1001}.$$

## Lucky Lines and Lucky Links

*Lucky Lines* and *Lucky Links* are two related games, from Oregon and Connecticut respectively, that involve eight numbers and rules that determine prizes depending on where any matched numbers fall in a $3 \times 3$ grid with a center "free" space.

### Lucky Lines

In Oregon, Lucky Lines combines lotto action with the geometry of tic-tac-toe and the center free square from bingo. Players choose one number from four for each of eight blocks in a tic-tac-toe grid, with the center square designated as a free square. The numbers assigned to each block are shown in Figure 4.9.

| | | |
|---|---|---|
| 1–4 | 5–8 | 9–12 |
| 13–16 | Free | 17–20 |
| 21–24 | 25–28 | 29–32 |

FIGURE 4.9: Lucky Lines number assignment.

The state draws its own set of 8 numbers nightly, and players win an amount (Table 4.47) based, not on the number of matches, but on the number of tic-tac-toe lines, from 1 to 8, that they complete on their $2 ticket. The top prize is a pari-mutuel jackpot that starts at $10,000. It should be noted that completing exactly 7 lines is impossible—matching 7 of 8 numbers completes either 5 or 6 lines, depending on whether or not the unmatched number is in a corner. Matching all 8 numbers completes all 8 lines, of course.

TABLE 4.47: Lucky Lines pay table [121]

| Lines | Prize |
|---|---|
| 1 | $2 |
| 2 | $4 |
| 3 | $7 |
| 4 | $25 |
| 5 | $100 |
| 6 | $500 |
| 7–8 | Jackpot |

The number of matches on a Lucky Lines ticket is a simple binomial random variable with $n = 8$ and $p = \frac{1}{4}$, hence a typical ticket will match 2 numbers. Since the center square functions as a free match for any line on which it falls, it is possible to win a prize in Lucky Lines by matching as few as 2 numbers. However, a prize is only guaranteed when a ticket matches 5 or more of the 8 numbers. For example, matching only the numbers 2, 7, 17 and 23 in Figure 4.9 would fail to complete a line, and so would win nothing.

**Example 4.47.** Say that a Lucky Lines ticket is *unlucky* if it matches 4 numbers but nonetheless wins no prizes. What is the probability of an unlucky ticket?

There are $4^8 = 65,536$ different Lucky Lines tickets, and ${}_8C_4 \cdot 3^4 = 5670$ ways to match 4 of the 8 numbers. We need to count the number of configurations that correspond to no completed lines. Consider a blank Lucky Lines ticket, with only the free center square filled in.

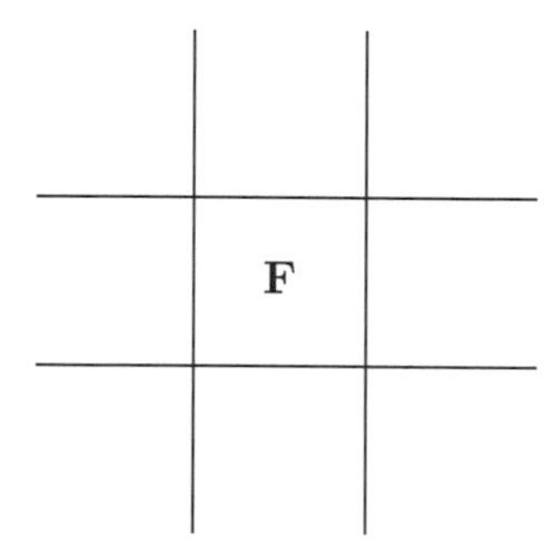


The unlucky match 4 combinations can be counted by following this algorithm:

1. The middle row must contain exactly one match besides the free square, because four matches in the top and bottom rows will complete at least one row, column, or diagonal, and two matches will complete the middle row. Pick one square from the middle row. There are 2 choices for that square.

2. Similarly, the middle column must contain one match, and again there are 2 ways to select it.

   These two choices identify a $2 \times 2$ corner of the board with 3 of 4 squares, including the center free square, filled in. In Figure 4.10, with • denoting a matched number and **F** the free center square, this is the upper right corner. Note that the fourth square, in the upper right, has not been matched. The number in this square must remain unmatched, for if it is drawn, *any* fourth match will complete a line.

3. Of the 5 squares remaining outside that $2 \times 2$ corner, we need to choose 2, which gives ${}_5C_2 = 10$ possibilities. We cannot pick the other square remaining in either the middle row or column; that eliminates 7 of the

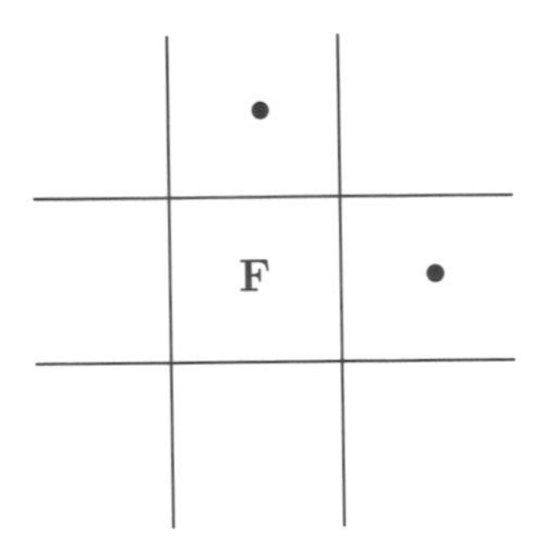


FIGURE 4.10: Lucky Lines: Configuration with two matches.

10 options. Nor can we choose the two that lie on opposite ends of a diagonal. This leaves only 2 valid choices: the square opposite the empty corner square identified above together with the square in its opposite vertical or horizontal corner.

Combining the choices here gives $2 \cdot 2 \cdot 2 = 8$ ways to match 4 of 8 numbers without completing a line. One such arrangement is shown in Figure 4.11.

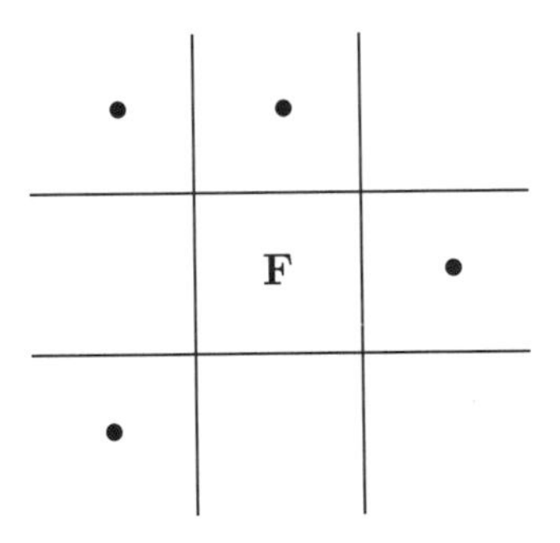


FIGURE 4.11: Lucky Lines: Unlucky configuration with 4 numbers matched but no lines complete.

An alternate argument relying on geometry confirms this conclusion: We seek to count the number of unlucky Lucky Line configurations by starting with one, as in Figure 4.11, and resorting to symmetries of the square layout to generate the others by rotating or reflecting this configuration.

1. Rotating the figure $90°, 180°$, or $270°$ counterclockwise (or, equivalently, $270°, 180°$, or $90°$ clockwise) around its center gives three new configurations.

2. Four additional configurations can be generated by reflecting the original design across any of four lines of symmetry: a horizontal line through the middle row, a vertical line through the middle column, and two diagonal lines passing through the center square, one from upper left to lower right and one from upper right to lower left.

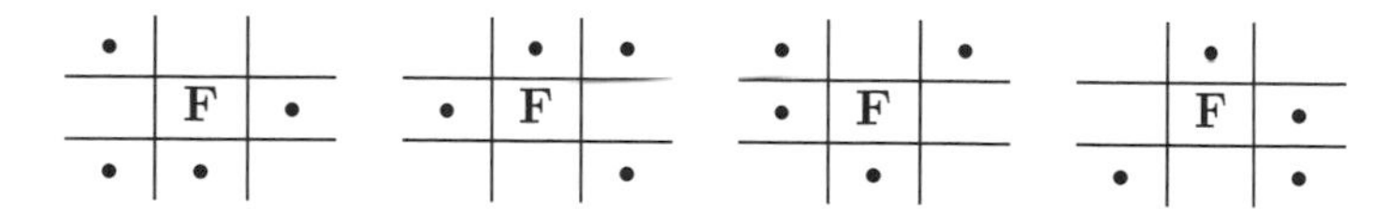

Combining the choices here gives 8 ways to match 4 numbers without completing a line.

Each of these 8 ways can be obtained by matching one of 4 numbers in each square, giving $4^4 = 256$ combinations of 4 numbers corresponding to each unlucky pattern, for a total of 2048 unlucky combinations. The probability of such an unlucky match is then

$$\frac{2048}{65,536} = \frac{1}{32}.$$

■

This does not take into account the difficulty of matching 4 numbers out of 8. A more accurate measure of the likelihood of crushing disappointment is the conditional probability of filling no lines, given that 4 numbers are matched. By reducing the sample space, we see that that probability is

$$P(\text{Fill 0 lines} \mid \text{Match 4 numbers}) = \frac{2048}{5670} \approx .3612,$$

or approximately 36.12%.

### Lucky Links

A game similar to Lucky Lines, Lucky Links, is offered by the Connecticut Lottery. Lucky Links is a passive 8/22 lottery; tickets are assigned randomly and players have no choice of numbers or positions.

The 8 numbers on a Lucky Links ticket are arranged in a $3 \times 3$ array of circles with the center circle, marked with a \$, as a free space (Figure 4.12). Like Lucky Lines, the state draws its numbers and payoffs are based on the number of completed tic-tac-toe lines.

Lucky Links tickets cost \$2. An optional wager, 2XPOWER, doubles the payoff for filling 3–6 lines for an additional dollar wagered. Table 4.48 gives the standard Lucky Links and 2XPOWER pay tables.

As with Lucky Lines, it is impossible to fill exactly 7 lines on a Lucky Links ticket. The \$50,000 payoff for filling 8 lines is paid from a prize pool capped at \$1,000,000. If 21 or more players fill all 8 lines, the million dollars is

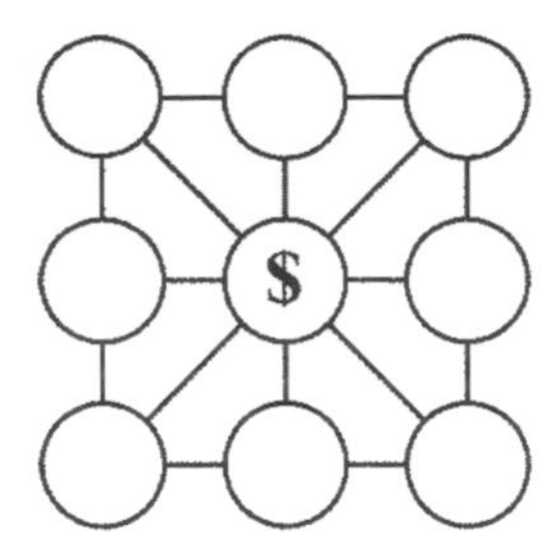


FIGURE 4.12: Lucky Links ticket.

TABLE 4.48: Lucky Links pay table [24]

| Lines | Standard | 2XPOWER |
|---|---|---|
| 2 | $5 | $5 |
| 3 | $10 | $20 |
| 4 | $50 | $100 |
| 5 | $100 | $200 |
| 6 | $1000 | $2000 |
| 7–8 | $50,000 | $50,000 |

divided among all such winning tickets; this is the only payoff that is a shared jackpot.

Since any of the 22 numbers may be placed in any of the eight circles, order seems like it should matter when considering the possibilities of a Lucky Links ticket. There are ${}_{22}P_8 = 12,893,126,400$ distinct tickets, far more than the 65,536 Lucky Lines tickets. However, it is not necessary to consider the ordered possibilities in computing the probability of filling any number of lines, since all we are interested in is how any matched numbers are distributed about the 8 circles.

**Example 4.48.** Just as we did with Lucky Lines, we consider the probability of matching 4 numbers while completing no lines. The probability of matching exactly 4 drawn numbers is

$$P(4) = \frac{{}_8C_4 \cdot {}_{14}C_4}{{}_{22}C_8} = \frac{70,070}{319,770} \approx .2191,$$

which is more than double the corresponding probability of .0865 in Lucky Lines.

The layout of the numbers on a ticket matters, and we take that into account next. There are ${}_8C_4 = 70$ ways to choose the 4 circles occupied by the matched numbers. Of these 70 selections, 8 contain no complete lines (as we saw in examining Lucky Lines), 44 contain 1 line, and 18 contain 2. It follows that the probability of matching 4 numbers and completing no lines is

$$P(\text{Fill 0 lines} \mid \text{Match 4 numbers}) = \frac{70,070}{319,770} \cdot \frac{8}{70} \approx .0250.$$

■

Since Lucky Links has no prize for filling in only 1 line, the probability of matching 4 numbers and not winning anything is higher than this: approximately .1628.

It is evident that an important part of the analysis of Lucky Links is determining, for each $k$, how many tic-tac-toe lines are associated with each of the ${}_8C_k$ ways to select $k$ matched circles. Figure 4.13 is a spreadsheet computing the probabilities and expected values associated with Lucky Links; columns E through M show how the arrangements of $k$ matched circles (column B) are distributed from 0 through 8 completed lines.

| | A | B | C | D | E | F | G | H | I | J | K | L | M | N | O | P | Q |
|---|---|---|---|---|---|---|---|---|---|---|---|---|---|---|---|---|---|
| 1 | Lucky Links, Connecticut Lottery | | | | Lines Filled | | | | | | | | | Ticket Price: | 2 | | |
| 2 | Pick | Match | Probability | | 0 | 1 | 2 | 3 | 4 | 5 | 6 | 7 | 8 | Total | | Expectation | 2XPOWER Ex. |
| 3 | 8 | 0 | 0.00939112 | | 1 | 0 | 0 | 0 | 0 | 0 | 0 | 0 | 0 | 1 | | 0 | 0 |
| 4 | | 1 | 0.08586171 | | 8 | 0 | 0 | 0 | 0 | 0 | 0 | 0 | 0 | 8 | | 0 | 0 |
| 5 | | 2 | 0.2629515 | | 24 | 4 | 0 | 0 | 0 | 0 | 0 | 0 | 0 | 28 | | 0 | 0 |
| 6 | | 3 | 0.350602 | | 31 | 25 | 0 | 0 | 0 | 0 | 0 | 0 | 0 | 56 | | 0 | 0 |
| 7 | | 4 | 0.21912625 | | 8 | 44 | 18 | 0 | 0 | 0 | 0 | 0 | 0 | 70 | | 0.281733746 | 0.281733746 |
| 8 | | 5 | 0.06374582 | | 0 | 8 | 35 | 13 | 0 | 0 | 0 | 0 | 0 | 56 | | 0.347187041 | 0.495168402 |
| 9 | | 6 | 0.00796823 | | 0 | 0 | 0 | 18 | 8 | 2 | 0 | 0 | 0 | 28 | | 0.221972042 | 0.443944085 |
| 10 | | 7 | 0.00035025 | | 0 | 0 | 0 | 0 | 0 | 4 | 4 | 0 | 0 | 8 | | 0.192638459 | 0.385276918 |
| 11 | | 8 | 3.1272E-06 | | 0 | 0 | 0 | 0 | 0 | 0 | 0 | 0 | 1 | 1 | | 0.156362385 | 0.156362385 |
| 12 | | | | | | | | | | | | | | | | | |
| 13 | | | Payoff | | 0 | 0 | 5 | 10 | 50 | 100 | 1000 | 50000 | 50000 | | E: | -0.80010633 | -1.237514464 |
| 14 | | | 2XPOWER | | 0 | 0 | 5 | 20 | 100 | 200 | 2000 | 50000 | 50000 | | HA: | 40.01% | 41.25% |

FIGURE 4.13: Spreadsheet calculations for Lucky Links.

Column C computes the probability of matching exactly $k$ numbers; this is

$$P(k \mid 8) = \frac{{}_8C_k \cdot {}_{14}C_{8-k}}{{}_{22}C_8}.$$

Finding the expected return on a ticket requires separating the ${}_8C_k$ ways to arrange the matched numbers (column N) by number of lines completed. For example, consider row 8, which covers the case $k = 5$: Of the 56 possible arrangements, 8 complete one line, 35 complete two, and 13 complete three. It follows that if 5 numbers have been matched, the probability distribution in Table 4.49 is in force.

Since Lucky Links includes no payoff for filling in only one line, it is possible to match 5 numbers out of 8 and still lose. Table 4.49 shows that

$$P(\text{Lose} \mid \text{Match 5}) = \frac{8}{56} = \frac{1}{7}.$$

Combining Table 4.49 with Figure 4.13 shows that the probability of matching

TABLE 4.49: Lucky Links probability distribution with 5 numbers matched

| Lines completed | Probability |
|---|---|
| 1 | 8/56 |
| 2 | 35/56 |
| 3 | 13/56 |

5 numbers and not winning anything is

$$\frac{{}_8C_5 \cdot {}_{14}C_3}{{}_{22}C_8} \cdot \frac{1}{7} = \frac{20,384}{2,238,390} \approx .0091,$$

or just less than 1%.

To compute the expected value of either the standard or 2XPOWER ticket, we use the Excel **SUMPRODUCT** function to multiply the elements of cells E through M in a given row by the corresponding payoff cells, row 13 or 14 depending on the wager, and add the products. This sum is then divided by ${}_8C_k$ (column N), to convert the frequencies of each count of filled lines to probabilities, and multiplied by the probability of matching $k$ numbers (column C). These values are displayed in columns P and Q; summing and subtracting the ticket price gives an house advantage of 40.01% for the $2 ticket and 41.25% for 2XPOWER.

For the 2XPOWER bet, doubling some of the payoffs while only multiplying the ticket price by 1.5 does not turn out the be the advantage it might seem. If the prize for filling in 2 lines was also doubled, to $10, the resulting ticket would have a HA of only 25.22%, and so would represent a better—though still not a good—wager.

## City Picks—Derangements

In example 2.50 (page 39), we introduced City Picks, a $1 game operated by the Wisconsin Lottery in 2002 and 2003. The object was to arrange a list of nine Wisconsin cities in order and to match the order determined by the state in its nightly drawing [162]. Payoffs, which are shown in Table 4.50, were based on how many of the player's ordered cities matched the state's order. We note immediately that it is impossible to match exactly 8 of the 9 cities, since if 8 cities match, the ninth must also be in the right place.

Computing the probabilities for City Picks involves a different approach to combinatorics than mere permutations and combinations. It is clear that order matters in selecting a ticket, so the number of possible tickets is

$${}_9P_9 = 9! = 362,880.$$

As with keno and other lotto games, counting the number of ways to match $k$ cities involves consideration of the cities that are matched and the ones that

TABLE 4.50: Pay table for City Picks [23]

| Match | Prize |
|---|---|
| 9 | $50,000 |
| 7 | $1000 |
| 6 | $200 |
| 5 | $20 |
| 4 | $4 |
| 3 | $1 |

are not matched. As an illustration, we shall consider the case $k = 5$. The number of ways to choose a ticket with exactly 5 cities in the correct slots is the product of two factors, denoted here by $M$ and $N$.

- $M$ = Number of ways to pick 5 matching cities from 9 (the *match* number). $M$ is easily seen to be ${}_9C_5$: once the cities are selected, there is no choice about the order in which to place them.

- $N$ = Number of ways to arrange the other 4 cities without matching any (the *non-match* number).

$N$ involves order: we have identified the four non-matching cities, but we need to count the number of ways to arrange them in order without putting any in the correct slot. This calls for permutations rather than combinations, but clearly $N < {}_4P_4$, since some permutations of four cities will leave at least one city unmoved. The type of permutation we seek here is called a *derangement.*

**Definition 4.1.** A *derangement* of a finite set is an ordered rearrangement of the elements of the set which has no fixed points: every element of the set is moved from its original position. The number of derangements of a set with $n$ elements is denoted $D_n$.

**Example 4.49.** Consider the set $\{1, 2, 3, 4\}$. If we rearrange the set into the order $\{2, 1, 4, 3\}$, we have a derangement. The order $\{4, 1, 3, 2\}$ is not a derangement, since the 3 has not moved. ■

In the City Picks problem, we see that $N = D_4$, because we want to count the ways to arrange four cities in order without placing any city in the "correct" location determined by the state's drawing. The number of tickets with 5 cities correctly placed is then $MN = {}_9C_5 \cdot D_4$. In general, the number of ways to match $k$ of 9 cities is ${}_9C_k \cdot D_{9-k}$.

For small values of $n$, $D_n$ can be found by simply listing all the possible cases. If $n = 1$, there are no derangements. If $n = 2$, there is one derangement: $\{1, 2\}$ can be deranged to $\{2, 1\}$. It is convenient to define $D_0 = 1$, much as we define $0! = 1$.

More generally, a simple recurrence relation defines $D_n$ in terms of $D_k$ for some values of $k < n$. We begin by showing that $D_3 = 2$. Since the number 1 cannot be first in a derangement, we have 2 choices for the first number: it can be either 2 or 3. Once that choice is made, the rest of the derangement is determined:

- If the first number is 2, 1 cannot be placed in the second position, since that would require that 3 fill in the only remaining position, and we would not have a derangement. The only derangement of 3 numbers starting with 2 is $\{2, 3, 1\}$.
- Similarly, if 3 is the first number, the derangement must be $\{3, 1, 2\}$.

At the $D_4$ level, we start deriving the recursion. Suppose that the number 4 is deranged into position $k$. The derangements of the set $\{1, 2, 3, 4\}$ all fall into one of two types.

1. The number $k$ is not in position 4. An example of this type would be $\{4, 3, 1, 2\}$, where 4 is in position 1 and 1 is not in position 4. Removing the 4 from consideration shows that the remaining three numbers must be deranged: 2 cannot be in position 2, 3 cannot be in position 3, and 1 cannot be in position 4 (by the definition of this type of derangement). Accordingly, there are $D_3$ ways to form a derangement with the remaining numbers. There are 3 ways to select where the number 4 goes in each derangement of $\{1, 2, 3, 4\}$, so derangements of this type number $3 \cdot D_3$. In general, there are $(n-1) \cdot D_{n-1}$ of these derangements.
2. The number $k$ is in position 4, as with $\{3, 4, 1, 2\}$. Removing the numbers 4 and $k$ from this derangement leaves a derangement of $n-2$ numbers. Once again, there are $n-1$ choices for $k$, which contributes $3 \cdot D_2$ new derangements, or $(n-1) \cdot D_{n-2}$ in general.

Combining these two types of derangements gives the following relationship for $D_n$ when $n \geqslant 3$:

$$D_n = (n-1) \cdot (D_{n-1} + D_{n-2}).$$

Using the initial conditions $D_1 = 0$ and $D_2 = 1$ allows us to generate $D_n$ for any value of $n$; the values pertinent to City Picks, where $0 \leqslant n \leqslant 9$, are shown in Table 4.51.

TABLE 4.51: Derangements

| $n$ | 0 | 1 | 2 | 3 | 4 | 5 | 6 | 7 | 8 | 9 |
|---|---|---|---|---|---|---|---|---|---|---|
| $D_n$ | 1 | 0 | 1 | 2 | 9 | 44 | 265 | 1854 | 14,833 | 133,496 |

We conclude that $D_4 = 9$ and thus that the number of winning tickets with 5 cities correctly matched is ${}_9C_5 \cdot 9 = 1134$.

An alternate derivation gives

$$D_n = \left\lfloor \frac{n!}{e} + \frac{1}{2} \right\rfloor,$$

where $e \approx 2.71828...$ is the base of the natural logarithm and $\lfloor x \rfloor$ denotes the *floor function*: the greatest integer less than or equal to $x$. This formula can be used to calculate $D_n$ for a given $n$ directly, without the need to compute all of the previous values.

A probability distribution for City Picks can then be derived:

$$P(\text{Match } k \text{ cities}) = \frac{{}_9C_k \cdot D_{9-k}}{9!}.$$

These values are tabulated in Table 4.52.

TABLE 4.52: Probability distribution for City Picks

| $k$ | $P$(Match $k$) |
|---|---|
| 0 | .3679 |
| 1 | .3679 |
| 2 | .1839 |
| 3 | .0611 |
| 4 | .0153 |
| 5 | .0028 |
| 6 | $4.630 \times 10^{-4}$ |
| 7 | $9.921 \times 10^{-5}$ |
| 8 | 0 |
| 9 | $2.756 \times 10^{-6}$ |

Multiplying the probabilities in Table 4.52 by the corresponding prizes in Table 4.50, adding the products together, and subtracting the $1 ticket price gives an expected value of –$.4855, and a house advantage of 48.55%.

Recall Alexander Hamilton's maxim that successful lotteries had both simple rules and inexpensive tickets. While few players were likely to be deterred by the cost of a $1 City Picks ticket, this game may well have fallen short on the simplicity metric, as it lasted only eleven months. In the time that City Picks was available for play, 16 people won the $50,000 jackpot for correctly ordering all 9 cities [162].

## 4.5 Powerball

*Powerball* is a popular American lotto game administered by the Multi-State Lottery Association. Tickets for this lottery are sold in 44 states and 3

U.S. territories: the District of Columbia, Puerto Rico, and the U.S. Virgin Islands. The structure of Powerball has remained unchanged through several adjustments to the parameters: players must choose five numbered white balls from one bin and one red numbered ball from a separate bin—this last ball is known as the Powerball. Payoffs are based on the numbers matched: matching the Powerball always wins a prize which rises with the number of white balls also matched. Matching at least 3 white balls without matching the Powerball also wins a prize. Matching all 5 white balls and the Powerball wins a progressive jackpot that starts at $40,000,000.

If there is no jackpot winner in a drawing, the jackpot rolls over and is increased for the next drawing. Multiple rollovers can lead to jackpots in the hundreds of millions of dollars.

The basic mechanics of the game have been constant since Powerball began. What has changed several times is the number of white and red balls. When Powerball was first launched in 1992, in 13 states and the District of Columbia, there were 45 white balls and 45 red balls, making the game a 5/45 + 1/45 proposition [23]. Order doesn't matter when choosing the white balls, so they may be chosen in ${}_{45}C_5 = 1,221,759$ different ways. By the Fundamental Counting Principle, we must multiply this number by the 45 choices for the Powerball; this gives 54,979,155 possible results for each drawing, and so the chance of winning the jackpot on a single ticket was originally 1 in 54,979,155.

While this probability is objectively very small, as more states signed on to Powerball, the top prize was won frequently enough that sales began to suffer from the lack of truly huge jackpots that garnered immense publicity for the game. In 1997, the game was changed to a 5/49 + 1/42 game [23]. Going from 90 to 91 balls may not seem like much of a difference, but changing the distribution between white and red balls in this manner took the number of possible tickets to 80,089,128—nearly a 50% increase.

A third change was instituted in 2002, when the number of white balls was increased again, to 53. This brought the number of possible tickets to 120,526,700. Changes continued, with each change increasing the number of combinations and thus increasing the likelihood of a jackpot rolling over. Table 4.53 tracks the number of white and red balls, and the corresponding number of possible tickets.

The most recent change occurred in 2015, when the New York Gaming Commission, overseers of Powerball, changed the number of white balls from 59 to 69 and decreased the number of red balls from 39 to 26. These changes were made in an effort to ensure the continued survival of Powerball, whose sales had been dropping in the face of smaller big jackpots [151]. Under the newest rules, the number of possible tickets is now 292,201,338, over 5 times the number of combinations in the game's original configuration.

**Example 4.50.** To put this number in perspective, consider Michigan Stadium at the University of Michigan-Ann Arbor, which is the largest stadium in the USA, with a capacity of 109,901. If we were to fill the stadium 2658 times,

TABLE 4.53: Powerball game history [129]

| Year | White balls | Red balls | Possible tickets |
|---|---|---|---|
| 1992 | 45 | 45 | 54,979,155 |
| 1997 | 49 | 42 | 80,089,128 |
| 2002 | 53 | 42 | 120,526,700 |
| 2005 | 55 | 42 | 146,107,962 |
| 2009 | 59 | 39 | 195,249,054 |
| 2015 | 69 | 26 | 292,201,338 |

we would have almost as many people as there are possible Powerball tickets. In the history of the stadium, which includes many years before its capacity was expanded to its current number, there have not been even half that many games. In short, fewer people have attended Michigan home football games than there are Powerball combinations in its current configuration. ■

Powerball's current pay table is given in Table 4.54.

TABLE 4.54: Powerball pay table

| Match | Prize |
|---|---|
| 5 + Powerball | Jackpot |
| 5 | \$1,000,000 |
| 4 + Powerball | \$50,000 |
| 4 | \$100 |
| 3 + Powerball | \$100 |
| 3 | \$7 |
| 2 + Powerball | \$7 |
| 1 + Powerball | \$4 |
| 0 + Powerball | \$4 |

In exchange for the decreased jackpot probability outlined in Table 4.53, the chance of winning certain other prizes was increased slightly in 2015, so something was given as something else was taken away. For example, a \$2 Powerball ticket pays a \$4 return when the Powerball is matched with either zero or one of the five white balls. Under the 2009 rules, the probability of winning \$4 is

$$\frac{{}_{54}C_5 + 5 \cdot {}_{54}C_4}{{}_{59}C_5} \cdot \frac{1}{35} = \frac{4,743,765}{175,223,510} \approx .0271 \approx \frac{1}{37}.$$

With the new ball distribution introduced in 2015, this probability rises, to

$$\frac{{}_{64}C_5 + 5 \cdot {}_{64}C_4}{{}_{69}C_5} \cdot \frac{1}{26} = \frac{10,801,392}{292,201,338} \approx .0370 \approx \frac{1}{27}.$$

On the other hand, the probability of turning \$2 into \$1,000,000 by matching all 5 white balls and not matching the Powerball dropped by more than half, from

$$\frac{1}{{}_{59}C_5} \cdot \frac{34}{35} \approx \frac{1}{5,153,633}$$

to

$$\frac{1}{{}_{69}C_5} \cdot \frac{25}{26} \approx \frac{1}{11,688,054}.$$

The change worked. The new rules went into effect in November 2015, and the jackpot rolled over until January 13, 2016 when, on the 21st drawing, the jackpot of \$1.586 billion was split among three winning tickets. The winning numbers were 4, 8, 19, 27, and 34; the Powerball was 10. In addition to the three jackpot winners, nearly \$273.9 million was paid out in lesser prizes.

## Some Financial Mathematics

The January 2016 Powerball jackpot was advertised as a 30-year annuity, or regular series of payments, totalling the jackpot amount. A single winner could have chosen instead to accept a single cash payment of about \$930 million. Where does this second value come from?

Since the first annuity payment is made immediately after the winning ticket is verified, this annuity is known as an *annuity due*, and we think of each payment as associated with the year following it, with payments to be made at the beginning of year 1 through the beginning of year 30, 29 years later. \$930 million represents the *present value* of the jackpot: the amount of money that the lottery agency needs to have on hand right now to generate the 30 annual payments, assuming that money can be invested at a return of $i\%$ annually. In order that a payment of $R$ dollars may be made $k$ years from now, a fixed amount less than $R$ can be invested for those $k$ years.

Specifically, let $i$ be the interest rate, compounded annually, expressed as a decimal number, so 3½% interest translates to $i = .035$. It follows that $R \cdot (1+i)^{-k}$—the discounted amount of the eventual payment $R$—set aside now to accumulate interest will accumulate in value to $R$ dollars at the end of $k$ years. Adding up the accumulated value of all 30 payments, which correspond to $k$ running from 0 to 29 years, gives the total present value $A$:

$$A = \sum_{k=0}^{29} R(1+i)^{-k} = R \cdot \sum_{n=0}^{29} (1+i)^{-k}.$$

For convenience, we define $v = (1+i)^{-1}$, giving

$$A = \sum_{k=0}^{29} Rv^k.$$

This sum is finite, and so may be evaluated in closed form for a given value of $n$:

$$\sum_{k=0}^{n-1} v^k = \frac{1 - v^n}{1 - v}.$$

This final expression is denoted $\ddot{a}_{\overline{n}|i}$ and pronounced "a dots angle $n$ at $i$":

$$\ddot{a}_{\overline{n}|i} = \sum_{k=0}^{n-1} (1+i)^{-k} = (1+i) \cdot \frac{1 - (1+i)^{-n}}{i} = \frac{1 - v^n}{1 - v} = \frac{1 - v^n}{iv},$$

and we have

$$A = R\ddot{a}_{\overline{n}|i}.$$

**Example 4.51.** In our example above, with $n = 30$ and $i = 3\frac{1}{2}\%$, we have

$$\ddot{a}_{\overline{30}|.035} = \frac{1 - \left(\frac{1}{1.035}\right)^{30}}{\left(\frac{.035}{1.035}\right)} \approx 19.0358.$$

■

Prior to the advent of powerful inexpensive computers and electronic calculators, performing calculations involving $\ddot{a}_{\overline{n}|i}$ involved working with extensive precomputed tables of these values for common interest rates and annuity lengths.

At the same time, the advertised jackpot amount $S$ is interpreted as the *accumulated amount* of the annuity. $S$ and $A$ are related by the equation

$$S = A(1+i)^n.$$

In words, $S$ is the amount that could be earned by investing $A$ at an annual interest rate of $i$ for $n$ years. Given $S$, this equation can be used to compute $A$, which can then be used to find $R$, the amount of the annual payments.

**Example 4.52.** For $S = \$1.586$ billion, $i = 3\frac{1}{2}\%$, and $n = 30$, we find that $A \approx \$565,057,559$, and then

$$R = \frac{A}{\ddot{a}_{\overline{30}|.035}} \approx \frac{565,057,559}{19.0358} \approx \$29,683,992.$$

■

This value for $A$—nearly \$350 million less than the announced lump sum payment—suggests that Powerball was not counting on a 3½% return. Solving for $i$ gives

$$i = \sqrt[n]{\frac{S}{A}} - 1.$$

Using the announced values of $S = \$1.586$ billion and $A = \$930$ million shows that the effective interest rate that Powerball management was expecting to earn was about 1.86%.

This simple analysis assumes an annuity with a fixed annual payment $R$. The Powerball annuity is actually a variable annuity whose payment increases by 5%—the *escalator rate*—each year; this represents some attempt to ensure that the prize maintains its value despite inflation. If the initial payment is denoted $R'$, the annual payment after $k$ years is then $R' \cdot (1.05)^k$, and so we have

$$A = \sum_{k=0}^{29} R' \cdot (1.05)^k \cdot (1+i)^{-k}.$$

We shall work with this formula in the general case where the escalator rate is denoted by $r$ and there are $n$ payments, so

$$A = \sum_{k=0}^{n-1} R' \cdot (1+r)^k \cdot (1+i)^{-k}.$$

.

We note that

$$A = \sum_{k=0}^{n-1} R' \cdot (1+r)^k \cdot (1+i)^{-k} = \sum_{k=0}^{n-1} R' \cdot \left(\frac{1+r}{1+i}\right)^k,$$

which has a form similar to the formula for $\ddot{a}_{\overline{n}|i}$. This can be simplified, yielding

$$A = R' \cdot \frac{1 - \left(\frac{1+r}{1+i}\right)^n}{1 - \frac{1+r}{1+i}} = R' \cdot \frac{1 - v_j^n}{1 - v_j} = R' \cdot \ddot{a}_{\overline{n}|j},$$

where

$$v_j = \frac{1+r}{1+i} = \frac{1}{1+j}.$$

This defines an interest rate

$$j = \frac{i-r}{1+r}$$

and allows us to evaluate the series of increasing payments as a single annuity due. It is understood here that $i > r$: the interest rate earned exceeds the escalator rate, so that $j > 0$ [19].

**Example 4.53.** If $i = 6\%$ above, and the escalator rate remains $r = 5\%$, then $j \approx .952\%$. The initial payment $R'$ would be

$$R' = \frac{A}{\ddot{a}_{\overline{30}|.00952}} \approx \frac{930,000,000}{26.2356} \approx \$35,447,962.$$

The final payment, 29 years later, would be $R' \cdot (1.05)^{29} \approx \$145,908,617$. ■

Of course, these windfall jackpots, like all gambling winnings, represent taxable income, and so the appropriate taxes must be deducted along the way.

**Example 4.54.** The Iowa Lottery runs a game called "Lucky for Life", in which the top prize is $1000 a day for the winner's lifetime. The prize is guaranteed for a minimum of 20 years, and has the option of a one-time payment of $5.75 million [60].

Consider the 20-year minimum prize, which is paid as 20 annual payments of $365,000. The effective annual interest rate that Iowa lottery officials are counting on is the solution to

$$5,750,000 = 365,000 \cdot \ddot{a}_{\overline{20}|i},$$

from which it follows that

$$\ddot{a}_{\overline{20}|i} = \frac{5,750,000}{365,000} = 15.7534.$$

Solving for $i$ with the computer algebra system Mathematica gives $i \approx 2.67\%$. ■

If we look instead at the advertised lifelong series of payments, rather than speculate on the expected lifespan of the winner, we can think of this as an annuity with an infinite number of payments, which is called a *perpetuity*. How much money must the state have on hand at the start of the payments to fund a perpetuity of $1000 per day?

We look to find the effect on the present value $A$ of letting $n \to \infty$ by taking a limit.

$$A = \lim_{n\to\infty} 1000\, \ddot{a}_{\overline{n}|i} = 1000 \cdot \lim_{n\to\infty} \left( (1+i) \cdot \frac{1-(1+i)^{-n}}{i} \right).$$

As $n \to \infty$, $(1+i)^{-n} \to 0$, and we have

$$A = 1000 \cdot \left( \frac{1+i}{i} \right) = 1000 \cdot \left( \frac{1}{i} + 1 \right) = 1000 \cdot \ddot{a}_{\overline{\infty}|i},$$

a value which depends only on the interest rate $i$, which we interpret as an interest rate compounded daily because payments are made daily. Using the annual interest rate of 2.67% computed above, we have

$$1.0267 = (1+i)^{365},$$

so

$$i = \sqrt[365]{1.0218} - 1 \approx 7.2152 \times 10^{-5},$$

or $i \approx .0072\%$ per day. The state's cash reserve must then be

$$1000 \cdot \ddot{a}_{\overline{\infty}|7.2152\times10^{-5}} = 1000 \cdot \left( \frac{1}{7.2152 \times 10^{-5}} + 1 \right) \approx \$13,860,569.70.$$

## Jackpots Revisited

Following the triple win of January 13, 2016, the Powerball jackpot was reset to $40 million paid over 30 years, with a lump-sum cash value of $24.8 million.

As with progressive keno (page 116), it is worthwhile to consider the question of an accumulated jackpot so large as to make a $2 Powerball ticket a positive-expectation venture. While simple mathematics can derive a minimum value for such a jackpot, that view of Powerball overlooks the fact that as the jackpot rises, the number of tickets sold also increases, as does the probability of sharing the prize—possibly dividing it below the level where the expectation is positive.

A back-of-the-envelope analysis of Powerball might suggest that when the jackpot exceeds double the number of tickets, or $584,402,676, then the expected value of a single ticket is positive. This disregards all prizes except for the jackpot, which may not be an unreasonable simplification. It also ignores tax implications on the prize, which is a more problematic omission that nonetheless can be accommodated by considering only the after-tax portion of the jackpot. However, this does not account for increased ticket demand; if 584,402,676 people each reasoned this way and bought a ticket with a positive expected value, very few of them would actually make money.

In 2010, a paper by Aaron Abrams and Skip Garibaldi [4] addressed the interplay between lottery jackpots and ticket demand, attempting to determine whether any lotto jackpot with a positive expectation could rise to the level of a good investment when the increase in ticket sales that accompanies a rising jackpot was taken into account. Their work led to two inequalities that were necessary to identify a good investment:

- $N < J$: The number of tickets sold, $N$, should be less than the jackpot $J$.
- $J > 2J_0$: The jackpot must exceed twice the *jackpot cutoff* $J_0$, which is the product of the number of possible tickets and the proportion of ticket sales allocated to the jackpot plus lottery expenses.

**Example 4.55.** For the 2015 Powerball rules, approximately 82.13% of sales were allocated to the jackpot and overhead, so

$$J_0 = ({}_{69}C_5 \cdot 26) \cdot .8213 \approx \$240,000,000.$$

Consequently, Powerball qualifies as a good bet if the jackpot is at least $480 million and the number of tickets sold is less than the jackpot amount. ■

A good bet is not necessarily a good investment; these conditions merely identify a lotto drawing with a positive expected rate of return. The authors went on to claim that no Powerball jackpot would ever achieve that second goal.

## 4.6 Lotto Strategies: Do They Work?

### Checking Claims

In Montana, a lotto game called *Big Sky Bonus* is a Powerball-like 4/28 + 1/28 game. The number of distinct tickets is then

$$_{28}C_4 \cdot 28 = 20,475 \cdot 28 = 573,300,$$

far smaller than for Powerball or similar drawings in other states.

The bonus number does not figure into the progressive jackpot, making Big Sky Bonus essentially two separate games on a single $2 ticket, as in Table 4.55 below.

TABLE 4.55: Pay table for Big Sky Bonus [101]

| Match | Payoff |
|---|---|
| Bonus | $10 |
| 2 of 4 | $2 |
| 3 of 4 | $20 |
| 4 of 4 | Progressive |

The progressive jackpot for matching 4 of 4 numbers starts at $2000.

**Example 4.56.** The expected value on the combined ticket, assuming the minimum progressive jackpot, is

$$\begin{aligned} E &= (2) \cdot \frac{_4C_2 \cdot {}_{24}C_2}{_{28}C_4} + (20) \cdot \frac{_4C_3 \cdot {}_{24}C_1}{_{28}C_4} + (2000) \cdot \frac{1}{_{28}C_4} + (10) \cdot \frac{1}{28} - 2 \\ &\approx -\$1.29. \end{aligned}$$

The house advantage is 64.5%. As is the case with progressive jackpots, increasing the jackpot decreases the HA; if the jackpot rises to $28,406, then a single winning ticket has a positive expected return. ■

A Web page at smartluck.com [144] purports to offer advice for picking numbers in Big Sky Bonus. While many of these claims stand up to mathematical scrutiny, they are not helpful in changing the game's odds.

- *Claim: All odd numbers or all even numbers are rarely drawn, so choose a mix of odd and even numbers.*

  The probability of getting 4 numbers of the same parity (odd or even) in the pick-4 game is

$$\frac{_{14}C_4}{_{28}C_4} = \frac{1001}{20,475} \approx .0489,$$

so the probability of getting either 4 odd or 4 even numbers is double that: just under 10%. All-odd or all-even number combinations are rarely drawn because there are far fewer of them than of mixed combinations. In a single drawing, all 4-number combinations are equally likely.

- *All high (15–28) or all low (1–14) numbers are rarely drawn. As with odd and even, it is better to choose a mix of different types of numbers.*

  The same probabilities calculated above for odd or even numbers apply to high and low numbers. Over a year's worth of daily drawings, we would expect only about 18 of 365 winning combinations to include only numbers from 1–14, for example. This tells us nothing about the numbers which will be drawn tomorrow.

  In fact, *any* division of the numbers from 1–28 into two disjoint sets of 14 numbers each will produce the same low probability, about 10%, that all 4 of the drawn numbers will fall into one set or the other. Odd/even, high/low, or something less mathematically meaningful—they all produce the same probability. The fact that this advice is correct doesn't make it terribly useful in picking numbers.

Both of these claims suffer from an inversion of the two events in conditional probabilities. The probability

$$P(\text{The combination is a winner} \mid \text{It contains both types of numbers})$$

is constant, regardless of the types of numbers (high/low, odd/even, etc.) involved. Using high/low or odd/even splits, this probability is

$$p = \frac{1}{{}_{28}C_4 - 2002} = \frac{1}{18,473}.$$

Choosing mixed combinations, as the Web site suggests, does nothing to change this.

The related probability

$$P(\text{The combination contains both types of numbers} \mid \text{It is a winner})$$

is

$$p = \frac{18,473}{20,475} \approx .9022.$$

While this does exceed 90%, this calculation assumes that the combination has won—which is not something that can be known before the numbers are drawn.

- *One number will be a repeat from the last drawing every other drawing, or 53% of the time.*

  The Complement Rule is useful in assessing this claim. The probability that none of the 4 numbers from a given drawing is repeated in the next drawing is

$$\frac{_{24}C_4}{_{28}C_4} = \frac{10,626}{20,475} \approx .5190,$$

so the probability of at least one repeated number is approximately $1 - .5190 = .4810$, which is in the vicinity of 53%.

However, the leap from "approximately 50%", whether 48% or 53%, to "every other drawing" suggests a regularity to repeated numbers that is inconsistent with randomly-generated winning combinations. The assumption of independence does not mean that every odd-numbered drawing will include a number from the previous drawing and every even-numbered drawing will not, for example.

Can you identify the drawings where a repeated number will be drawn? Given that, can you correctly identify which number will be repeated? Those are the challenges you face if taking this advice.

## How to Avoid Sharing

In 2005, Powerball officials were very surprised when 110 people submitted winning tickets matching 5 numbers, omitting only the Powerball, for the March 30 drawing [42]. The numbers drawn that day were 22, 28, 32, 33, and 39, with 42 as the Powerball. The 110 players had all chosen the five regular numbers, but each had selected 40 as the Powerball rather than 42. After some investigation of this highly unlikely event, lottery officials discovered that most of the winners had chosen their numbers from a fortune found in a fortune cookie. The cookies were traced to Wonton Food, a bakery in New York City, where officials reported that the numbers were chosen by drawing numbers out of a bowl. Wonton Food made as many as 4 million fortune cookies a day, which were distributed nationally; 26 of the 29 Powerball states at the time were represented among the match-5 winners.

Had the Powerball been 40, the 110 players would have split the $13.8 million jackpot among themselves, with each winner qualifying for a jackpot of $227,272.73—far less than the multi-million dollar jackpot a Powerball player hopes for. As it turned out, the sole winner of the jackpot in the March 30 drawing had used Quick Pick to select her numbers [42].

A lesson to be taken away from this story is that duplicating other players' numbers when playing Powerball and similar lotteries is to be avoided, if at all possible. Not choosing numbers based on fortune cookie messages is one bit of useful advice; this section will describe another. While it is not possible to change the probability of winning a multi-million dollar lotto jackpot by choosing numbers in accordance with any strategy, it *is* possible to minimize the likelihood of sharing a jackpot by following the ideas presented here. Any set of six numbers from 1 to 47 has the same probability of winning a 6/47 jackpot, whether it is a combination like $\{1, 2, 3, 4, 5, 6\}$ or something more random like $\{7, 20, 25, 35, 43, 45\}$. A life-changing lottery jackpot in Powerball or a similar lottery, when it's won, will be split among all tickets bearing the winning numbers, so your goal in picking numbers is to choose numbers that are unlikely to be chosen by other players. You will still have the same (tiny) probability of winning the jackpot, but if you do win, you will be far less likely to have to share your winnings with another player or players.

The basic idea behind the strategy outlined here is very simple: Find out what other people are likely to do, and *don't do that.* For purposes of illustration, we shall consider a Classic Lotto 47 game, where players pick six numbers in the range from 1 to 47. Studies have shown that many lottery players tend to pick their lotto numbers using some combination of the following methods [54]:

1. Choose numbers with personal significance, such as birthdays, anniversaries, or other important dates.
2. Choose numbers that make interesting patterns on the bet slip.
3. Choose numbers toward the center of the bet slip.
4. Choose numbers in arithmetic progression, such as the set {7, 14, 21, 28, 35, 42}. Seven, of course, is considered by many to be a lucky number, and so this progression starts with 7 and counts by 7s.

There is some overlap among these methods, of course—numbers in arithmetic progression often make interesting patterns on the bet slip, for example. Avoiding the numbers that these methods tend to choose is the key to minimizing the likelihood that you will have to share the jackpot in the unlikely event that you win it.

The first criterion is the easiest to avoid. Lottery numbers based on birthdays will not be greater than 31, so a player seeking to pick a unique set of numbers should primarily choose high numbers, with no more than one under 31. In practice, we would seek to make the sum of the chosen numbers large.

Consider the general $r/s$ lottery (page 233), where players select $r$ numbers in the range from 1 to $s$. Classic Lotto 47, then, is a $6/47$ game. In an $r/s$ lottery, the sum of the numbers in any possible combination will have (see [54]) mean

$$\mu = \frac{r \cdot (s+1)}{2}$$

and standard deviation

$$\sigma = \sqrt{\frac{(s-r) \cdot r \cdot (s+1)}{12}}.$$

Your goal in choosing numbers should be to have a sum higher than about 75% of all possible sums. Since the sums of the $r$ numbers chosen in an $r/s$

lottery are bell-shaped and symmetrically distributed about the mean, the Empirical Rule can be used here. In a 6/47 lottery such as Classic Lotto 47, we find that the sum of a player's numbers has mean $\mu = 144$ and SD $\sigma \approx 31.37$. The Empirical Rule states that about 68% of the sums will fall between 113 and 175, and since the data set is bell-shaped, 50% of the values lie below the mean of 144. If we choose numbers that add up to 166 or greater, we will have a combination whose sum exceeds 75% of all possible sums, and we will avoid most popular date combinations.

Of course, it is necessary to take the other factors listed above and explained below into account, for while the combination $\{42, 43, 44, 45, 46, 47\}$ may have a high sum, it is actually a relatively popular choice and fails our other tests of suitability [54].

Since lottery bet slips are optically scanned, interesting patterns on the bet slip that are not associated with arithmetic progressions tend to be clusters of adjacent numbers. Once again, we look at common practice and then do something else. This factor depends on the exact layout of the betting slip, which in turn may depend on $s$. A general rule is to avoid too many clusters, which may correspond to arithmetic progressions, and also to avoid too few, which result from bettors making nice-looking patterns. Toward that end, we define the *cluster number* of a wager as the number of adjacent—either edge-adjacent or diagonally adjacent—blocks of numbers. A single number in isolation comprises a cluster, as does any string of bet squares connected at their edges or corners.

**Example 4.57.** The bet slip shown in Figure 4.14 for a 6/47 lottery has a cluster number of 4. ■

| 1 | 2 | 3 | 4 | 5 |
|---|---|---|---|---|
| 6 | 7 | 8 | 9 | 10 |
| 11 | 12 | 13 | 14 | 15 |
| 16 | 17 | 18 | 19 | 20 |
| 21 | 22 | 23 | 24 | 25 |
| 26 | 27 | 28 | 29 | 30 |
| 31 | 32 | 33 | 34 | 35 |
| 36 | 37 | 38 | 39 | 40 |
| 41 | 42 | 43 | 44 | 45 |
| 46 | 47 | *Lotto 6/47* | | |

FIGURE 4.14: 6/47 lotto ticket with cluster number 4.

For an $r/s$ lottery, cluster numbers of 1 or $s$ are to be avoided, as they tend to correlate with smaller overall payouts at the high levels [54, p. 92].

*Edge numbers* are numbers that appear on an edge of the bet slip, and their exact number and values depend on the format of the slip, as with clusters. Players trying to be "random" in their selections often choose few numbers near the edges, so if you don't want to share your prize, look to the edge numbers in selecting your combination [54]. From a practical perspective, you should select combinations with at least four edge numbers—keeping in mind, of course, the need for a suitably large sum and an appropriate cluster number.

Finally, we address arithmetic progressions. A combination such as {1, 2, 3, 4, 5, 6} is very simple, very popular, and no more or less likely to win than any other [54]. In no small part, this is because the rule for determining this combination is very simple and easily discovered by many people. The same objection may be raised against a combination like {5, 10, 15, 20, 25, 30} or even one like {8, 15, 22, 29, 36, 43}. To avoid such popular methods of choosing numbers, we define the *arithmetic complexity* of a combination:

**Definition 4.2.** The *arithmetic complexity* (AC) of a set of $r$ numbers is the number of positive differences among all of the numbers in that set, minus $r - 1$.

**Example 4.58.** The set {23, 31, 34, 35, 41, 47} shown on the ticket in Example 4.57 leads to the following positive differences:

$$
\begin{array}{lllll}
47-23=24 & 47-31=16 & 47-34=13 & 47-35=12 & 47-41=6 \\
41-23=18 & 41-31=10 & 41-34=\ 7 & 41-35=\ 6 & \\
35-23=12 & 35-31=\ 4 & 35-34=\ 1 & & \\
34-23=11 & 34-31=\ 3 & & & \\
31-23=\ 8 & & & &
\end{array}
$$

Note that each number is subtracted from every other number in such a way that the difference is positive. There are 13 different differences—6 and 12 are repeated—so the AC of this set is $13 - 5 = 8$. ■

**Example 4.59.** For the set {1, 2, 3, 4, 5, 6}, the numbers give rise to the following positive differences:

$$
\begin{array}{lllll}
6-1=5 & 6-2=4 & 6-3=3 & 6-4=2 & 6-5=1 \\
5-1=4 & 5-2=3 & 5-3=2 & 5-4=1 & \\
4-1=3 & 4-2=2 & 4-3=1 & & \\
3-1=2 & 3-2=1 & & & \\
2-1=1 & & & &
\end{array}
$$

We count five different differences, so the AC of this set is $5 - 5 = 0$. ■

When a set of $r$ different numbers is in an arithmetic progression like those above, the AC will always be 0—we subtract $r-1$ from the number of different differences in order to make this happen. An AC of 0 suggests that the set in question is not very complex, and that certainly applies to an arithmetic sequence. The maximum AC is

$$\frac{r(r-1)}{2} - (r-1) = \frac{(r-1)(r-2)}{2},$$

when all of the differences are different.

Since lottery players tend toward patterns and simple combinations, the gambler seeking an edge should avoid those and choose more complex sets: anything with an AC of at least 6. In a 6/49 lottery, over 96% of combinations have $AC > 6$ [54]), so this rule excludes few combinations—but they are popular combinations.

Taking everything together, we have arrived at one possible strategy for choosing numbers that will decrease the likelihood of choosing someone else's numbers and thus having to split the prize if you should win: For a 6/47 lottery, play only combinations fitting all four of the following criteria:

- The sum of the numbers is at least 166.
- The cluster number is between 2 and 5.
- The edge number is at least 4.
- The AC is at least 6.

In addition, any combination that satisfies these requirements should nonetheless be discarded if it is a recent winning combination in this lottery or a neighboring state's lottery, for people are known to favor those combinations as well, or if it has a recognizable pattern—for example, {32, 33, 35, 38, 42, 47}.

Another option when buying a lottery ticket is to let a computer select your numbers randomly, an option called "Quick Pick" or something similar, as we have seen in keno drawings (p. 113). This makes it easier to buy a ticket and may actually be an option worth using if you're trying not to share a prize—provided that your random selection isn't a collection of possible birthday numbers, for example. If you have the chance to examine your Quick Pick numbers before committing to the ticket, this can be an easy way to risk money on a longshot without much likelihood that your numbers will be chosen, nonrandomly, by another gambler.

## Random Strategies Investments

On its face, Cash Winfall, a game run by the Massachusetts Lottery beginning in 2004, was quite simple: a \$2 twice-weekly 6/46 lotto game with a progressive jackpot that rolled over after each drawing without a jackpot winner. The pay table (Table 4.56) was generous, paying off for as few as two matches.

The progressive jackpot had an unusual condition that the New Zealand Bullseye lottery would later use: While it accumulated value from its starting amount of \$500,000 with each jackpot-free drawing, it was capped at \$2 million. If a rollover meant that the jackpot would exceed \$2 million, the excess amount was rolled down and used to augment the prize pools for fewer than 6 matched numbers: 26% to the Match-5 pool, 47% to Match-4, and 27%

TABLE 4.56: Massachusetts Lottery Cash WinFall pay table [32]

| Match | Prize |
|---|---|
| 2 | Free ticket |
| 3 | $5 |
| 4 | $150 |
| 5 | $4000 |
| 6 | Progressive |

to Match-3 [147]. This procedure was instituted as a response to a previous game, Mass Millions, which had a traditional progressive jackpot but which failed when no jackpots were won in 2003 despite a year of rollovers [147]. Cash Winfall was designed as a game where more numerous smaller jackpots might prove a greater attraction to players than infrequent large ones.

This nontraditional feature led to an exploitable opportunity for investors: if it was possible to pinpoint the drawing when the jackpot would roll down, a player buying many tickets—as in hundreds of thousands—would be nearly certain to make a significant profit. It would not be necessary to hit the jackpot at its maximum; all that would be necessary would be to match 5 out of 6 numbers on as many tickets as possible. The probability of matching 5 numbers out of 6 in a 6/46 lottery is

$$\frac{{}_6C_5 \cdot {}_{40}C_1}{{}_{46}C_6} = \frac{240}{9,366,819} \approx \frac{1}{39,029},$$

suggesting that if the Match-5 prize were increased to more than about $40,000 by a rolldown, the expected return on a single ticket would be positive. To take full advantage of this positive expectation, though, would require buying many tickets, as the probability of any one ticket winning was still very small.

In 2005, a group of students at the Massachusetts Institute of Technology deduced that the new game was exploitable and set out to identify rolldowns in advance and buy up thousands of tickets in anticipation of inflated payoffs for matching 5 numbers. The group dubbed their enterprise "Random Strategies Investments LLC" after Random Hall, their dormitory at MIT.

The MIT team was not the only one to recognize this opportunity, of course. They were joined in their quest by two other teams: one based at Northeastern University and one from the state of Michigan. The Michigan team had run a similar scheme on a Michigan Lottery game called Winfall, a 6/49 lotto game with a $5 million rolldown jackpot [147].

The challenges to Random Strategies Investments and its competitors were primarily logistical rather than mathematical:

- Identify when a drawing was likely to roll down. Since the rolldown was triggered when the top jackpot hit $2 million, a previous jackpot close to, but not exceeding, $1.6 million was seen as a lucrative target.

Ideally, this identification was done in such a way that the state's lottery computers would not predict a rolldown for the upcoming drawing. If the previous game's jackpot had gone unwon at $1.95 million, for example, a rolldown was virtually guaranteed for the next drawing. Sales would increase, as was always the case when a rolldown was expected, and the opportunity for profit would be diminished.

- Trigger a rolldown by buying as many tickets as possible for that drawing, making sure to minimize duplicated number combinations. Forcing Cash Winfall into a rolldown involved sending group members out to multiple lottery agents and then tying up lottery terminals for the hours needed to make the necessary purchases. Fortunately for the MIT team, bet slips, once filled out, could be saved and reused from one drawing to another.

  The Michigan team primarily used Quick Pick to select its numbers. While this made the mechanics of buying thousands of tickets much easier, it meant that the tickets they bought had less-than-optimal distributions of numbers, including possibly duplicating some number combinations.

- Sort through hundreds of thousands of paper tickets, once the drawing ended, to find the winners. This was close to a full-time occupation in the days following a rolldown.

Additionally, there was a very small risk that some other player would win the top jackpot by matching all six drawn numbers, thus eliminating the rolldown. Over the 8-year lifetime of Cash Winfall, though, only one of 45 rolldown drawings produced a jackpot winner [147].

**Example 4.60.** Buying 200,000 tickets for a drawing with a rolldown sets up a binomial random variable $X$ counting the number of wins at a specified level. For the Match-5 prize, $n = 200,000$ and $p = \dfrac{240}{9,366,819}$. We have

$$P(X = x) = {}_{200,000}C_x \cdot p^x \cdot (1-p)^{200,000-x},$$

leading to the probability distribution in Table 4.57.

The most likely outcome is 5 Match-5 winners. (The mean outcome is $np \approx 5.12$ winning tickets.)

Similar calculations will show that the most likely outcomes for other prizes are 250 Match-4 winners and 4219 Match-3 winners. For the drawing on February 8, 2010, the total amount won from these tickets would have been $425,640, leaving $25,640 as profit after deducting the cost of the tickets, a 6.41% profit on a very short-term (3 or 4 days) investment [147].

This drawing would also be likely to yield 29,283 free tickets from matching 2 numbers, which could be used to offset ticket expenses for the next rolldown drawing. ■

TABLE 4.57: Binomial probability distribution for 200,000 Mass Winfall tickets, Match 5 payoff

| $x$ | $P(X=x)$ |
|---|---|
| 0 | .0059 |
| 1 | .0305 |
| 2 | .0781 |
| 3 | .1334 |
| 4 | .1710 |
| 5 | .1752 |
| 6 | .1496 |
| 7 | .1095 |
| 8 | .0702 |
| 9 | .0400 |
| 10 | .0205 |
| $> 10$ | .0161 |

Random Strategies Investments reportedly applied this approach to Cash Winfall over several years, winning about $3.5 million before taxes, before the actions of the three teams started raising questions among lottery officials [88]. The fallout from the assault on Cash Winfall was perhaps not what one might have expected. Since no rules were broken, and since the actions taken by the three teams did not change the odds of winning or affect the probability of winning for any other lottery patron, there were no grounds for lottery officials to act against the teams. The investigation did turn up a few minor violations of lottery ticket-selling procedures, but no action was taken against team members. The money that was collected by this approach would have been awarded to someone even if there had been no coordinated action, and the Commonwealth of Massachusetts received its share of the proceeds for public works projects. Upon a thorough review of the game, the Massachusetts Inspector General concluded that there had been no wrongdoing and that Cash Winfall had been a financial success [147].

---

## 4.7 Exercises

Solutions begin on page 304.

**4.1.** Assuming 120,000 eligible tickets and a constant probability of a ticket being drawn for any one prize, find the amount that the million-dollar drawing adds to the expected value of a Michigan Green Game ticket.

**4.2.** Figure 4.15 shows a Michigan Jackpot passive ticket which dates from 1974.

FIGURE 4.15: Michigan Lottery's Michigan Jackpot ticket.

There are 11 ways to win on a single ticket. The state would draw one 6-digit number, one 5-digit number, and one 3-digit number in each week's drawing. The "Jackpot Number" at the top of the ticket qualified the ticket holder for entry into a separate drawing for cash prizes ranging from $20,000 to $333,333. With the understanding that the numbers on each ticket were different, so no ticket could win more than three prizes, find the expected return on this $1 ticket using the minimum $20,000 payoff for the jackpot number.

**4.3.** Confirm the assertion on page 247 that there are 155 different ways to match 8 numbers on an Atlantic Bucko ticket.

**4.4.** In Lotto :D's evening drawing, compare the house advantage for a $2 bet to the HA for a $5 bet.

**4.5.** In Iowa and Minnesota, All or Nothing (page 253) is played as a 12/24 game. The state draws 12 numbers, and the pay table is shown in Table 4.58.

TABLE 4.58: Pay table for All or Nothing: Iowa and Minnesota [59]

| Catch | Prize |
|---|---|
| 0 or 12 | $100,000 |
| 1 or 11 | $1000 |
| 2 or 10 | $20 |
| 3 or 9 | $5 |
| 4 or 8 | $1 |

Find the probability of losing money on a $1 All or Nothing ticket and the house advantage.

**4.6.** In the late 1970's, when casino gambling in the USA was limited to Nevada and New Jersey, the New York state lottery offered a stripped-down version of keno on a lottery ticket [69]. Players picked four numbers in the range from 1–40, and the lottery drew its seven numbers once weekly. A 50¢ wager paid off $500 (odds of 1000 for 1) if the player's four numbers were among the state's seven. How does the house advantage of this game compare to a 70¢ Pick 4 keno ticket paying $80 when all four numbers are caught?

**4.7.** Find the probability of filling exactly 1 line on a Lucky Links ticket.

**4.8.** The Illinois Lottery debuted in 1974, with the ticket shown in Figure 4.16.

FIGURE 4.16: Illinois Lottery's first ticket.

This passive ticket combined a 5/49 Weekly Lotto game with a pair of drawings involving three-digit numbers. For the lotto game, prizes were $20 for matching three of the state's five numbers, $100 for matching four, and $5000 for matching all five.

The state drew 3 three-digit numbers, which applied to both the Weekly Bonanza and Millionaire Game drawings. Matching two of the state's three numbers in the Weekly Bonanza section won entry into a separate drawing, with prizes ranging from a guaranteed $1000 to $300,000 paid in a 15-year annuity. Matching two numbers in the Millionaire Game section won $500 plus entry into a separate drawing, with prizes from $1000 to $1,000,000 in a 20-year annuity.

a. Find the probability of winning on the Weekly Lotto portion of the ticket.

b. How much does the Weekly Lotto game contribute to the expected value of a ticket?

c. Find the probability of winning on either the Weekly Bonanza or Millionaire Game drawings.

**4.9.** Another Illinois Lottery passive game was the baseball-themed Super

Slam (Figure 4.17), which offered 22 weekly ways to win on a single ticket, and matched the Pennsylvania Lottery's 23 For A Dollar ticket by including a number for a one-time bonus drawing on the right–side ticket stub.

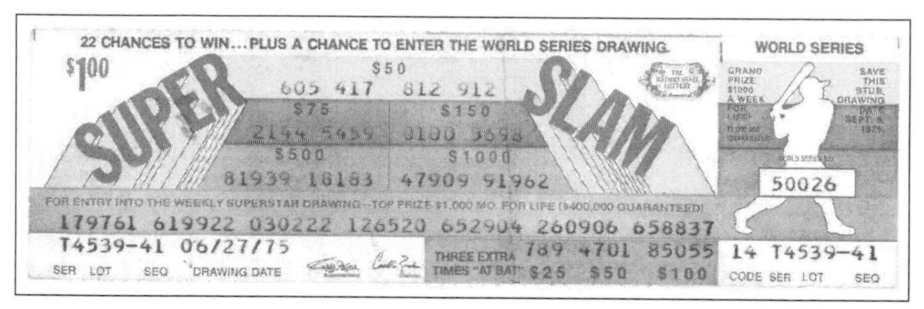


FIGURE 4.17: Illinois Lottery's Super Slam ticket.

A Super Slam ticket contained the following numbers:

- Four 3-digit numbers, which carried a \$50 prize.
- Four 4-digit numbers. Two were worth \$75 when matched; the other two had a \$150 prize attached.
- Four 5-digit numbers. Two could win \$500 and two could win \$1000.
- Seven 6-digit numbers, which were eligible for entry into a weekly drawing. The prizes available in this drawing ranged from \$1000 to a top prize of \$1000 a month for life, with a minimum guarantee of \$400,000.
- In a separate bonus area of the ticket (at the lower right): a 3-digit number paying \$25, a 4-digit number paying \$50, and a 5-digit number paying \$100.
- The right-hand stub bore a 5-digit number offering \$100 a week for life as the top prize in a single drawing, with \$1 million guaranteed. Other prizes in this drawing were \$2000, \$15,000, and \$50,000.

The state drew one number of each length every week for the 12 weeks that the game was active. The bonus drawing was conducted once, on October 11, 1975. Leo Puralewski Sr., of Chicago, was the grand prize winner [62].

a. How do the chances of winning at Super Slam compare to 23 For A Dollar?

b. Assuming that the various numbers on a single ticket were all distinct, find the probability that a Super Slam ticket wins at least one prize.

**4.10.** A 1991 US patent application described an additional Daily 3 game bet called *Add-Em-Up Lotto* that offered players a chance to wager on the sum of the digits of the 3-digit number drawn [165]. As proposed, this bet could be made on its own or in addition to a traditional bet on a chosen 3-digit number.

Since three-digit lottery numbers range from 000 to 999, the possible digit sums range from 0 to 27, with an average sum of $3 \times 4.5 = 13.5$. Table 4.59 shows a hypothetical pay table for this bet. Each of the 10 lines of the table is intended as a separate $1 wagering option: for example, betting $1 on the option "3 or 24" pays off $25 (25 for 1) if the sum is 3 or 24, and nothing otherwise.

TABLE 4.59: Add-Em-Up betting options [165]

| Sum | Prize |
|---|---|
| 0 or 27 | $250.00 |
| 1 or 26 | $82.50 |
| 2 or 25 | $41.50 |
| 3 or 24 | $25.00 |
| 4 or 23 | $16.50 |
| 5 or 22 | $12.00 |
| 6, 7, 20, or 21 | $4.00 |
| 8, 9, 18, or 19 | $2.50 |
| 10, 11, 16, or 17 | $2.00 |
| 12, 13, 14, or 15 | $1.75 |

a. Find the probability of winning a bet on the combination "2 or 25".

b. Find the expected value of a bet on "1 or 26".

c. By placing a $1 bet on each of the ten combinations in the table, it is possible to guarantee a win. Find the probability that this wager will wind up as a net winner.

**4.11.** In Aruba, the *Zodiac* lottery game combines choosing an astrological sign, as in Quebec's Astro game, with a Daily 4 drawing. Players choose a 4-digit number and a sign of the zodiac, and the lottery agency draws its number and sign twice daily [7]. There are four winning combinations:

A: Matching the four-digit number and astrological sign pays off 25,000 for 1.

B: Matching the four-digit number, but not the sign, pays 2500 for 1.

C: Matching the last three digits of the number and the zodiac sign pays 500 for 1.

D: Matching the last three digits without matching the sign pays 50 for 1.

The cost of a ticket, in Aruban florins (1.77 Afl. = $1 US), is determined by what the player wishes to spend. The minimum wager is 0.50 florins; the maximum is effectively 25 florins, as lottery computers are programmed to reduce the government's exposure by allowing only 25 florins of action on any single four-digit number.

a. Show that, despite the impression given by the payoffs listed above, event B has a higher probability than event C.

b. Find the probability of winning on a Zodiac ticket.

c. Compute the house advantage on a 1-florin ticket.

**4.12.** *EuroMillions* is a 5/50 + 2/12 lottery game administered in the United Kingdom and available in 12 European countries. The numbers in the second set, from 1–12, are called "Lucky Stars". Taken together, the two pools of numbers result in ${}_{50}C_5 \cdot {}_{12}C_2 = 139,838,160$ possible tickets. EuroMillions prizes are pari-mutuel, based on the number of tickets sold and the number of tickets bearing each winning combination. The top prize, awarded for matching 5 of 5 numbers and both Lucky Stars, averages €46,596,455.43 [86].

In September 2016, the size of the Lucky Stars pool was changed from 11 to 12, and the price of a ticket rose from £2 to £2.50 (or €2.50 or 3.50 Swiss francs), both in an effort to generate larger jackpots as was done with Powerball.

a. Show that under the new rules, the probability of winning the smallest EuroMillions prize by matching 2 numbers and no Lucky Stars has increased.

b. Which event has the smallest probability: matching 5 numbers and no stars, matching 4 numbers and 1 star, or matching 3 numbers and 2 stars?

c. In Austria, lottery players may add the *Austria Joker* game to their EuroMillions tickets. For an additional €1.30, players receive a 6-digit number, and are paid off if the last 1–6 digits of the number drawn by lottery officials match theirs, in accordance with Table 4.60. The top jackpot is also a pari-mutuel prize.

TABLE 4.60: Austria Joker pay table [85]

| Match | Prize |
|---|---|
| All 6 digits | Jackpot |
| Last 5 digits | €7777 |
| Last 4 digits | €777 |
| Last 3 digits | €77 |
| Last 2 digits | €7 |
| Last digit | €1.50 |

Find the probability of winning a prize in Austria Joker.

d. How large, to the nearest euro, must the jackpot be for Austria Joker to have a positive expectation?

**4.13.** A predecessor to Amigo, *Rapido* was a lottery that ran in France from 1998–2014. Rapido tickets cost €1, half the price of Amigo tickets, and the drawings were held every 2½ minutes, twice as often as Amigo.

Rapido was an 8/20 + 1/4 lottery. Payoffs began when a player matched 4 of the 8 main numbers (Grid A) together with the 1 bonus number from Grid B, and are shown in Table 4.61.

TABLE 4.61: Rapido pay table

| **Numbers Matched** | | |
|---|---|---|
| **Grid A** | **Grid B** | **Prize** |
| 4 | 1 | 1 |
| 5 | 0 | 2 |
| 5 | 1 | 6 |
| 6 | 0 | 10 |
| 6 | 1 | 30 |
| 7 | 0 | 50 |
| 7 | 1 | 50 |
| 8 | 0 | 1000 |
| 8 | 1 | 10,000 |

a. How many different Rapido tickets are possible?

b. Suppose that a ticket matches $a$ numbers from Grid A and $b$ numbers from Grid B, where $0 \leqslant a \leqslant 8$ and $b = 0$ or 1. Derive the probability distribution function $p(a, b)$.

**4.14.** Suppose that the bonus numbers were eliminated from Amigo and the game drawn as a standard 7/28 lottery, with the ticket price lowered to €1. Payoffs for matching $k$ numbers are to be taken from Table 4.24 on page 244, using the "Match $k$ blue numbers and 0 bonus numbers" lines. Payoffs should not be decreased to account for the lower ticket price.

a. What is the probability of winning a prize in the revised Amigo?

b. Find the house edge of this new game.

**4.15.** SCRATCHEZZ is an online variation on instant lottery tickets that allows a player to choose his or her own lucky numbers or words. Developed by Craig Sedoris and Gary Kadlec, SCRATCHEZZ takes a player's 6-character phrase and embeds it in a 6 by 6 grid. 40 characters—26 letters, 10 digits, and the characters $, @, #, and ♡—are available; only one character may be used twice in a player's phrase [152].

For example, if the player chooses the phrase "♡MONEY", the virtual ticket appears as shown in Figure 4.18. Each of the 6 characters is hidden in each column under the "ZZ" logo.

| ♡ | M | O | N | E | Y |
|---|---|---|---|---|---|
| ZZ | ZZ | ZZ | ZZ | ZZ | ZZ |
| ZZ | ZZ | ZZ | ZZ | ZZ | ZZ |
| ZZ | ZZ | ZZ | ZZ | ZZ | ZZ |
| ZZ | ZZ | ZZ | ZZ | ZZ | ZZ |
| ZZ | ZZ | ZZ | ZZ | ZZ | ZZ |
| ZZ | ZZ | ZZ | ZZ | ZZ | ZZ |

FIGURE 4.18: SCRATCHEZZ virtual ticket.

To play the game, players "scratch" off one cell in each column. The game pays off if 2 or more of the 6 characters are revealed in the column which they head. While the game is not yet playable for money, the proposed ticket price is $5; the pay table is shown in Table 4.62.

TABLE 4.62: SCRATCHEZZ pay table [128]

| Match | Prize |
|---|---|
| 6 | $25,000 |
| 5 | $2500 |
| 4 | $1500 |
| 3 | $150 |
| 2 | $50 |

a. How many phrases consisting of 6 different characters are possible?

b. How many phrases may be created with one character appearing twice?

c. Find the probability distribution function for a phrase with 6 different characters and construct the probability distribution.

d. Find the probability distribution function for a phrase with 5 different characters, one appearing twice, and construct the probability distribution.

e. What is the expected value of a $5 ticket in each phrase configuration?

# *Answers to Selected Exercises*

## Chapter 2

Exercises begin on page 55.

**2.1.** 18.
**2.2.** 55.
**2.3.** 24.
**2.4.** $\frac{1}{45}$.
**2.5a.** $\frac{1}{29,370}$.
**2.5b.** $\frac{43}{29,370}$.
**2.5c.** $\frac{24}{n(n-1)(n-2)}$.
**2.6.** 20%.
**2.7.** 30%.
**2.8.** 78.
**2.9.** 816.
**2.10.** 10.
**2.12a.** $7.3930 \times 10^{-4}$.
**2.12b.** .0067.
**2.12c.** .0077.
**2.13.** 50.
**2.14.** $46.50.
**2.15a.** 12,103,014.
**2.15b.** 302,575,350.
**2.15c.** 20,800.
**2.15d.** $6.8743 \times 10^{-5}$.
**2.16a.** 2,598,960.
**2.16b.** 1/25,989,600.
**2.16c.** .5312.

## Chapter 3

Exercises begin on page 192.

**3.1.** $2.

**3.2.** .3672.
**3.3.** Between the Hardway 6 or 8 bet and the baccarat Tie bet.
**3.4.** $P$(Match 5) is greater.
**3.5a.** .0089.
**3.5b.** $5.328 \times 10^{-4}$.
**3.5c.** .0023.
**3.6.** 10%.
**3.7.** .0729.
**3.8.** 71.44%.
**3.9.** 50.98%.
**3.10.** 1164.
**3.12a.** $7.7259 \times 10^{-10}$.
**3.12b.** $5.0859 \times 10^{-11}$.
**3.12c.** $3.5556 \times 10^{-5}$.
**3.12d.** $1.3936 \times 10^{-10}$.
**3.13a.** 170.
**3.13b.** $F(m,n) = .55n + 1.25m^2 - 1.25m + 11.50mn - 1.70$.
**3.13c.** \$212.90.
**3.14a.** 110.
**3.14b.** $F(m,n,j) = 2.5nj + 1.25m^2 - 1.25m + 11.50mn + 5.75m^2j - 5.75mj - 1.10$.
**3.14c.** \$318.90.
**3.14d.** \$197.90.
**3.15a.** 28.42%.
**3.16c.** $k \leqslant 4$.
**3.18a.** –\$.2730.
**3.18b.** \$112,826.
**3.20.** 23.66%.
**3.21.** \$10,350, \$10,344.
**3.22.** \$2903.50.
**3.23a.** 88.
**3.23b.** .0057.
**3.24.** \$311.910.
**3.28b.** $P(6 \mid 7)$.
**3.29a.** $P(0) \approx .0017$.
**3.29b.** $P(19) \approx 9.0119 \times 10^{-15}$.
**3.29c.** $P(k) = \dfrac{{}_{18}C_{k-2} \cdot {}_{61}C_{21-k}}{{}_{80}C_{20}} + \dfrac{{}_{18}C_{k} \cdot {}_{61}C_{20-k}}{{}_{80}C_{20}}$.
**3.29d.** 7.89%.

## Chapter 4

Exercises begin on page 294.
**4.1.** $\$2.58 \times 10^{-5}$.
**4.2.** –68¢.

**4.4.** 91.4% vs. 91.7%.
**4.5.** .7797, 39.49%.
**4.6.** New York: 61.70%. Keno: 64.99%.
**4.7.** .3409.
**4.8a.** .0051.
**4.8b.** 11.34¢.
**4.8c.** $6.006 \times 10^{-6}$.
**4.9b.** .0056.
**4.10a.** .012.
**4.10b.** –50.5¢.
**4.10c.** .112.
**4.11b.** .001.
**4.11c.** 48.375%.
**4.12c.** .1.
**4.12d.** €892,777.
**4.13a.** 503,880.
**4.13b.** $p(a, b) = \dfrac{{}_{8}C_a \cdot {}_{12}C_{8-a}}{{}_{20}C_8} \cdot \dfrac{3-2b}{4}$.
**4.14a.** .0432.
**4.14b.** 57.13%.
**4.15a.** 2,763,633,600.
**4.15b.** 1,184,414,400.
**4.15c.** $P(\text{Match } k) = {}_{6}C_k \cdot \left(\frac{1}{6}\right)^k \cdot \left(\frac{5}{6}\right)^{6-k}$.
**4.15d.** $P(\text{Match } k) = \sum_{a+b=k,\ 0 \leqslant a \leqslant 2,\ 0 \leqslant b \leqslant 4}$
$\left[ {}_{2}C_a \left(\frac{1}{3}\right)^a \left(\frac{2}{3}\right)^{2-a} \cdot {}_{4}C_2 \left(\frac{1}{6}\right)^b \left(\frac{5}{6}\right)^{4-b} \right]$.
**4.15e.** \$27.28, \$61.38.

# References


[1] 23 For A Dollar (advertisement), *The Daily News*, Huntingdon, PA, 11 May 1979, p. 26. Online at https://www.newspapers.com/newspage/17377596/. Accessed 28 April 2016.

[2] 25 cent lottery tickets offered, Las Vegas *Sun*, 19 September 2002. Online at http://lasvegassun.com/news/2002/sep/19/25-cent-lottery-tickets-offered/. Accessed 27 April 2016.

[3] 2by2—Prizes and Odds. Online at http://www.powerball.com/2by2/2by2_prizes.asp. Accessed 21 May 2016.

[4] Abrams, Aaron, and Skip Garibaldi, Finding Good Bets in the Lottery, and Why You Shouldn't Take Them. *The American Mathematical Monthly* **117**, January 2010, p. 3–26.

[5] Alwes, Berthold C., *The History of the Louisiana State Lottery Company*. Master's thesis in history, Louisiana State University, Baton Rouge, 1929.

[6] Arizona Lottery, 5 Card Cash. Online at https://www.arizonalottery.com/en/play/draw-games/5-card-cash, Accessed 22 April 2016.

[7] Aruba Lottery, How to play Zodiac. Online at http://www.lottoaruba.com/zodiac. Accessed 4 August 2016.

[8] Australian Soccer Pools. Online at http://www.australiangambling.com.au/lotto/soccer-pools/. Accessed 8 October 2016.

[9] Big Red Keno. Online at http://www.bigredkeno.com. Accessed 26 March 2016.

[10] Big Red Keno, Lincoln Playbook. Online at http://bigredkeno.com/Content/Media/file/document/Locations/lincoln_paybook.pdf. Accessed 26 March 2016.

[11] Biggest Keno Jackpots Ever, 25 September 2008. Online at http://pro360.com/keno/biggest-keno-jackpots-ever.html. Accessed 16 May 2016.

[12] Blayney, Richard D., *Win More At Keno*. Blayney Enterprises, Garden Grove, CA, 1971.

[13] Boddy, Gillian, and Jacqueline D. Matthews, *Disputed Ground: Robin Hyde, Journalist*, Victoria University Press, Wellington, New Zealand, 1991.

[14] Bollman, Mark, *Basic Gambling Mathematics: The Numbers Behind The Neon*. CRC Press/Taylor & Francis, Boca Raton, FL, 2014.

[15] Bradley, Robert C., Euler and the Genoese Lottery. Online at http://citeseerx.ist.psu.edu/viewdoc/download?doi=10.1.1.24.5959&rep=rep1&type=pdf. Accessed 2 September 2017.

[16] *A brief description of the manner by which the Chinese count, combine, and establish the different tickets in the Chinese lottery*, San Francisco: s.n., 1891.

[17] British Columbia Lottery Corporation, About Keno. Online at http://lotto.bclc.com/keno-and-keno-bonus/how-to-play.html. Accessed 26 March 2016.

[18] Brooks, Henry M., *The Olden Time Series: Curiosities of the Old Lottery*. Ticknor and Company, Boston, 1886.

[19] Broverman, Samuel A., *Mathematics of Investment and Credit*, 5th edition. ACTEX Publications, Inc., Winsted, CT, 2010.

[20] California Lottery, Daily Derby: How to Play. Online at http://www.calottery.com/play/draw-games/daily-derby/how-to-play. Accessed 21 April 2016.

[21] California Lottery, Regulations. Online at http://static.www.calottery.com/~/media/Publications/Lottery_Regulations/Approved Regulations 4-25-13.pdf. Accessed 21 April 2016

[22] Carradine, B., *The Louisiana State Lottery Company Examined and Exposed*. D.L. Mitchel, Publisher and Printer, New Orleans, 1889.

[23] Catlin, Don, *The Lottery Book: The Truth Behind The Numbers*. Bonus Books, Chicago, 2003.

[24] Connecticut Lottery Lucky Links. Online at https://www.ctlottery.org/Modules/Games/default.aspx?id=8. Accessed 2 April 2017.

[25] Continental Casino has keno specials, *Los Angeles Times*. 9 February 1992, p. F70.

[26] Cowles, David, *Complete Guide to Winning Keno*, 3rd edition. Cardoza Publishing, New York, 2003.

[27] Culin, Stewart, *The Gambling Games of the Chinese in America*. University of Pennsylvania Press, Philadelphia, 1891.

[28] Curtin, Lawrence, and Karen Bernardo, *The History of Sweepstakes*. Sweepstakes News, Key Biscayne, FL, 1997.

[29] Daston, Lorraine, *Classical Probability in the Enlightenment*. Princeton University Press, Princeton, NJ, 1988.

[30] D.C. Lottery, DC-5. Online at http://dclottery.com/games/dc5/default.aspx. Accessed 28 May 2016.

[31] Elwood, Bryan, and Hal Tennant, *The Whole World Lottery Guide*, 2nd edition. Million$ Magazine, Toronto, 1981.

[32] Estes, Andrea, and Scott Allen, A game with a windfall for a knowing few. Boston.com, 31 July 2011. Online at http://archive.boston.com/news/local/massachusetts/articles/2011/07/31/a_lottery_game_with_a_windfall_for_a_knowing_few/?page=full. Accessed 30 April 2016.

[33] Euler, Leonhard, Reflections on a singular Kind of Lottery named the Genoise Lottery. Translated by Richard J. Pulskamp. Online at http://cerebro.xu.edu/math/Sources/Euler/E812.pdf. Accessed 2 September 2017.

[34] Everett, George, How Keno was Born in Butte, Montana. Online at http://www.butteamerica.com/keno.htm. Accessed 20 March 2016.

[35] Ezell, John Samuel, *Fortune's Merry Wheel: The Lottery In America*. Harvard University Press, Cambridge, MA, 1960.

[36] Farkas, Karen, Ohio Lottery's first drawing 40 years ago drew thousands to a mall parking lot. Cleveland.com, 22 August 2014. Online at http://www.cleveland.com/metro/index.ssf/2014/08/ohio_lotterys_first_drawing_40.html. Accessed 5 November 2016.

[37] Ferguson, Tom W., The Curse of the Big Lottery Win. Online at http://www.lastchaser.com/BugleTwo.html#anchor_141, 30 September 2012. Accessed 4 October 2012.

[38] Flynn, Kevin, *American Sweepstakes: How One Small State Bucked the Church, the Feds, and the Mob to Usher in the Lottery Age*. University Press of New England, Lebanon, NH, 2015.

[39] The Football Pools, online at http://www.footballpools.com/cust. Accessed 18 June 2012.

[40] Fredella, Brad, Two video keno strategies that work. Online at https://www.gamingtoday.com/articles/article/29263-Two_video_keno_strategies_that_work#.V1t7JzXbBKk, 1 February 2011. Accessed 10 June 2016.

[41] Fredella, Brad, Video keno Volatility. Online at http://www.gamingtoday.com/articles/article/28738-Video_keno_Volatility#.VvdJYEfbBKk, 22 December 2010. Accessed 26 March 2016.

[42] Garcia, Michelle, Fortune Cookie Has Got Their Numbers. Washington *Post*, 12 May 2005. Online at http://www.washingtonpost.com/wp-dyn/content/article/2005/05/11/AR2005051101772.html

[43] Georgia Lottery Corporation Annual Report 2003. Online at https://gas-cdn.lotterycrs.com/content/dam/portal/pdfs/about-us/FY04AR.pdf. Accessed 27 April 2016.

[44] Georgia lottery starts new Change Game. *Savannah Daily News*, Savannah, GA, 30 July 2002. Online at http://savannahnow.com/stories/073002/LOCchangegame.shtml#.VyF-7kfbBKk.

[45] Glasgow, Kathy, Bolita in Havana. *Miami New Times*, 19 December 2002.

[46] Goren, Charles, *Go With The Odds: A Guide To Successful Gambling.* Macmillan, Toronto, 1969.

[47] Griffin, Peter, Griffin Comments on 8-Spot Casino Story. *Casino & Sports* **15**, January 1983, p. 22.

[48] Grutzner, Charles, Dimes Make Millions for Numbers Racket. *New York Times*, 26 June 1964, p. 16–17.

[49] Gutterz Fun Center, Keno Play Book. Online at http://www.gutterzfuncenter.com/wp-content/uploads/keno_play_book_rules.pdf. Accessed 3 April 2016.

[50] Haigh, John, *Taking Chances: Winning With Probability*, 2nd edition. Oxford University Press, Oxford, 2003.

[51] Hamilton, Alexander, Idea Concerning A Lottery [January 1793]. *Founders Online*, National Archives, last modified 29 June 2017, http://founders.archives.gov/documents/Hamilton/01-13-02-0291. [Original source: *The Papers of Alexander Hamilton*, volume 13, *November 1792-February 1793*, ed. Harold C. Syrett. New York, Columbia University Press, 1967, p. 518–521.] Accessed 13 July 2017.

[52] John Hamilton Collection of Keno Pay Charts, 1958–1991. MS-00340. Special Collections, University Libraries, University of Nevada, Las Vegas. Las Vegas, NV.

[53] Hannum, Robert C., and Anthony N. Cabot, *Practical Casino Math*, 2nd edition. Institute for the Study of Gambling and Commercial Gaming, Reno, NV, 2005.

[54] Henze, Norbert, and Hans Riedwyl, *How to Win More: Strategies for Intcreasing a Lottery Win.* A K Peters Ltd., Natick, MA, 1998.

[55] "A History of Central Florida, Episode 30: Bolita." RICHES of Central Florida. Online at https://richesmi.cah.ucf.edu/omeka2/items/show/4575. Accessed 2 August 2017.

[56] Hoosier Lottery, Max 5: How To Play. Online at http:web.archive.org/web/20020803162808/http://www.in.gov/hoosierlottery/yes.htm. Accessed 23 May 2016.

[57] Hoosier Lottery, Poker Lotto: How To Play. Online at https://www.hoosierlottery.com/games/poker-lotto/how-to-play. Accessed 23 April 2016.

[58] International Casino News. *Casino & Sports* **1**, 32–35, 1977.

[59] Iowa Lottery, About All or Nothing™. Online at http://www.ialottery.com/Games/Online/AON.asp. Accessed 10 May 2016.

[60] Iowa Lottery Blog, How Long DO Those Lucky For Life Prizes Really Last? Online at http://www.ialotteryblog.com/2016/04/how-long-do-those-lucky-for-life-prizes-really-last.html. Accessed 30 April 2016.

[61] Ip, Manying. *Unfolding History, Evolving Identity: The Chinese in New Zealand.* Auckland University Press, New Zealand, 2003.

[62] Jackson, James A., Machinist, 54, wins $1,000 a week for life. *Chicago Tribune*, 12 October 1975, Section 1, page 4. Online at http://archives.chicagotribune.com/1975/10/12/page/4/article/machinist-54-wins-1-000-a-week-for-life/index.html. Accessed 15 June 2016.

[63] Jerzes Sports Bar and Keno, Pay tables and Rules. Online at http://jerzes.com/wp-content/uploads/2014/04/PlayersKeno_booklet.pdf. Accessed 30 March 2016.

[64] Kaplan, Lawrence J., and James M. Maher, The Economics of the Numbers Game. *The American Journal of Economics and Sociology* **29**, # 4 (October 1970), p. 391–408.

[65] Keno Casino—Your Best Bet. Online at http://www.playkenotoday.com/pay tables.php. Accessed 31 March 2016.

[66] Keno Gamerule. Online at http://kn1.12macau.com/KenoGame/keno_rule_E.html. Accessed 24 September 2016.

[67] Keno House Percentage Guide, *Keno Newsletter* **3**, #12, December 1991, p. 3.

[68] Kentucky Lottery, How to Play 5 Card Cash. Online at https://kylottery.com/apps/draw_games/5cardcash/howtoplay.html. Accessed 23 April 2016.

[69] Keno in New York, *Casino & Sports* **8**, 4–5, 1978.

[70] The Kirby Keno System. *Gambling Times* **1**, #11, January 1978, p. 44–6.

[71] Kremen, Pesach, In Keno, play the Pop 80 rate to go home with a decent win. *Gaming Today*, 11 November 2014. Online at https://www.gamingtoday.com/articles/article/50303-In_Keno_play_the_Pop_80_rate_to_go_home_with_a_decent_win#.VwsE_UfbBKk. Accessed 10 April 2016.

[72] Kremen, Pesach, Look to Oregon for top keno lottery deals. *Gaming Today*, 18 February 2014. Online at http://www.gamingtoday.com/casino_games/keno_bingo/article/45711-Look_to_Oregon_for_top_keno_lottery_deals#.Vv_KxUfbBKl.

[73] Kremen, Pesach, The D Deano rate still the best keno in Vegas. *Gaming Today*, 4 August 2015. Online at https://www.gamingtoday.com/articles/article/55197-The_D_Deano_rate_still_the_best_keno_in_Vegas#.VwsII0fbBKk. Accessed 10 April 2016.

[74] La Vista Keno Pay Book and Rules. Online at http://www.lavistakeno.com/uploads/LaVista-Keno-Pay-Book-and-Rules.pdf. Accessed 3 April 2016.

[75] Landmark Hotel featuring Bonus Ball Keno in casino, *Los Angeles Times*. 14 July 1985, p. O86.

[76] Lawson, Edward, The New Jersey Lottery. *Gambling Times* **2**, #2, April 1978, p 20, 87.

[77] Legislature of the Virgin Islands Committee on Finance, Fiscal Year 2016 Budget Hearings, Virgin Islands Lottery. Online at http://tinyurl.com/VILottery2016. Accessed 13 February 2016.

[78] Liggett, Byron, New games give boost to keno and baccarat. *Reno Gazette-Journal*, 15 July 1993, p. 58.

[79] Lo Giuoco del Lotto. Online at http://www.logiuocodellotto.com/. Accessed 29 August 2017.

[80] Loto Québec, Astro. Online at https://loteries.lotoquebec.com/en/lotteries/astro. Accessed 8 April 2016.

[81] Loto Québec, Lotto :D. Online at https://loteries.lotoquebec.com/en/lotteries/lottod. Accessed 17 February 2016.

[82] Loto Québec, Lotto Poker. Online at https://loteries.lotoquebec.com/en/lotteries/lotto-poker. Accessed 23 April 2016.

[83] Lotteria Nacional [Ecuador], Pozo. Online at http://www.loteria.com.ec/compra-tus-juegos/conoce-nuestros-productos/pozo. Accessed 21 December 2016.

[84] Lottery Canada, Atlantic Bucko. Online at http://www.lotterycanada.com/atlantic-bucko. Accessed 27 April 2016.

[85] The Lottery Company Limited (GB), Austria Joker. Online at https://www.euro-millions.com/austria-joker. Accessed 13 August 2016.

[86] The Lottery Company Limited (GB), EuroMillions Odds of Winning. Online at https://www.euro-millions.com/odds-of-winning. Accessed 13 August 2016.

[87] Lottoland, KeNow Results & Winning Numbers. Online at https://www.lottoland.com/en/kenow/results-winning-numbers. Accessed 2 October 2016.

[88] Lyons, Casey, All Hail the MIT Lottery Scammers. *Boston*, 9 August 2012. Online at http://www.bostonmagazine.com/news/blog/2012/08/09/hail-lottery-scammers/. Accessed 3 May 2016.

[89] Maryland Lottery, 5 Card Cash Prize Structure. Online at http://www.mdlottery.com/games/5-card-cash/payouts/. Accessed 23 April 2016.

[90] McClure, Wayne, *Keno Winning Ways*. Gamblers Book Club, Las Vegas, 1979.

[91] McClure, Wayne, *Lottery and Keno Winning Strategies*. Gamblers Book Club, Las Vegas, 1991.

[92] Mechigian, John, *Keno Encyclopedia*. Funtime Enterprises, Inc., Fresno, CA, 1972.

[93] Michigan Lottery Starts Nov. 13, *Milwaukee Journal*, 19 October 1972, part 1, p. 17. Online at https://news.google.com/newspapers?nid=1499&dat=19721019&id=yJ8oAAAAIBAJ&sjid=qygEAAAAIBAJ&pg=7379,6177214&hl=en.

[94] Michigan Lottery, Holly Jolly Jackpot. Online at https://www.michiganlottery.com/instant_games_info?. Accessed 20 March 2016.

[95] Michigan Lottery, Poker Lotto: How to Play. Online at https://www.michiganlottery.com/pokerlotto_info?#how_to_play. Accessed 23 April 2016.

[96] Michigan Lottery, *Where The Money Goes*. Online at http://www.michiganlottery.com/where_the_money_goes?cid=p1core-tbx09.f.1800/b59a4/a8/ddbfb883.1d0e9ebe75e0486e4a4a4791fba1d1b5. Accessed 5 June 2012.

[97] Mifal Hapais, The KENO Draw. Online at http://www.pais.co.il/sites/en/Keno/Pages/howtoplay.aspx. Accessed 25 November 2016.

[98] Mifal Hapais, Pais 777—Even if you Don't have Luck. Online at http://www.pais.co.il/sites/en/777/Pages/hottoplay.aspx. Accessed 26 November 2016.

[99] Miller, Jonathan, Players Bid Lotzee Goodbye, But Say It Won't Be Missed. *New York Times*, 12 September 2003. Online at http://www.nytimes.com/2003/09/12/nyregion/players-bid-lotzee-goodbye-but-say-it-won-t-be-missed.html. Accessed 22 April 2017.

[100] Minnesota Millionaire Lottery 2016. Online at https://www.mnlottery.com/games/raffle_2016/.

[101] Montana Lottery, Big Sky Bonus Progressive. Online at http://montanalottery.com/bigskybonus.

[102] National Savings and Investments, *Happy Days!*. Online at http://www.nsandi.com/files/published_files/asset/pdf/premium-bonds-brochure.pdf. Accessed 30 September 2016.

[103] National Savings and Investments, Prize Checker. Online at http://www.nsandi.com/prize-checker. Accessed 30 September 2016.

[104] Nebraska Department of Revenue, Charitable Gaming Press Kit—County/City Lottery (Keno). Online at http://www.revenue.nebraska.gov/gaming/presskit.html#kenodesc. Accessed 20 March 2016.

[105] Nebraska Lottery, Nebraska MyDaY. Online at https://www.nelottery.com/homeapp/lotto/33/gamedetail. Accessed 8 April 2016.

[106] Nevada Numbers. Online at https://en.wikipedia.org/wiki/Nevada_Numbers. Accessed 22 August 2015.

[107] Nevada Numbers Lite. Online at https://web.archive.org/web/20140518021626/http://nevadanumberslite.com/. Accessed 22 August 2015.

[108] New Jersey Lottery, Dollar Throwdown. Online at https://www.njlottery.com/en-us/fastplay/29.html. Accessed 22 August 2017.

[109] New Jersey Lottery, Lotzee Game Rules (4 out of 100 Lottery) Amended August 17, 2000. Online at https://web.archive.org/web/20010417072128/http://www.state.nj.us:80/lottery/lotzeerule.htm. Accessed 22 April 2017.

[110] New Mexico Lottery Unveils '4 This Way'. *Lottery Post*, 20 September 2004. Online at https://www.lotterypost.com/news/96240. Accessed 29 July 2017.

[111] New Zealand Lotteries Commission, How to play Bullseye. Online at https://mylotto.co.nz/assets/bullseye/bullseyehow-to-play.pdf. Accessed 8 October 2016.

[112] Nolan, Walter I., *The Facts of Keno*. Gamblers Book Club, Las Vegas, 1978.

[113] Nordell, Philip G., The Rutgers Lotteries. *The Journal of the Rutgers University Library* **16**, p. 1–12, December 1952.

[114] The Numbers Game | Massachusetts State Lottery. Online at http://www.masslottery.com/games/lottery/numbers-game.html. Accessed 28 February 2016.

[115] Oklahoma Lottery, How to play Poker Pick. Online at http://www.lottery.ok.gov/howtoplay_pokerpick.asp. Accessed 23 April 2016.

[116] Ohio's Great State Lottery Starts Today! (advertisement), *News-Journal*, Mansfield, OH, 12 August 1974, p. 14. Online at https://www.newspapers.com/newspage/47534381/. Accessed 5 November 2016.

[117] Ontario Lottery and Gaming Corporation, Encore. Online at http://www.olg.ca/lotteries/games/encore.jsp. Accessed 11 August 2016.

[118] Ontario Lottery and Gaming Corporation, Megadice Lotto. Online at http://www.olg.ca/lotteries/games/howtoplay.do?game=megadice. Accessed 16 February 2016.

[119] Ontario Lottery and Gaming Corporation, NHL Lotto Prize Structure. Online at http://www.olg.ca/lotteries/games/howtoplay.do?game=nhllotto#prizeStructure. Accessed 10 August 2016.

[120] Oregon Lottery Keno. Online at http://oregonlottery.org/games/draw-games/keno. Accessed 2 April 2016.

[121] Oregon Lottery Lucky Lines. Online at http://www.oregonlottery.org/games/draw-games/lucky-lines. Accessed 11 May 2016.

[122] Osman, Magda, *Future-Minded: The Psychology of Agency and Control.* Palgrave/Macmillan, London, 2014.

[123] Over-10 Keno Proposition At Royal Americana Lounge. *Casino & Sports* **12**, 20, 1979.

[124] Oz Lotteries, Play The Pools. Online at https://www.ozlotteries.com/the-pools#play_lottery_the_pools. Accessed 8 October 2016.

[125] Parker, Robert E., Las Vegas: Casino Gambling and Local Culture, in *The Tourist City*, Dennis R. Judd & Susan S. Feinstein, editors. Yale University Press, New Haven, CT, 1999, p. 107-123.

[126] Penn. Lottery introduces new 'Mix & Match' game. *Lottery Post*, 19 January 2007. Online at https://www.lotterypost.com/news/149136. Accessed 18 April 2017.

[127] Pettus, Emily Wagster, Mississippi group will study lottery but not take a stand, *CDC Gaming Reports*, 25 May 2017. Online at http://www.cdcgamingreports.com/mississippi-group-will-study-lottery-but-not-take-a-stand/. Accessed 25 May 2017.

[128] PHRAZZE$: The next new word in gaming. Online at http://phrazzeslottery.com/. Accessed 12 November 2016.

[129] Powerball History. Online at http://www.powerball.com/pb_history.asp. Accessed 11 May 2016.

[130] Progressive Jackpot Keno Games, *Keno Newsletter*, **3** #12, December 1991, p. 2–4.

[131] Resultados Pozo Millonario Sorteo 563, *Ecuador Noticias*. Online at http://www.ecuadornoticias.com/2014/05/resultados-pozo-millonario-sorteo-563.html. Accessed 21 December 2016.

[132] Richmond, Kelly, Reno Hotel Offers Million Dollar Keno for an Extra Quarter. Online at http://www.prweb.com/releases/2004/04/prweb116974.htm, 8 April 2004.

[133] Roberts, Francis, An Arithmetical Paradox, Concerning the Chances of Lotteries. *Philosophical Transactions* **17**, 677-681, 1 January 1693.

[134] Royer, Victor M., *Powerful Profits From Keno.* Kensington Publishing Corp., New York, 2004.

[135] Scarne, John, *Scarne's Complete Guide to Gambling.* Simon and Schuster, New York, 1961.

[136] Schumpeter, Against the odds—Informal betting in Myanmar. *The Economist*, 8 October 2012. Online at https://www.economist.com/blogs/schumpeter/2012/10/informal-betting-myanmar.

[137] Seville, Adrian, The Italian Roots of the Lottery. *History Today*, **49**, #3, March 1999. Online at http://www.historytoday.com/adrian-seville/italian-roots-lottery. Accessed 29 August 2017.

[138] Shackleford, Michael, Cleopatra Keno. Online at https://wizardofodds.com/games/cleopatra-keno/, 3 January 2009. Accessed 4 September 2016.

[139] Shackleford, Michael, Harrah's. Online at http://wizardofvegas.com/hotels/harrahs/, 1 February 2010. Accessed 20 August 2016.

[140] Shackleford, Michael, Power Keno. Online at http://wizardofodds.com/games/keno/appendix/6/. Accessed 14 October 2015.

[141] Silberstang, Edwin, Keno Anyone?. *Gambling Times* **1**, #11, January 1978, p. 22–24.

[142] Sklansky, David, Getting the Best of It: Casinos and Their Mistakes. *Gambling Times* **4**, #8, December 1980, p. 56–58.

[143] Sklansky, David, and Alan E. Schoonmaker, *DUCY?: Exploits, Advice, and Ideas of the Renowned Strategist.* Two Plus Two Publishing LLC, Henderson, NV, 2010.

[144] SmartLuck, Inc., How to Win Montana Big Sky Bonus 4/28 + 1/28. Online at http://www.smartluck.com/free-lottery-tips/montana-big-sky-bonus-428-pb.htm. Accessed 4 May 2016.

[145] South Dakota Lottery, Wild Card Game Page. Online at http://lottery.sd.gov/games/online/wildcard/. Accessed 28 February 2016.

[146] Sullivan, George, *By Chance A Winner: The History of Lotteries.* Dodd, Mead, & Company, New York, 1972.

[147] Sullivan, Gregory W., Letter to Massachusetts State Treasurer Steven Grossman, 27 July 2012. Online at http://www.mass.gov/ig/publications/reports-and-recommendations/2012/lottery-cash-winfall-letter-july-2012.pdf. Accessed 30 April 2016.

[148] Supreme Ventures, How To Play Lotto. Online at http://www.supremeventures.com/content/svlgames/lotto.html. Accessed 17 September 2016.

[149] Sweeney, Matthew, *The Lottery Wars.* Bloomsbury USA, New York, 2009.

[150] The POS Game, AMIGO, Continues Its National Deployment Replacing RAPIDO Game. *Lottery Insider News* **63**, #13, 1 July 2013. Online at http://www.lotteryinsider.com/vol63/no13.htm#1311. Accessed 9 October 2016.

[151] Transcript of the 6 July 2015 meeting of the New York Gaming Commission. Online at http://www.yorkmedia.com/nysgc/transcripts/2015/07/06/New_York_State_Gaming_Comission_07-06-2015.pdf. Accessed 11 July 2015.

[152] United States Patent Approved for a Revolutionary Instant Online Alpha Numeric Lottery Game, *Lottery Insider: The Week In Review*, **74**, #1, 4 January 2015.
Online at http://www.lotteryinsider.com/vol74/no1.htm.

[153] Virgin Islands Lottery Extra Ordinary Prospectus. Online at http://www.winusvilottery.com/vi-lottery-game/extra-ordinary-game-prospectus. Accessed 13 February 2016.

[154] Virgin Islands Lottery—How To Play. Online at http://www.winusvilottery.com/vi-lottery-game/how-to-play. Accessed 13 February 2016.

[155] Virgin Islands Lottery Ordinary Prospectus. Online at http://www.winusvilottery.com/vi-lottery-game/ordinary-game-prospectus. Accessed 13 February 2016.

[156] Vogel, Ed, Meet 'Mr. Keno' Harry Hooser—King of Nevada Ticket Writers. *Casino & Sports* **21**, 78–79, 1981.

[157] Waverly Keno Specials. Online at http://www.waverlykeno.com/Keno_Specials.html. Accessed 21 April 2016.

[158] Ways To Win—Virgin Islands Lottery. Online at http://www.winusvilottery.com/vi-lottery-game/ways-to-win-the-vi-lottery.

[159] Weisenberg, Michael, The Kings of Keno. *Gambling Times* **8**, #11, March 1985, p. 74–77.

[160] Weiss, Harry B., and Grace M. Weiss, *The Early Lotteries of New Jersey.* Trenton, NJ, The Past Times Press, 1966.

[161] Weisstein, Eric W "Sphere Packing." From MathWorld—A Wolfram Web Resource. Online at http://mathworld.wolfram.com/SpherePacking.html. Accessed 25 May 2017.

[162] Wisconsin Lottery, About the Wisconsin Lottery. Online at https://www.wilottery.com/about.aspx. Accessed 8 May 2016.

[163] Wong, Stanford, *Blackjack Secrets.* Pi Yee Press, Las Vegas, 1993.

[164] Wong, Stanford, *Casino Tournament Strategy.* Pi Yee Press, Las Vegas, 1993.

[165] Wood, M.W, Lottery summing game, United States Patent #5,106,089 A. Online at http://www.google.com/patents/US5106089. Accessed 20 August 2016.

[166] XpertX, Inc., MegaKeno® Statewide Linked Progressives. Online at http://www.xpertx.com/assets/pdfs/XpertX_MegaKeno_13.pdf. Accessed 30 March 2016.

[167] Yin and Yang (Over-and-Under)—A Way to Shoot for $50,000 in Keno at the Marina Casino. *Casino & Sports* **9**, 65–68, 1979.

# Index

Abrams, Aaron, 284, 307
Add-Em-Up Lotto, 297–298
Aggregate limit, 81–84, 89, 90, 116, 117, 122, 149, 175, 181, 189–191
Alamo Casino (Las Vegas), 159
All or Nothing, 253, 295
Allen, Scott, 309
Alwes, Berthold C., 307
American roulette, 154
Amigo, 243–245, 300
Angel of the Winds Casino, 74, 85, 108, 109, 186
Annuity, 280–283, 296
Aquarius Casino, 134
Aqueduct Racetrack, 14
Arithmetic complexity, 290–291
Arithmetic progression, 288–290
Arizona, 229, 253
Arizona Charlie's Decatur, 64, 69
Arlington, WA, 85
Aruba, 298–299
Astro, 298
Atlantic Bucko, 245–249, 313
Atlantic City, 103
Augustus, 9
Australia, 3, 199, 237
Austria Joker, 299–300
Average, 28
Avi Casino, 82

Baccarat, 144, 149, 155, 156
Bally's Casino, 78, 133, 143, 144
Bank Shot, 196
Barbary Coast Casino, 143–144
Barney's Casino, 150
Barton's 93 Club, 192
Battle Creek, MI, 68, 122
Beatrice, NE, 196
Belize, 56
Bellevue, NE, 60, 124, 125, 130, 176, 192
Bernardo, Karen, 309
Big 'Q', 125
Big Bucks Keno, 144
Big Red Keno, 64, 65, 68, 69, 87, 94–95, 109, 112–113, 193, 196, 197, 307
Big Sky Bonus, 285–287
Bingo, 268
Binion's Horseshoe Club, 57, 146, 193
Binomial distribution, 50
Binomial experiment, 50
Binomial formula, 51
Binomial random variable, 50, 168, 181, 269, 293
Blackjack, 24, 27, 65, 123, 155, 156, 159, 166, 168
Blair, NE, 68, 69, 193
Blayney, Richard D., 308
Bob Cashell's Horseshoe Club, 187
Boddy, Gillian, 308
Boledo, 56
Bolita, 15, 56
Bonanza Club, 47, 90
Bond, James, 155
Bourbon Street Casino, 106
Box bet, 205, 207
Bradley, Robert C., 308
British Columbia, 113, 308
Brooklyn number, 14
Brooks, Henry M., 308
Broverman, Samuel A., 308

Buckeye 300, 218–220
Bulls-eye keno, 113, 172–174
Bullseye lottery, 210–211, 291
Butte, MT, 4

Cabot, Anthony N., 311
Caesars Entertainment, 78
Caesars Palace, 88, 91, 92, 123, 144
Caesars Tahoe, 76, 120
Cal-Neva Casino, 118–120
California Casino, 65, 68, 159, 199
California Club, 7, 90
California Lottery, 266, 308
Canada, 245
Caribbean Keno, 31, 172
Carousel Casino, 57
Carradine, B., 308
Carson City, NV, 68
Carson Station Casino, 68
Cash Winfall, 291–294
Catch 22, 120–121
Catch All, 69–71, 73, 74, 77, 88, 143, 144, 193, 203
Catlin, Don, 308
Caveman Keno, ix, 162–166
Central Limit Theorem, 66
Change Game, 228
Change Play, 228
Chicken Keno, 190–192
Chinese lottery, 4–5, 55–56, 87, 93
City Picks, ix, 39–277
Civil War, 12
Classic Lotto 47, 288, 289
Cleopatra Keno, 180–181
Club Keno, 74, 172
Cluster number, 289, 291
Coin tossing, 30
Combination, 40
Combination ticket, 45
Complement, 19
Complement Rule, 67, 101, 168, 216, 223, 262, 286
Conditional probability, 26–173
Connecticut Lottery, 271
Continental Casino, 111, 147
Coupons, 148–150
Cover All, 84–86, 103
Coverall Keno, 115
Cowles, David, 309
Craps, 65, 94, 144, 155, 156, 168
Crazy Corners, 113
Cromwell Casino, 143
Crossroads Keno, 200–201
Crown Cigar Store, 3
Crystal Bay Casino, 107
Crystal Bay, NV, 46, 107
Cuba, 15
Culin, Stewart, 309
Curtin, Lawrence, 309

D Casino, 7, 43, 60, 65, 85, 86, 127
Daily 2 game, 51, 55, 204–205
Daily 3 game, 205–206
Daily 4 game, 206–208, 265
Daily Derby, 266–268
Daily Double, 124
Daily Keno, 74
Daston, Lorraine, 309
DC Lottery, 208, 309
DC5, 208–210, 212, 213
Deano rate, 127
Delaware Sports Lottery, 34–36
Derangement, 275–277
Desert Inn Casino, 144
Detroit *Free Press*, 203
Diamond Daze, 113
Diamond Lil's Special, 106
Diamond's Casino, 199
Dice, 18, 27
Dick Graves' Nugget Casino, 59
Diggin's, The, 86
Disjoint sets, 19
District of Columbia, 10
Dollar Throwdown, 19, 25
Dotty's Casino, 158
Double-Up Keno, 128–130, 197
Doublette, 110–111
Dunes Casino, 142

Eagle, NE, 64

Early Bird, 34
Ecuador, 250–251, 313
Edge number, 289, 291
Egg multiplier, 163–165
El Cortez Casino, 63, 68, 71–73, 87, 117, 132, 163
Elwood, Bryan, 309
Empirical Rule, 54–55, 289
Encore, 211–213
ERNIE, 225
Estes, Andrea, 309
Euler, Leonhard, 56, 234–235, 309
EuroMillions, 299–300
European roulette, 27, 130
Event, 18
Everett, George, 309
Exacta Keno, 175–178
Excalibur Casino, 43, 81, 99–101, 121, 122
Expectation, 28–36, 52, 59, 63, 64, 68, 70, 71, 85, 88, 89, 95, 97, 109, 115–117, 120, 122, 124, 126, 133, 147, 150–152, 154, 160, 163, 164, 167, 169, 170, 172, 174–179, 182–185, 197, 198, 205, 206, 208, 209, 211, 220, 223, 230, 236, 241, 245, 249–251, 259, 274, 277, 284, 285, 292, 298, 300
Experiment, 17–18
Experimental probability, 20–21
eXtra Million, 181–183
Ezell, John Samuel, 309

Factorial, 38
Fair game, 30
Farkas, Karen, 309
Fast Play, 19
Feasible point, 138–140
Ferguson, Tom W., 309
Fiesta Henderson, 60
Fire Bet, 168
FireKeepers Casino, 68, 122, 135, 154, 159, 193
First Addition Rule, 22
Fitzgerald's Casino (Las Vegas), 42
5 Card Cash, 229–230
Five Ranges, 185
Floor function, 277
Florence, 9
Florida, 15
Flynn, Kevin, 309
Football pools, 48–49
Four Corners Keno, 71
Four Queens Casino, 81, 144, 145
4 This Way, 264–265
Foxwoods Casino, 8, 67, 68, 70, 199
France, 243, 300
Fredella, Brad, 158, 310
Fremont Casino, 90, 142
Frontier Casino, 144
Fundamental Counting Principle, 37, 41, 58, 106, 278

Gambler's Fallacy, 24
*Gambling Times*, 153, 189
Game King, 157, 158
Gaming Arts, 166, 174, 175
Garcia, Michelle, 310
Garibaldi, Skip, 284, 307
General Multiplication Rule, 27
Genoa, 10, 56
Georgia Lottery, 228
Germany, 9
Gibraltar, 199
Glasgow, Kathy, 310
Gold Coast Casino, 142
Golden Keno Ball, 174–175
Golden Nugget, 76
Golden Nugget Casino, 57, 90, 145
Goose, 8
Goren, Charles, 310
Griffin, Peter, 310
Grutzner, Charles, 310
Gutterz Fun Center, 64, 65, 310

Hacienda Casino, 153
Haigh, John, 310
Hamilton, Alexander, 249, 277, 310
Hannum, Robert C., 311

Harolds Club, 84, 90, 148, 149
Harrah's Casino (Las Vegas), 78, 88, 89, 144
Harrah's Casino (Reno), 68
Harvey's Casino, 68, 187
Hastings, NE, 109
Helena, MT, 86
Henderson, NV, 60, 68, 158
Henze, Norbert, 311
Hi-Low Keno, 93, 95–101, 161
High Frequency Pay, 76
High Roller Special, 143
Hold, 123
Holdrege, NE, 64
Holly Jolly Jackpot, 227
Honduras, 13
Hoosier Lottery, 208, 263, 311
Horseshoe Casino, 90
Hot $10 Ticket, 142
House advantage, ix, 30, 31, 34–36, 43, 47, 56, 59, 61, 63–70, 72–74, 76–79, 82, 84, 88–91, 94, 95, 98, 107, 112, 115–117, 120, 122–124, 126, 127, 129, 134, 135, 142, 146, 148, 150, 154–158, 160, 161, 165–169, 172, 174, 175, 177, 178, 180, 182, 184–186, 192, 193, 196, 197, 199–201, 204, 205, 207–209, 216, 230, 235, 236, 241, 250, 255, 258, 274, 277, 285, 295, 296, 300

*Idea Concerning A Lottery*, 249
Illinois Lottery, 296–297
Inclusion/exclusion, 105
Independent events, 24
Indiana, 230, 263
Instant tickets, 226–228
Integer programming, 138
Internal Revenue Service, 73
International Game Technology (IGT), 157, 162
Intersection number, 200
Iowa Lottery, 283, 311
Ip, Manying, 311
Island Special, 199
Israel, 79–80, 194

Jackpot, NV, 192
Jackson, James A., 311
Jamaica, 237
Japanese King 12-Way, 47
Java, 184
Jefferson, Thomas, 10
Jerry's Nugget Casino, 47, 64, 117, 134, 141, 176, 177
Jerzes Sports Bar, 112, 198
Jumbo Keno Progressive, 196–197
Just 7777s, 78

Kadlec, Gary, 300
Kansas, 172, 253
Kansas Lottery, 64
Kaplan, Lawrence J., 311
Keno, 1–9, 28, 56–57, 59–201, 234, 237, 243, 253, 284, 291
Keno Booster, 184
Keno Casino (Nebraska), 82, 124
Keno insurance, 168–171
Keno Millions, 182–183
Keno Millions Mini, 182–183
*Keno Newsletter*, 31, 312, 316
Keno On The Go, 60
KeNow, 200
Kentucky, 230
King, 87, 100, 128, 195
King Me Eights, 198
King ticket, 46
Kingo, 130–132
Kirby Keno System, 189–190
Korner Keno, 109–110
Kremen, Pesach, 312
Krugerrand, 148

La Lotto di Firenze, 9
La Vista Keno, 65, 69, 116, 198
La Vista, NE, 65
Lagniappe Keno, 125–127
Lake Tahoe, 118
Landmark Casino, 152

Las Vegas, 6, 7, 43, 57, 63–65, 68, 70, 71, 76–78, 81, 87, 90, 103, 111, 117, 127, 132–134, 142, 146, 150, 152, 153, 159, 196, 199
Las Vegas Club, 6, 90
Las Vegas, NV, 60
Laughlin, NV, 82, 133, 134, 170
Law of Large Numbers, 21, 24, 147, 189
Lawson, Edward, 312
Liggett, Byron, 313
Line Drive Keno, 198
Littlewoods, 48
Lo Giuoco del Lotto, 10, 234, 313
Loto Québec, 313
Lottery, 9–16, 29–30, 47–48, 74–75, 133, 203–301
Lottery Canada, 313
Lotto (Jamaica), 237
Lotto :D, 230–233, 258, 261, 295
Lotto America, 58
Lotto Nebraska, 190, 192
Lotto Poker, 230
Lottoland, 199
Lotzee, 249–250
Louisiana Lottery, 12–14
Lucky for Life, 283
Lucky Lines, ix, 268–272
Lucky Links, 268, 271–274, 296
Lucky-N-Wild 7–11, 94, 95
Luxor Casino, 60
Lyden, Francis, 4, 6
Lyden, Joseph, 4, 7
Lyons, Casey, 313

Macau, 171, 193
MacMillan, Harold, 224
Maher, James M., 311
Manhattan way, 14
Marina Casino, 76
Martingale, 190
Maryland, 230
Maryland Lottery, 24, 313
Mashantucket, CT, 8, 68
Mass Millions, 292
Massachusetts Lottery, 291
Massachusetts Numbers Game, 204, 228
Mathematica, 283
Matthews, Jacqueline D., 308
Max 5, 263–264
McCarran Airport, 159
McClure, Wayne, 313
Mean, 28
Mechigian, John, 313
Mega Millions, 12, 57–58
Mega10 Keno, 117
Megadice Lotto, 258–261
Merlin's Magical Keno Kube, 121–123
Mermaids Casino, 57
Meskwaki Casino, 28, 62, 68, 111, 115
Mesquite, NV, 92, 171
MGM Grand Casino, 76
Michigan, 230
Michigan Jackpot, 294–295
Michigan Lottery, 24, 32, 205–208, 217–218, 220–221, 227, 228, 292, 294, 295, 314
Microsoft Excel, 49–51
Miller, Jonathan, 314
Millionaire Pool, 219
Minnesota Millionaire Raffle, 223–224
Mint Casino, 8, 90
Mirage Casino, 103, 170
Mississippi, 16
Missouri, 172
Missouri Lottery, 172, 206
MIT, 292
Mix & Match, 39, 265–266
Montana, 34, 285
MontBleu Casino, 76
Monte Carlo Casino (Reno), 199
Mount Pleasant, MI, 159
Multi-State Lottery Association, 251, 253, 277
Multiplication Rule, 25, 124, 263

Mutually exclusive events, 21, 24
Myanmar, 15

Nebraska, 60, 65, 69, 82, 94, 190, 197, 253, 314
Nebraska City, NE, 60
Nebraska Lottery, 236
Nebraska MyDaY, 255–257, 315
Netherlands, 9
Nevada, 34, 91, 117
Nevada Club, 168, 169
Nevada Gaming Commission, 91
Nevada Numbers, 133–135, 193
Nevada Numbers Lite, 134–135
New Hampshire Sweepstakes, 216–217
New Jersey, 10, 249, 250
New Jersey Lottery, 19
New Mexico, 233
New Mexico Lottery, 264
New Orleans, 13, 125
New York City, 15
New York Gaming Commission, 278
New York Lottery, 296
New York, NY, 14
New Zealand, 3, 210, 291
Newbury, MA, 10, 30
Nhit-lone, 15
NHL Lotto, 261–263
No Loser 20 Spot, 109
No Tax Pick 5, 73
Nolan, Walter I., 116, 315
Nonlinear programming, 138
Nordell, Philip G., 315
Norfolk, NE, 65, 112–113
North Dakota, 254
North Las Vegas, NV, 47, 64, 68, 90, 134
Northeastern University, 292
Nugget Casino, 59
Numbers Game, 14–15, 27, 32–34, 203–204, 213

Odds, 23
Ohio, 184, 237
Ohio Lottery, 64, 218–220
Oklahoma, 230
Oklahoma Lottery, 315
Olympia, WA, 107
Omaha, NE, 68, 87, 109
100 Dimes keno, 132–133, 193
$100,000 Parlay Card, 36
1-Off, 206
Ontario, 211, 258, 261
Opera House Casino, 68
Oregon, ix, 34, 183, 237
Oregon Keno Multiplier, 183–184, 237
Orleans Casino, 6, 63, 68, 69, 79, 92, 93, 125, 126, 178
Osman, Magda, 316
Over-10, 115

Pais 777, 194
Pák kòp piú, 1–5, 9, 88
Pakapoo, 3, 88
Palace Club, 6
Palmetto Cash 5, 236–237
Papillion, NE, 73, 112
PAR Sheets, 261
Pari-mutuel, 203, 204, 228
Parker, Robert E., 316
Parlay card, 34
Passive ticket, 213
Pattern Play, 113–115
Pawns, 47
Pennsylvania, 39, 55, 242
Pennsylvania Lottery, 15, 204, 221–223, 265, 297
Penny Keno, ix, 135–142, 150, 194–196
Peppermill Casino (Mesquite), 171
Peppermill Casino (Reno), 75, 110, 181
Perkins, William, 16
Permutations, 38
Perpetuity, 283
Pettus, Emily Wagster, 316
Pick-It, 203–204
Pioneer Club (Las Vegas), 90

Pioneer Club (Reno), 87, 90
Players Keno, 73, 112
Playing cards, 17
Plaza Casino, 87, 117
Poker, 27
Poker Lotto, 230
Poker Pick, 230
Policy game, 14
Pools, The (Australia), 237
Pop 80, 127–128
Power Keno, 178–180, 196
Power-Up, 237
Powerball, ix, 12, 15, 16, 24, 40, 116, 133, 203, 204, 213, 228, 237, 253, 277–285, 287, 288, 299
Pozo Millonario, 250–251
Premium Bonds, 224–226
Present value, 280
Primadonna Casino, 93
Probability distribution, 27–28
Probability distribution function, 27, 79, 265
Probability, defined, 20
Professional and Amateur Sports Protection Act, 34
Progressive jackpots, 116–117, 171
Puralewski, Leo, 297

Quebec, 230, 258, 298
Quebec Astro, 257–258
Queen's College Literature Lottery, 11, 41, 57
Quick Pick, 8, 113, 130, 249, 291, 293
Quinela (Keno), 171–172, 178, 193

Racehorse keno, 6, 171
Raffle, 10, 223
Random, 289
Random Hall, 292
Random Strategies Investments LLC, 291–294
Random variable, 27–28
Rapido, 300
Red Hot Keno Special, 81
Red Rock Resorts, 117, 196
Red Wind Casino, 107
Reno, NV, 6, 68, 75, 84, 87, 90, 93, 133, 168, 181, 199
Reserve Casino, 60, 68
Riedwyl, Hans, 311
Ring Of Fire, 122
Rio Casino, 78
Riverside Casino, 170
Riviera Casino, 124, 125, 153
Roadrunner Cash, 233
Roberts, Francis, 29, 317
Rock/paper/scissors, 19, 25
Rockingham Park, 216
Roman Brother, 217
Roulette, 24, 155
Royal Americana Casino, 115
Royal Casino, 57
Royer, Victor M., 317
Rutgers University, 11

Sahara Casino, 7, 83, 90, 153
Sahara Tahoe Casino, 148
Salem, NH, 216
Sam's Town, 77
Sample space, 18
Sands Casino, 91
Scarne, John, 317
Schoonmaker, Alan E., 317
SCRATCHEZZ, 301
Second Addition Rule, 22
Sedoris, Craig, 300
7 Kings All Ways, 46–47
7–11 Keno, 151–152
Seville, Adrian, 317
Shackleford, Michael, 317
Shortcake Keno, 146, 193
Silberstang, Edwin, 317
Silver Sevens Casino, 111
Sklansky, David, 317
Slot machines, 71, 156, 261
SLS Casino, 7
Soaring Eagle Casino, 159
South Carolina Education Lottery, 236–237
South Lake Tahoe, NV, 68, 187

Sparks, NV, 59, 196
Speed Keno, 166–168, 174
Spotlight Keno, 60, 193
Standard deviation, 53–55
Star number, 193
Stardust Casino, 91
Stateline, NV, 76, 148, 150
Stetson's Casino, 158
Stratosphere Casino, 150
Subset, 37
Sullivan, George, 318
Sullivan, Gregory W., 318
Super 31, 168
Super 7, 242–243
Super Jackpot Keno, 75–76
Super Keno, 178, 196
Super Slam, 297
Superball, 208
Superplay, 220–221
Sweeney, Matthew, 318
Sweet Sixteen, 108, 120
Synthetic lottery, 199
Systematic ticket, 79, 194

Tahoe Biltmore Lodge, 46
Tama, IA, 28, 68
Tennant, Hal, 309
Tetris, 85
Texas Station, 68, 69, 156
Texas Triple Chance, 240–242
Theoretical probability, 20–21
*Thousand Character Classic*, 2
321 Boom, 193
369 Way Keno, 160–161
Tic-tac-toe, 268, 271
Ticket To Ride, 199
Top and Bottom, 111–116
Traditional Extra Ordinary game, 213, 215, 216
Traditional Ordinary game, 213–216
Treasure Island Casino, 70, 73, 75, 99, 100, 134
*Ts'in Tsz' Man*, 2
20 Spot Special, 108–109
23 For A Dollar, 220–223, 297
2by2, 253–255
2-Game Parlay Keno, 124, 176, 192

U.S. Virgin Islands, 57, 213
Union Plaza Casino, 87, 98, 102, 125
University of Michigan, 278
Urn of Fortune, 9

Valley, NE, 193
Vannini's Patent Lotteries, 10
Vegas Golden Knights, 261
Vegas World Casino, 150–153
Video keno, 156–168
Virgin Islands Lottery, 213–216
Virgin River Casino, 92
Virginia Company, 10
Vogel, Ed, 318

Waverly, NE, 65
Way ticket, 42–47, 57, 84, 95, 131, 135, 136, 141, 160, 167, 237
Weisenberg, Michael, 319
Weiss, Harry B. and Grace M., 319
Western Village Casino, 196, 198
Wheel bet, 207
Whiskey Pete's Casino, 59
White Pigeon Ticket, 1
Wild Card, 251
Wild Card 2, 251–253
Winfall, 292
Wisconsin, ix
Wisconsin Lottery, 39, 274
Wong, Stanford, 319
Wonton Food, 287

XpertX, 117, 319

You Pick 'Em Treasury Ticket, 14, 37, 203

Zayacher, Martin, 216
Zodiac, 298